D0876279

Multiphoton Spectroscopy of Molecules

QUANTUM ELECTRONICS — PRINCIPLES AND APPLICATIONS

A Series of Monographs

EDITED BY

PAUL F. LIAO
AT&T Bell Telephone Laboratories
Murray Hill, New Jersey

PAUL KELLEY
Lincoln Laboratory
Massachusetts Institute of Technology
Lexington, Massachusetts

N. S. Kapany and J. J. Burke. OPTICAL WAVEGUIDES, 1972

Dietrich Marcuse. THEORY OF DIELECTRIC OPTICAL WAVEGUIDES, 1974

Benjamin Chu. LASER LIGHT SCATTERING, 1974

Bruno Crosignani, Paolo Di Porto, and Mario Bertolotti. STATISTICAL PROPERTIES OF SCATTERED LIGHT, 1975

John D. Anderson, Jr. GASDYNAMIC LASERS: AN INTRODUCTION, 1976

W. W. Duley. CO_2 LASERS: EFFECTS AND APPLICATIONS, 1976

Henry Kressel and J. K. Butler. SEMICONDUCTOR LASERS AND HETEROJUNCTION LEDs, 1977

H. C. Casey and M. B. Panish. HETEROSTRUCTURE LASERS: PART A, FUNDAMENTAL PRINCIPLES; PART B, MATERIALS AND OPERATING CHARACTERISTICS, 1978

Robert K. Erf (Ed.). SPECKLE METROLOGY, 1979

Marc D. Levenson. INTRODUCTION TO NONLINEAR LASER SPECTROSCOPY, 1982

David S. Kliger (Ed.). ULTRASENSITIVE LASER SPECTROSCOPY, 1983

Robert A. Fisher (Ed.). OPTICAL PHASE CONJUGATION, 1983

John F. Reintjes. NONLINEAR OPTICAL PARAMETRIC PROCESSES IN LIQUIDS AND GASES, 1984

S. H. Lin, Y. Fujimura, H. J. Neusser, and E. W. Schlag. MULTIPHOTON SPECTROSCOPY OF MOLECULES, 1984

YOH-HAN PAO
Case Western Reserve University
Cleveland, Ohio

Founding Editor
1972–1979

MULTIPHOTON SPECTROSCOPY OF MOLECULES

S. H. Lin

Department of Chemistry
Arizona State University
Tempe, Arizona

Y. Fujimura

Department of Chemistry
Tohoku University
Sendai, Japan

H. J. Neusser
E. W. Schlag

Institute for Physical and Theoretical Chemistry
Technical University of Munich
Munich, Federal Republic of Germany

1984

ACADEMIC PRESS, INC.

(Harcourt Brace Jovanovich, Publishers)

Orlando San Diego San Francisco New York London

Toronto Montreal Sydney Tokyo São Paulo

ACADEMIC PRESS, INC.
Orlando, Florida 32887

United Kingdom Edition published by
ACADEMIC PRESS, INC. (LONDON) LTD.
24/28 Oval Road, London NW1 7DX

Library of Congress Cataloging in Publication Data
Main entry under title:

Multiphoton spectroscopy of molecules.

   (Quantum electronics--principles and applications)
   Includes index.
   1. Molecular spectra. 2. Photons. I. Lin, S. H.
(Sheng Hsien), Date. II. Title. III. Series.
QD96.M65M84   1983   543'.0858   83-2584
ISBN 0−12−450520−1

PRINTED IN THE UNITED STATES OF AMERICA

84 85 86 87   9 8 7 6 5 4 3 2 1

Contents

Appendix II

Appendix III

Appendix IV

Preface

In the past decade there has been very rapid growth in multiphoton spectroscopy, one of the most interesting fields of research made possible by the development of powerful lasers. Multiphoton spectroscopy has been widely used in biology, chemistry, materials science, physics, and other disciplines. We realize that it is perhaps foolhardy to write a monograph about multiphoton spectroscopy at this time, when new results are still published at a high rate in scientific journals. Nevertheless, it is hoped that this volume may have lasting value. Since the field of multiphoton spectroscopy is still in a stage of rapid development, no effort has been made to give a complete bibliography or to achieve complete coverage of all experimental data. The fundamental theory and methods and the basic experimental results are stressed. This volume is intended for all who have an active interest in the field of multiphoton spectroscopy and those who want to enter this field. Thus the presentation is elementary and self-contained wherever possible.

After an introductory chapter giving a general survey of the progress of multiphoton spectroscopy and the features of the multiphoton processes, in Chapter 2 we present several theoretical methods (the time-dependent perturbation, Green's function, density matrix, and susceptibility methods) for treating the multiphoton transitions in a molecular system, starting from first principles. In Chapter 3 the development of two-photon spectroscopy is reviewed, and then various experimental methods are given; multiphoton ionization mass spectroscopy is considered in some detail. Characteristics of molecular multiphoton spectroscopy are presented in Chapter 4, and intensity dependence, saturation phenomena, and polarization dependence are also discussed. In Chapter 5 spectroscopic properties of molecular multiphoton transitions in both nonresonant and resonant cases are presented. Mechanisms of vibronic coupling in forbidden two-photon transitions are discussed, and it is shown that some rovibronic bands appearing in the two-photon excitation spectra of benzene in vapor can be analyzed in terms of intramolecular vibrational relaxation. Visible/UV multiphoton spectroscopic results of the Rydberg and valence states of molecules of interest from a static or a dynamical point of view are discussed in Chapter 6. Interesting new applications of multiphoton spectroscopy, such as multiphoton ionization mass spectroscopy, multiphoton circular dichroism, and ion dip spectroscopy, are introduced in Chapter 7.

We are grateful to many colleagues and friends for encouragement, help, and suggestions, and thank especially Professors T. Nakajima, M. A. El-Sayed, K. F. Freed, I. Tanaka, M. Ito, A. R. Ziv, E. C. Lim, E. K. C. Lee, M. Kawasaki, T. Azumi, and N. Mikami; Doctors Y. Mikami, H. L. Selzle, K. H. Fung, U. Boesl, A. A. Villaeys, H. Kono, and H. P. Lin; and C. Fujimura, N. Ito, W. Dietz, Z. Z. Ho, Y. L. Lee, W. E. Henke, and E. Riedle.

We should like to thank the authors and publishers of the referenced papers, journals, and books for allowing us to reproduce their results in this volume. Financial support from the Alexander von Humboldt Foundation and the National Science Foundation (US–Germany Program) helped to make this book possible. We should also like to thank our families for their patience and understanding.

Multiphoton Spectroscopy of Molecules

CHAPTER
1

Introduction

1.1 GENERAL SURVEY OF THE PROGRESS OF VISIBLE/UV MULTIPHOTON SPECTROSCOPY

Visible/UV multiphoton spectroscopy has made a great contribution to molecular spectroscopy. This spectroscopy basically consists in the excitation of molecules by two or more photons of visible or UV frequency. New vibronic (vibration–electron) and electronically excited states, which were not found in ordinary single-photon spectroscopy because of their different selection rules, can be observed in a wide range from lower excited states to ionized continua.

Though the possibility of simultaneous two-photon absorption or emission (the lowest order multiphoton processes) in molecules was pointed out in 1931 by Goeppert-Mayer, who applied Dirac's dispersion theory, experimental observation of two-photon absorption in the optical region was made possible only after lasers were developed as an intense incident light source, especially after the late 1960s, when tunable dye lasers appeared (a detailed discussion of two-photon spectroscopy will be given in Chapter 3). In fact, compared with a one-photon cross section for a typical molecule ($\sim 10^{-17}$ cm^2), the cross sections of multiphoton transitions are extremely low at the intensity of conventional light sources: for example, $\sim 10^{-51}$ cm^4 s and $\sim 10^{-82}$ cm^6 s^2 for two- and three-photon transitions, respectively. Today we can measure multiphoton spectra associated with two or more photons in the visible/UV region with high-power pulsed lasers.

The development of visible/UV multiphoton spectroscopy since the late 1960s can be roughly divided into two stages. The first stage began with the advent of dye lasers, which opened up a new field in molecular spectroscopy. The tunability of dye lasers is particularly important for multiphoton processes because one can obtain an excitation source by using only a single-frequency beam rather than two or more lasers of different frequencies. During the first stage, two-photon absorption and excitation spectra of organic molecules in solid, liquid, and vapor phases were measured. *Gerade–gerade* (g–g) transitions of molecules with inversion symmetry, which are forbidden in one-photon spectroscopy, were observed. Low-lying excited ^1A_g states of linear polyenes were identified; we shall present these in Chapter 6. Mechanisms of vibronic coupling in nonresonant two-photon transitions were studied; we shall discuss these in detail in Chapter 5. Monson and McClain (1970) and McClain (1971) have reported polarization effects in the two-photon transition of randomly oriented nonrotating molecules, and polarization effects in the two-photon excitation of crystals have been theoretically investigated by Inoue and Toyozawa (1965). The polarization effects for freely rotating molecules in the gas phase have been theoretically described by Metz *et al.* (1978) and experimentally investigated by Hampf *et al.* (1977) and Wunsch *et al.* (1977). The two-photon transition probability, after averaging over the orientation, depends on the polarization of the incident laser beams (linear or circular, and so on). The one-photon transition probability in randomly oriented molecules, on the other hand, is independent of the polarization. Polarization measurements are a very convenient method for determining the symmetry of the excited states. The polarization effects of rotating and nonrotating molecules are presented in Chapter 4. Meanwhile, progress with dye lasers that generate tunable radiation of narrow bandwidth (less than 0.01 cm^{-1}) has made it possible to measure the rotational structure of the vibronic bands. Two-photon spectroscopy was somewhat restricted to the observation of low-lying excited states because the method of detecting signals from excited molecules was limited.

The second stage of development began in 1975, when a multiphoton ionization technique for detecting information from excited molecules was developed independently by Johnson and Dalby (see Johnson, 1975; Johnson *et al.*, 1975; Petty *et al.*, 1975). This method consists in collecting electrons released from the molecules after irradiation by a tunable laser pulse, amplifying the pulse, and recording the signal as a function of the laser frequency. The experimental method is presented in Chapter 3. The sensitivity of ion detection is high compared with that of absorption. The sensitivity even exceeds that of fluorescence detection if high-lying states close to the ionization threshold, which only weakly fluoresce, are investigated. However, this method is more sophisticated in interpretation because resonances of different absorption steps may overlap, giving rise to overlapping spectra. It is possible to observe Rydberg states as well as valence states. Spectroscopic properties of the excited states of a large number of molecules in various conditions have thus been clarified, and this method has become an increasingly popular area of molecular spectroscopy.

A new branch of multiphoton ionization called multiphoton ionization mass spectroscopy (MPIMS), in which a mass spectrometer is used to identify the ionization products, is now being developed (Boesl *et al.*, 1978; Zandee *et al.*, 1978; for a review see Bernstein, 1982; Schlag and Neusser, 1983). MPIMS yields important information on the dynamic behavior involving both energy relaxations and fragmentation reactions that take place in the excited states of neutral and ionic molecules. The mechanisms of MPIMS are discussed in Chapter 7. This may be regarded as the third stage in the development of multiphoton molecular spectroscopy.

1.2   CHARACTERISTIC PROPERTIES OF MULTIPHOTON TRANSITIONS

Visible/UV multiphoton transitions have several characteristic features such as laser intensity dependence, resonance enhancement, and polarization dependence; we shall study these properties in this volume. As preliminaries, intensity dependence and the resonance effect on molecular multiphoton spectroscopy are briefly outlined in this section.

1.2.1   Intensity Dependence

The multiphoton transition probability is formulated according to time-dependent perturbation theory, as can be seen in Chapter 2. As an example, let us consider a two-photon absorption from states a to n, as shown in Fig. 1.1a. The transition probability $W^{(2)}$, taking into account only the lowest

Fig. 1.1   The two-photon absorption processes of a molecule: (a) nonresonant, (b) resonant.

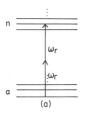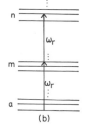

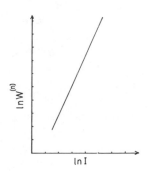

Fig. 1.2 The formal intensity law for multiphoton processes. Here I and $W^{(n)}$ denote the laser intensity and the transition probability of an n-photon process, respectively.

order term of the radiation–molecule interaction, is given as

$$W^{(2)} \propto I^2 \left| \sum_m \frac{\langle n|\boldsymbol{\mu}|m\rangle\langle m|\boldsymbol{\mu}|a\rangle}{\Delta E_{ma} - \hbar\omega_r} \right|^2, \tag{1.2.1}$$

where I is the intensity of the laser, m the virtual intermediate states, ΔE_{ma} the energy difference between the intermediate and initial states, $\boldsymbol{\mu}$ the dipole moment, and ω_r the laser frequency. Equation (1.2.1) shows that the two-photon transition probability is proportional to the square of the laser intensity. More generally, the n-photon transition probability is proportional to I^n. This is called the formal intensity law for the multiphoton transition. If no saturation occurs, one can determine the order of the multiphoton transition from the slope of the log–log plot of the transition probability as a function of laser intensity,

$$\ln W^{(n)} = n \ln I + C, \tag{1.2.2}$$

as shown in Fig. 1.2. The intensity law holds for multiphoton transitions of molecules irradiated just above the detection threshold by light from a moderately high-power laser. In multiphoton experiments in which a strong laser beam brings about saturation of the population between the relevant states, one can often see a deviation from the intensity law. This will be discussed in detail in Chapter 4.

1.2.2 Resonant Effect

When the laser is tuned and its frequency approaches a real intermediate electronic state (Fig. 1.1b), we can see a drastic increase in the two-photon absorption signal (resonance enhancement). This process is called a resonant two-photon transition. If a rigorous resonance condition were satisfied, that is, $\Delta E_{ma} = \hbar\omega_r$ in Eq. (1.2.1), then the magnitude of the transition probability would go to infinity. However, the energy levels of the intermediate states are not infinitely sharp but have widths Γ_{ma}, and the divergence of the transition can be avoided. The width originates from intra- and inter-molecular perturbations and from the higher order radiation–molecule interaction. In order to take the resonant effect into account phenomenologically, we replace the real energy denominator in Eq. (1.2.1) by a complex energy denominator with the term $i\Gamma_{ma}$. If we neglect the higher order radiation–molecule interaction, Γ_{ma} is called the dephasing constant—it describes the rate of phase loss between the m and a states associated with the transition, and it may be expressed as (see Chapter 2)

$$\Gamma_{ma} = \tfrac{1}{2}(\Gamma_{mm} + \Gamma_{aa}) + \Gamma_{ma}^{(d)}, \tag{1.2.3}$$

where Γ_{mm} and Γ_{aa} are the population decay constants of states m and a, respectively, and $\Gamma_{ma}^{(d)}$ is the pure dephasing constant that originates from a molecule–perturber elastic scattering process.

It is interesting to note that the vibronic structure appearing in the resonant multiphoton transition is generally different from that in the non-resonant transition: In the former case the vibronic structure reflects the potential differences between the initial, resonant, and final states or between these states, and in the latter case the vibronic structure is mainly determined by the Franck–Condon vibrational overlap integral between the initial and final states, since energy mismatch to the intermediate states is so large that the vibronic structure of the intermediate state $|m\rangle$ may be neglected in Eq. (1.2.1). A detailed discussion of the vibronic structure in multiphoton spectra may be found in Chapter 5.

REFERENCES

Bernstein, R. B. (1982). *J. Phys. Chem.* **86**, 1178.

Boesl, U., Neusser, H. J., and Schlag, E. W. (1978). *Z. Naturforsch., A* **334**, 1546.

Goeppert-Mayer, M. (1931). *Ann. Phys. (Leipzig)* [5] **9**, 273.

Hampf. W., Neusser, H. J., and Schlag, E. W. (1977). *Chem. Phys. Lett.* **46**, 406.

Inoue, M., and Toyozawa, Y. (1965). *J. Phys. Soc. Jpn.* **20**, 363.

Johnson, P. M. (1975). *J. Chem. Phys.* **62**, 4562.

Johnson, P. M., Berman, M. R., and Zakheim, D. (1975). *J. Chem. Phys.* **62**, 2500.

McClain, W. M. (1971). *J. Chem. Phys.* **55**, 2789.

Metz, F., Howard, W. E., Wunsch, L., Neusser, H. J., and Schlag, E. W. (1978). *Proc. R. Soc. London, Ser. A* **363**, 381.

Monson, P. R., and McClain, W. M. (1970). *J. Chem. Phys.* **53**, 29.

Petty, G., Tai, C., and Dalby, F. W. (1975). *Phys. Rev. Lett.* **34**, 1207.

Wunsch, L., Metz, F., Neusser, H. J., and Schlag, E. W. (1977). *J. Chem. Phys.* **66**, 386.

Zandee, L., Bernstein, R. B., and Lichten, D. A. (1978). *J. Chem. Phys.* **69**, 3427.

Supplementary reference books and review articles on multiphoton processes of atoms or molecules

Bernstein, R. B. (1982). *J. Phys. Chem.* **86**, 1178.
Bloembergen, N. (1979). In "Nonlinear Behaviour of Molecules, Atoms and Ions in Electric, Magnetic or Electromagnetic Fields" (L. Neel, ed.) p. 1. Elsevier, Amsterdam.
Bloembergen, N., and Levenson, M. D. (1976). *Appl. Phys.* **13**, 315.
Eberly, J., and Lambropoulos, P. (eds.) (1978). "Multiphoton Processes." Wiley, New York.
Johnson, P. M. (1980). *Acc. Chem. Res.* **13**, 20.
Johnson, P. M., and Otis, C. E. (1981). *Annu. Rev. Phys. Chem.* **32**, 139.
Lambropoulos, P. (1976). *Adv. At. Mol. Phys.* **12**, 87.
McClain, W. M. (1974). *Acc. Chem. Res.* **7**, 129.
McClain, W. M., and Harris, R. A. (1977). *Excited States* **3**, 2.
Parker, D. H., Berg, J. O., and El-Sayed, M. A. (1978). *Springer Ser. Chem. Phys.* **3**, 320.
Peticolas, W. L. (1967). *Annu. Rev. Phys. Chem.* **18**, 233.
Schlag, E. W., and Neusser, H. J. (1983). *Acc. Chem. Res.* **16**, 355.
Swofford, R. L., and Albrecht, A. C. (1978). *Annu. Rev. Phys. Chem.* **29**, 421.
Walther, H. (1976). *Top. Appl. Phys.* **2**, 1.

CHAPTER

2

Theory of Multiphoton Absorption and Ionization

In this chapter we are concerned with the derivation of expressions for the transition probabilities of multiphoton absorption and ionization processes. For this purpose several theoretical methods will be presented. Readers who are already familiar with this type of derivation or who are mainly concerned with the applications of the theory may skip this chapter.

2.1 THE PERTURBATION METHOD

In this section we shall present a systematic formulation of the theory of multiphoton processes by using the time-dependent perturbation method. This method is well known and can be found in standard textbooks (Schiff, 1955) and reference books (Heitler, 1954; Louisell, 1973; Sargent et al., 1974). First we shall briefly outline the time-dependent perturbation method, and then we shall show how to apply this method to multiphoton processes.

2.1.1   Time-Dependent Perturbation Theory

To solve time-dependent problems we need to be able to calculate the time evolution of the system caused by a perturbation. Thus we have to solve the wave equation, which expresses the manner in which the complete wave function Ψ changes with time,

$$\hat{H}\Psi = i\hbar(\partial\Psi/\partial t). \tag{2.1.1}$$

Let us now write the Hamiltonian operator as

$$\hat{H} = \hat{H}_0 + \lambda V, \tag{2.1.2}$$

where V is the perturbation and λ denotes the perturbation parameter. The unperturbed eigenfunctions Ψ_n^0 satisfy the equation

$$\hat{H}_0\Psi_n^0 = i\hbar(\partial\Psi_n^0/\partial t) \tag{2.1.3}$$

and are of the form

$$\Psi_n^0 = \psi_n \exp(-itE_n/\hbar), \tag{2.1.4}$$

where

$$\hat{H}_0\psi_n = E_n\psi_n. \tag{2.1.5}$$

In order to obtain a solution of Eq. (2.1.1) we expand the wave function Ψ in terms of the unperturbed basis set Ψ_n^0 (i.e., using the expansion theorem):

$$\Psi = \sum_n C_n(t)\Psi_n^0. \tag{2.1.6}$$

Substituting Eq. (2.1.6) into Eq. (2.1.1) gives (Eyring *et al.*, 1944)

$$i\hbar \frac{\partial C_n}{\partial t} = \lambda \sum_m C_m \langle \Psi_n^0 | V | \Psi_m^0 \rangle. \tag{2.1.7}$$

According to the usual practice of perturbation theory, we set

$$C_n = C_n^{(0)} + \lambda C_n^{(1)} + \lambda^2 C_n^{(2)} + \cdots. \tag{2.1.8}$$

Substituting Eq. (2.1.8) into Eq. (2.1.7) and equating the coefficients of λ^n, we find

$$\frac{dC_n^{(0)}}{dt} = 0, \tag{2.1.9}$$

$$i\hbar \frac{dC_n^{(1)}}{dt} = \sum_m C_m^{(0)} \langle \Psi_n^0 | V | \Psi_m^0 \rangle, \tag{2.1.10}$$

$$i\hbar \frac{dC_n^{(2)}}{dt} = \sum_m C_m^{(1)} \langle \Psi_n^0 | V | \Psi_m^0 \rangle, \tag{2.1.11}$$

$$ih\frac{dC_n^{(3)}}{dt} = \sum_m C_m^{(2)}\langle\Psi_n^0|V|\Psi_m^0\rangle, \tag{2.1.12}$$

and so forth.

Equation (2.1.7) indicates that in any particular problem we shall have a set of simultaneous differential equations that can be solved to give explicit expressions for the coefficients C_n. From Eqs. (2.1.8)–(2.1.12) we can see that owing to the use of the perturbation method, these simultaneous equations are decoupled; we can solve Eqs. (2.1.9)–(2.1.12) successively. As will be shown later, the use of the ordinary perturbation method cannot take into account the damping effect (Heitler, 1954). The damping effect can be treated by using the so-called singular perturbation method (Lee *et al.* 1973; Lee and Lee, 1973) and the Green's function method described in Section 2.2.

To solve the zeroth-order equation given by Eq. (2.1.9) we need the initial condition. Suppose that the system is initially in the k state. Then Eq. (2.1.9) yields

$$C_k^{(0)} = 1, \qquad C_m^{(0)} = 0. \tag{2.1.13}$$

By substituting Eq. (2.1.13) into the first-order equation given by Eq. (2.1.10), we find

$$ih\,(dC_n^{(1)}/dt) = \langle\Psi_n^0|V|\Psi_k^0\rangle. \tag{2.1.14}$$

For the case in which V is time independent, Eq. (2.1.14) can be integrated as

$$C_n^{(1)} = (V_{nk}/\hbar\omega_{nk})(1 - e^{it\omega_{nk}}), \tag{2.1.15}$$

where $\omega_{nk} = (E_n - E_k)/\hbar$.

Substituting Eq. (2.1.15) into Eq. (2.1.11) and carrying out the integration give

$$C_n^{(2)} = \sum_m \frac{V_{nm}V_{mk}}{\hbar\omega_{mk}}\left(\frac{1 - e^{it\omega_{nm}}}{\hbar\omega_{nm}} - \frac{1 - e^{it\omega_{nk}}}{\hbar\omega_{nk}}\right). \tag{2.1.16}$$

Similarly, we obtain the third-order result as

$$C_n^{(3)} = \sum_l \sum_m \frac{V_{nm}V_{ml}V_{lk}}{\hbar\omega_{lk}}\left[\frac{1}{\hbar\omega_{ml}}\left(\frac{1 - e^{it\omega_{nm}}}{\hbar\omega_{nm}} - \frac{1 - e^{it\omega_{nl}}}{\hbar\omega_{nl}}\right)\right.$$
$$\left. - \frac{1}{\hbar\omega_{mk}}\left(\frac{1 - e^{it\omega_{nm}}}{\hbar\omega_{nm}} - \frac{1 - e^{it\omega_{nk}}}{\hbar\omega_{nk}}\right)\right]. \tag{2.1.17}$$

Let us now derive the expressions for the transition probabilities. In the first-order approximation, for the transition $k \to n$ we have

$$|C_n^{(1)}|^2 = (|V_{nk}|^2/(\hbar\omega_{nk})^2)(2 - 2\cos\omega_{nk}t). \tag{2.1.18}$$

In the long-time limit ($t \to \infty$) we may use the relation (Heitler, 1954)

$$\delta(\omega_{nk}) = \frac{1}{\pi} \lim_{t \to \infty} \frac{1}{t} \frac{1 - \cos \omega_{nk} t}{\omega_{nk}^2}, \tag{2.1.19}$$

where $\delta(\omega_{nk})$ is the delta function, to rewrite Eq. (2.1.18) as

$$|C_n^{(1)}|^2 = \frac{2\pi t}{\hbar^2} |V_{nk}|^2 \delta(\omega_{nk}) = \frac{2\pi t}{\hbar} |V_{nk}|^2 \delta(E_n - E_k). \tag{2.1.20}$$

The transition probability per unit time is then given by

$$W_{k \to n}^{(1)} = \frac{|C_n^{(1)}|^2}{t} = \frac{d}{dt} |C_n^{(1)}|^2 = \frac{2\pi}{\hbar^2} |V_{nk}|^2 \delta(\omega_{nk})$$

$$= \frac{2\pi}{\hbar} |V_{nk}|^2 \delta(E_n - E_k) \tag{2.1.21}$$

for the transition $k \to n$. Normally, what one observes is not the detailed transition probability $W_{k \to n}^{(1)}$, but either the so-called single-level rate constant

$$W_k^{(1)} = \sum_n W_{k \to n}^{(1)} = \frac{2\pi}{\hbar} \sum_n |V_{nk}|^2 \delta(E_n - E_k) \tag{2.1.22}$$

or the average rate constant

$$W^{(1)} = \sum_k P_k W_k^{(1)} = \frac{2\pi}{\hbar} \sum_n \sum_k P_k |V_{nk}|^2 \delta(E_n - E_k), \tag{2.1.23}$$

where P_k represents the weighting factor of the initial state k. For a thermal system P_k is the Boltzmann factor, and for an isolated system P_k represents the microcanonical distribution; detailed discussions can be found in the literature (Lin, 1972, 1973). When a continuum is involved, the summations that appear in Eqs. (2.1.22) and (2.1.23) should be replaced by integrals; for example,

$$\sum_n \to \int \rho(E)\, dE, \tag{2.1.24}$$

where $\rho(E)$ denotes the density of states. Each of the transition probability expressions $W_{k \to n}^{(1)}$, $W_k^{(1)}$, and $W^{(1)}$ is usually called Fermi's golden rule.

Next we shall derive an expression for the transition probability in the second-order approximation. In the long-time limit, only the second term (i.e., the term with ω_{nk}) makes a dominant contribution:

$$|C_n^{(2)}|^2 = \frac{2 - 2\cos \omega_{nk} t}{(\hbar \omega_{nk})^2} \left| \sum_m \frac{V_{nm} V_{mk}}{\hbar \omega_{nk}} \right|^2. \tag{2.1.25}$$

It follows that

$$W^{(2)}_{k \to n} = \frac{d}{dt} |C^{(2)}_n|^2 = \frac{2\pi}{\hbar^2} \left| \sum_m \frac{V_{nm} V_{mk}}{\hbar \omega_{mk}} \right|^2 \delta(\omega_{nk})$$

$$= \frac{2\pi}{\hbar} \left| \sum_m \frac{V_{nm} V_{mk}}{\hbar \omega_{mk}} \right|^2 \delta(E_n - E_k), \tag{2.1.26}$$

$$W^{(2)}_k = \frac{2\pi}{\hbar} \sum_n \left| \sum_m \frac{V_{nm} V_{mk}}{\hbar \omega_{mk}} \right|^2 \delta(E_n - E_k), \tag{2.1.27}$$

and

$$W^{(2)} = \frac{2\pi}{\hbar} \sum_k \sum_n P_k \left| \sum_m \frac{V_{nm} V_{mk}}{\hbar \omega_{mk}} \right|^2 \delta(E_n - E_k). \tag{2.1.28}$$

The transition probability in the third-order approximation can be treated similarly:

$$|C^{(3)}_n|^2 = \frac{2 - 2 \cos \omega_{nk} t}{(\hbar \omega_{nk})^2} \left| \sum_l \sum_m \frac{V_{nm} V_{ml} V_{lk}}{(\hbar \omega_{lk})(\hbar \omega_{mk})} \right|^2, \tag{2.1.29}$$

$$W^{(3)}_{k \to n} = \frac{2\pi}{\hbar} \left| \sum_l \sum_m \frac{V_{nm} V_{ml} V_{lk}}{(\hbar \omega_{lk})(\hbar \omega_{mk})} \right|^2 \delta(E_n - E_k), \tag{2.1.30}$$

$$W^{(3)}_k = \frac{2\pi}{\hbar} \sum_k \left| \sum_l \sum_m \frac{V_{nm} V_{ml} V_{lk}}{(\hbar \omega_{lk})(\hbar \omega_{nk})} \right|^2 \delta(E_n - E_k), \tag{2.1.31}$$

and

$$W^{(3)} = \frac{2\pi}{\hbar} \sum_k \sum_n P_k \left| \sum_l \sum_m \frac{V_{nm} V_{ml} V_{lk}}{(\hbar \omega_{lk})(\hbar \omega_{mk})} \right|^2 \delta(E_n - E_k). \tag{2.1.32}$$

Although these expressions for the transition probabilities will be mainly applied in this volume to the absorption and emission of radiation by molecules, they are general and can be used to calculate the rate constants of various relaxation processes (Levine, 1969; Fong, 1975).

2.1.2 Interaction between a Molecule and the Radiation Field

In Section 2.1.1 we derived expressions for calculating the transition probabilities of rate processes. To apply them to the absorption and emission of radiation by molecules (or atoms), we need to know the interaction V between the molecule (or atom) and the radiation field; this will be discussed in this section. We shall present a simple treatment; for a rigorous treatment the specialized references should be consulted (Heitler, 1954; Louisell, 1973; Sargent et al., 1974).

In deriving the classical Hamiltonian, it is more convenient to use the vector potential **A** and the scalar potential ϕ rather than the electric and magnetic fields **E** and **H**. The relations among these quantities are given by

$$\mathbf{H} = \nabla \times \mathbf{A} \tag{2.1.33}$$

and

$$\mathbf{E} = -\frac{1}{c}\frac{\partial \mathbf{A}}{\partial t} - \nabla\phi, \tag{2.1.34}$$

where c is the velocity of light. A particle of mass m and charge e moving with a velocity **V** in an electromagnetic field is subjected to a force

$$\mathbf{F} = e[\mathbf{E} + 1/c(\mathbf{V} \times \mathbf{H})]. \tag{2.1.35}$$

The equations of motion are then given by

$$m\ddot{x} = m\frac{d^2x}{dt^2} = -e\frac{\partial\phi}{\partial x} - \frac{e}{c}\frac{\partial A_x}{\partial t} + \frac{e}{c}(\dot{y}H_z - \dot{z}H_y), \tag{2.1.36a}$$

with similar expressions for $\ddot{y}$ and $\ddot{z}$.

Using Eq. (2.1.33), Eq. (2.1.36) becomes

$$m\ddot{x} = -e\frac{\partial\phi}{\partial x} - \frac{e}{c}\frac{\partial A_x}{\partial t} + \frac{e}{c}\left[\dot{y}\left(\frac{\partial A_y}{\partial x} - \frac{\partial A_x}{\partial y}\right) + \dot{z}\left(\frac{\partial A_z}{\partial x} - \frac{\partial A_x}{\partial z}\right)\right].$$
$$\tag{2.1.36b}$$

It is not difficult to verify that these equations of motion can be derived from the Lagrangian

$$L = \tfrac{1}{2}m(\dot{x}^2 + \dot{y}^2 + \dot{z}^2) + (e/c)(\mathbf{V} \cdot \mathbf{A}) - e\phi \tag{2.1.37}$$

and Lagrange's equations of motion

$$\frac{d}{dt}\frac{\partial L}{\partial\dot{x}} - \frac{\partial L}{\partial x} = 0. \tag{2.1.38}$$

From the definition of generalized momentum, $p_i = \partial L/\partial\dot{q}_i$, we obtain

$$p_x = m\dot{x} + (e/c)A_x, \tag{2.1.39}$$

with analogous values for p_y and p_z. The Hamiltonian is therefore

$$H = p_x\dot{x} + p_y\dot{y} + p_z\dot{z} - L = \tfrac{1}{2}m(\dot{x}^2 + \dot{y}^2 + \dot{z}^2) + e\phi, \tag{2.1.40}$$

or in terms of coordinates and momenta,

$$H = \frac{1}{2m}\left[\left(p_x - \frac{e}{c}A_x\right)^2 + \left(p_y - \frac{e}{c}A_y\right)^2 + \left(p_z - \frac{e}{c}A_z\right)^2\right] + e\phi. \tag{2.1.41}$$

The procedure for constructing the quantum-mechanical Hamiltonian operator is to replace the momentum p_x in the classical Hamiltonian function by $-i\hbar(\partial/\partial x)$, and so on. After collecting terms and expressing the results in vector notation, we see that the Hamiltonian operator is given by

$$\hat{H} = \frac{1}{2m}\left(-\hbar^2\,\nabla^2 + \frac{i\hbar e}{c}\,\nabla\cdot\mathbf{A} + \frac{2i\hbar e}{c}\,\mathbf{A}\cdot\nabla + \frac{e^2}{c^2}\,|\mathbf{A}|^2\right) + e\phi. \quad (2.1.42)$$

For an electromagnetic field like that associated with a light wave $\nabla\cdot\mathbf{A} = 0$ and $\phi = 0$, and therefore for a system of charged particles with an internal potential energy u, we shall have the Hamiltonian operator

$$\hat{H} = \hat{H}_0 + V, \quad (2.1.43)$$

where

$$\hat{H}_0 = -\sum_j \frac{\hbar^2}{2m_j}\,\nabla_j^2 + u, \quad (2.1.44)$$

$$V = -\sum_j \frac{e_j}{m_j c}\,\mathbf{A}_j\cdot\mathbf{p}_j + \sum_j \frac{e_j^2}{2m_j c^2}\,|\mathbf{A}_j|^2, \quad (2.1.45)$$

where $\mathbf{p}_j$ is the momentum operator of the ith particle.

In a semiclassical treatment of the interaction between the molecule and the radiation field one regards $\mathbf{A}$ as a function of position and time. For example, for a plane-polarized wave propagating along the z direction with its polarization in the x direction, we have

$$\mathbf{A} = \hat{i}A_0\cos\omega\left(t - \frac{z}{c}\right). \quad (2.1.46)$$

The associated electric and magnetic fields are then

$$\mathbf{E} = -(1/c)\partial\mathbf{A}/\partial t = \hat{i}\frac{\omega}{c}\,A_0\sin\omega\left(t - \frac{z}{c}\right), \quad (2.1.47)$$

$$\mathbf{H} = \nabla\times\mathbf{A} = \hat{j}\frac{\omega}{c}\,A_0\sin\omega\left(t - \frac{z}{c}\right). \quad (2.1.48)$$

The electric and magnetic fields therefore have equal amplitudes and are at right angles to each other, as well as to the propagation direction, and hence represent a plane-polarized light wave moving along the z axis with velocity c. In the dipole approximation the term z/c in Eqs. (2.1.46)–(2.1.48) can be ignored. This term is important in magnetic and quadrupole radiation and in treating natural optical activity, and in such problems the vector potential $\mathbf{A}$ has to be expanded in a power series in z/c (Eyring et al., 1944).

In the quantum-mechanical treatment of radiation (Heitler, 1954; Louisell, 1973; Sargent et $al.$, 1974) the electromagnetic field is regarded as an ensemble of noninteracting harmonic oscillator modes, with the kth mode containing n_k photons, all of energy $\hbar\omega_k$, polarization $\hat{e}_k$, and propagation $\pm\mathbf{k}_k$. The propagation vector has magnitude

$$|\mathbf{k}_k| = \omega_k/c \qquad (2.1.49)$$

and is perpendicular to the polarization vector

$$\hat{e}_k \cdot \mathbf{k}_k = 0. \qquad (2.1.50)$$

The Hamiltonian of the ensemble for the radiation field is (Heitler, 1954; Lin, 1967)

$$\hat{H}_{\text{rad}} = \sum_k \hat{H}_k, \qquad (2.1.51)$$

where

$$\hat{H}_k = 2\omega_k^2 q_k q_k^* \qquad (2.1.52)$$

and q_k and q_k^* are the time-dependent complex oscillator amplitudes. The eigenfunctions for the unperturbed field are written in the product form

$$\Psi_{\text{rad}}(q_1, q_2, \ldots) = U_{n_1}(q_1)U_{n_2}(q_2)\cdots, \qquad (2.1.53)$$

where there are n_1 photons of frequency ω_1, polarization $\hat{e}_1$, and propagation $\pm\mathbf{k}_1$, and so on. The individual oscillator functions satisfy

$$\hat{H}_k U_{n_k}(q_k) = n_k\hbar\omega_k U_{n_k}(q_k), \qquad (2.1.54)$$

and the total energy is then

$$E_{\text{rad}} = \sum_k n_k\hbar\omega_k, \qquad (2.1.55)$$

where the sum is over all modes in the ensemble. The eigenfunction of the radiation field should be symmetric under exchange of any two coordinates q_k and q_l, since photons obey Bose–Einstein statistics. The functions $U_n(q)$ are the well-known harmonic oscillator eigenfunctions having the properties

$$\langle U_n | U_m \rangle = \delta_{nm}, \qquad (2.1.56)$$

$$qU_n = (n\hbar/2\omega)^{1/2}U_{n-1}, \qquad (2.1.57)$$

$$q^*U_n = ((n+1)\hbar/2\omega)^{1/2}U_{n+1}. \qquad (2.1.58)$$

It is convenient to use photon annihilation and creation operators defined by

$$\hat{a} = (2\omega/\hbar)^{1/2}q \qquad (2.1.59)$$

and

$$\hat{a}^\dagger = (2\omega/\hbar)^{1/2}q^*, \qquad (2.1.60)$$

respectively. In terms of $\hat{a}$ and $\hat{a}^{\dagger}$, we have

$$\hat{a}U_n = \sqrt{n}\, U_{n-1}, \tag{2.1.61}$$

$$\hat{a}^{\dagger}U_n = \sqrt{n+1}\, U_{n+1}, \tag{2.1.62}$$

$$\hat{H}_k = \hbar\omega_k \hat{a}_k^{\dagger}\hat{a}_k. \tag{2.1.63}$$

The properties of the electromagnetic field are described in terms of the vector potential $\mathbf{A}$, which satisfies the equations (Heitler, 1954; Louisell, 1973; Sargent *et al.*, 1974; Lin, 1967)

$$\nabla^2\mathbf{A} - \frac{1}{c^2}\frac{\partial^2\mathbf{A}}{\partial t^2} = 0 \tag{2.1.64}$$

and

$$\nabla \cdot \mathbf{A} = 0. \tag{2.1.65}$$

These can easily be obtained from Maxwell's equations. The reader may verify that the expression for $\mathbf{A}$ given by Eq. (2.1.46) indeed satisfies Eqs. (2.1.64) and (2.1.65). With the normalization condition

$$\langle \mathbf{A}_k | \mathbf{A}_l \rangle = (4\pi c^2/L^3)\,\delta_{kl}, \tag{2.1.66}$$

where L is the length of a cubic box, we have

$$\mathbf{A}_k = (4\pi c^2/L^3)^{1/2}\hat{e}_k \exp(i\mathbf{k}_k \cdot \mathbf{r}) \tag{2.1.67}$$

and

$$
\begin{aligned}
\mathbf{A} &= \sum_k (q_k\mathbf{A}_k + q_k^*\mathbf{A}_k^*) \\
&= \left(\frac{4\pi c^2}{L^3}\right)^{1/2} \sum_k \hat{e}_k(q_k e^{i\mathbf{k}_k\mathbf{r}} + q_k^* e^{-i\mathbf{k}_k\cdot\mathbf{r}}) \\
&= \sum_k e_k \left(\frac{2\pi\hbar c^2}{\omega_k L^3}\right)^{1/2}(\hat{a}_k e^{i\mathbf{k}_k\cdot\mathbf{r}} + \hat{a}_k^{\dagger} e^{-\mathbf{k}_k\cdot\mathbf{r}}),
\end{aligned}
\tag{2.1.68}
$$

where the sum is over all modes in the radiation field. The interaction between the molecule and the electromagnetic field is due to the coupling of the vector potential $\mathbf{A}$ with the moving charged particles (electrons and nuclei) as given by Eq. (2.1.45). In practice, the interaction operator is usually simplified by making the following approximation: The nuclear term in Eq. (2.1.45) is generally neglected altogether and the electric dipole approximation is made. The electric dipole approximation, as mentioned before, requires that $|\mathbf{k}||\mathbf{r}|$ be much less than unity so that the terms $\exp(\pm i\mathbf{k}\cdot\mathbf{r})$ in $\mathbf{A}$ can be approximated by the first term of the series

$$\exp(\pm i\mathbf{k}\cdot\mathbf{r}) = 1 + (\pm i\mathbf{k}\cdot\mathbf{r})^2 + \tfrac{1}{2}(\pm i\mathbf{k}\cdot\mathbf{r})^2 + \cdots. \tag{2.1.69}$$

In this approximation, the A^2 terms in Eq. (2.1.45) do not contribute to such inelastic processes as Raman scattering and multiphoton absorption owing to orthogonality within the set of zeroth-order molecular eigenstates.

By substituting Eq. (2.1.68) into Eq. (2.1.45), we obtain

$$V = -\sum_j \sum_k (\hat{e}_k \cdot \mathbf{p}_j) \frac{e_j}{m_j} \left(\frac{2\pi\hbar}{\omega_k L^3} \right)^{1/2} (\hat{a}_k e^{i\mathbf{k}_k \cdot \mathbf{r}_j} + \hat{a}_k^\dagger e^{-i\mathbf{k}_k \cdot \mathbf{r}_j}). \quad (2.1.70)$$

Here the A^2 terms have been ignored. Using the approximation given by Eq. (2.1.69), Eq. (2.1.70) becomes

$$V = -\sum_j \sum_k (\hat{e}_k \cdot \mathbf{p}_j) \frac{e_j}{m_j} \left(\frac{2\pi\hbar}{\omega_k L^3} \right)^{1/2} (\hat{a}_k + \hat{a}_k^\dagger). \quad (2.1.71)$$

This equation for V includes the contributions from both electrons and nuclei; in practice, the nuclear terms are neglected altogether, that is,

$$V = -\sum_k (\hat{e}_k \cdot \mathbf{P}) \frac{e}{m} \left(\frac{2\pi\hbar}{\omega_k L^3} \right)^{1/2} (\hat{a}_k^\dagger + \hat{a}_k), \quad (2.1.72)$$

where

$$\mathbf{P} = \sum_j \mathbf{P}_j, \quad (2.1.73)$$

the total linear momentum operator of the electrons, and e and m represent the charge and mass of an electron, respectively.

2.1.3   Expressions for Transition Probabilities I

We are now ready to derive the expressions for photon absorption. We first consider one-photon absorption. In this case the initial I and final F states are given by

$$|I\rangle \equiv |\varepsilon_i, n_l \hbar \omega_l\rangle \quad (2.1.74)$$

and

$$|F\rangle \equiv |\varepsilon_f, (n_l - 1)\hbar \omega_l\rangle. \quad (2.1.75)$$

Using Eqs. (2.1.72)–(2.1.75) and (2.1.21), we obtain

$$W^{(1)}_{I \to F} = (2\pi/\hbar) |V_{FI}|^2 \delta(E_F - E_I)$$
$$= (2\pi/\hbar^2) |V_{FI}|^2 \delta(\omega_F - \omega_I) \quad (2.1.76)$$

or

$$W^{(1)}_{i \to f} = (4\pi^2 e^2 n_l / m^2 \hbar \omega_l L^3) |\hat{e}_l \cdot \mathbf{P}_{fi}|^2 \delta(\omega_{fi} - \omega_l), \quad (2.1.77)$$

where $\omega_{fi} = (\varepsilon_f - \varepsilon_i)/\hbar$.

Using the relation (Eyring *et al.*, 1944)

$$\mathbf{P}_{fi} = (im\omega_{fi}/e)\boldsymbol{\mu}_{fi} = im\omega_{fi}\mathbf{R}_{fi}, \qquad (2.1.78)$$

where $\boldsymbol{\mu}_{fi}$ denotes the dipole transition moment, Eq. (2.1.77) becomes

$$W^{(1)}_{i\to f} = (4\pi^2 n_l\omega_l/\hbar L^3)|\hat{e}_l \cdot \boldsymbol{\mu}_{fi}|^2\,\delta(\omega_{fi} - \omega_l). \qquad (2.1.79)$$

This gives the probability per second for the absorption of a photon by the molecule when the radiation field is perfectly monochromatic and plane polarized. We have assumed that the molecular energy levels have no linewidth (no damping), so that only if $\omega_l = \omega_{fi}$ (energy conservation) will the molecule be able to absorb a photon and be excited. If $n_l = 0$, the molecule will not be excited. The probability is linearly proportional to $\omega_{fi} = \omega_l$. The experimentally observed transition probability is obtained by multiplying $W^{(1)}_{i\to f}$ by the initial distribution P_i and summing over all possible final states as

$$W^{(1)} = \frac{4\pi^2 n_l\omega_l}{\hbar L^3}\sum_i\sum_f P_i|\hat{e}_l \cdot \boldsymbol{\mu}_{fi}|^2\,\delta(\omega_{fi} - \omega_l). \qquad (2.1.80)$$

The corresponding absorption coefficient, with the light intensity measured in number of photons $n_l c/L^3$, is given by

$$k^{(1)}_{\text{abs}} = \frac{4\pi^2\omega_l}{\hbar c}\sum_i\sum_f P_i|\hat{e}_l \cdot \boldsymbol{\mu}_{fi}|^2\,\delta(\omega_{fi} - \omega_l). \qquad (2.1.81)$$

If the molecules are randomly oriented, Eq. (2.1.81) should be averaged over the molecular orientation. We find

$$k^{(1)}_{\text{abs}} = \frac{4\pi^2\omega_l}{3\hbar c}\sum_i\sum_f P_i|\boldsymbol{\mu}_{fi}|^2\,\delta(\omega_{fi} - \omega_l). \qquad (2.1.82)$$

Here the relation $\langle|\hat{e}_l \cdot \boldsymbol{\mu}_{fi}|^2\rangle_{\text{av}} = \frac{1}{3}|\boldsymbol{\mu}_{fi}|^2$ has been used.

One-photon emission can be treated similarly. In this case the initial and final states are given by

$$|I\rangle \equiv |\varepsilon_i, n_l\hbar\omega_l\rangle \qquad (2.1.74)$$

and

$$|F\rangle \equiv |\varepsilon_f, (n_l + 1)\hbar\omega_l\rangle. \qquad (2.1.83)$$

Substituting Eqs. (2.1.74) and (2.1.83) into Eq. (2.1.76) or (2.1.21) yields

$$W^{(1)}_{i\to f} = \frac{4\pi^2 e^2(n_l + 1)}{m^2\hbar\omega_l L^3}|\hat{e}_l \cdot \mathbf{P}_{fi}|^2\,(\omega_{if} - \omega_l). \qquad (2.1.84)$$

The n_l term represents the induced emission. For spontaneous emission we may set $n_l = 0$; since it may be emitted in all directions and in all polarizations, we must sum over all modes and both polarizations:

$$W^{(1)}_{i \to f} = \sum_l \frac{4\pi^2 e^2}{m^2 \hbar \omega_l L^3} |\hat{e}_l \cdot \mathbf{P}_{fi}|^2 \, \delta(\omega_{if} - \omega_l) \qquad (2.1.85)$$

or

$$W^{(1)}_{i \to f} = \sum_l \frac{4\pi^2 \omega_l}{\hbar L^3} |\hat{e}_l \cdot \boldsymbol{\mu}_{fi}|^2 \, \delta(\omega_{if} - \omega_l). \qquad (2.1.86)$$

Using the relation (Appendix I, Section I.A)

$$\sum_l \to \sum_{k_l} \to \frac{L^3}{8\pi^3 c^3} \int_0^\infty d\omega_k \, \omega_k^2 \int_{\Omega_k} d\Omega_k, \qquad (2.1.87)$$

Eq. (2.1.86) becomes

$$W^{(1)}_{i \to f} = (4\omega_{if}^3 / 3\hbar c^3) |\boldsymbol{\mu}_{fi}|^2. \qquad (2.1.88)$$

To obtain the observed radiative rate constant from Eq. (2.1.88), we need to multiply Eq. (2.1.88) by the initial distribution and sum over the final states as

$$W^{(1)} = \sum_i \sum_f \frac{4\omega_{if}^3}{3\hbar c^3} P_i |\boldsymbol{\mu}_{fi}|^2. \qquad (2.1.89)$$

In general, including the induced (or stimulated) and spontaneous emissions, we have

$$W^{(1)}_{i \to f} = \sum_l \frac{4\pi^2 \omega_l}{\hbar L^3} (n_l + 1) |\hat{e}_l \cdot \boldsymbol{\mu}_{fi}|^2 \, \delta(\omega_{if} - \omega_l)$$

$$= \frac{4\pi^2 \omega_l}{\hbar c} |\hat{e}_l \cdot \boldsymbol{\mu}_{fi}|^2 \left(I(\omega_l) + \frac{\omega_l^2}{2\pi^2 c^2} \right) \qquad (2.1.90)$$

for each polarization, where $\omega_l = \omega_{if}$. Here Eq. (2.1.87) and the following relation have been used:

$$I(\omega_l) = \frac{\omega_l^2}{8\pi^3 c^2} \int_{\Omega_k} d\Omega_k \, n(\omega_l), \qquad (2.1.91)$$

where $I(\omega_l)$ represents the intensity distribution. The total intensity in units of photon number is given by

$$I_{\text{tot}} = \int_0^\infty d\omega_l \, I(\omega_l). \qquad (2.1.92)$$

In Eq. (2.1.90), the term involving $I(\omega_l)$ represents the stimulated emission transition probability.

In the theory of the molecular emission and absorption of radiation presented in this section and Section 2.1.4, we assume that the molecular energy levels are infinitely sharp (i.e., delta functions), whereas we know from experiment that the observed emission and absorption lines have a finite width. This so-called damping effect will be included in the derivation of absorption and emission transition probabilities when we present the Green's function and density matrix methods.

2.1.4 Expressions for Transition Probabilities II

In Section 2.1.3 we derived the expressions of the transition probabilities for one-photon absorption and emission processes. In this section we shall be concerned with multiphoton processes. We first consider a two-photon process. For this purpose, we use Eq. (2.1.26). For two-photon absorption, the initial and final states are given by

$$|I\rangle \equiv |\varepsilon_i, n_l\hbar\omega_l, n_{l'}\hbar\omega_{l'}\rangle \tag{2.1.93}$$

and

$$|F\rangle \equiv |\varepsilon_f, (n_l - 1)\hbar\omega_l, (n_{l'} - 1)\hbar\omega_{l'}\rangle. \tag{2.1.94}$$

There are two possible intermediate states,

$$|M\rangle \equiv |\varepsilon_m, (n_l - 1)\hbar\omega_l, n_{l'}\hbar\omega_{l'}\rangle \tag{2.1.95}$$

and

$$|M\rangle \equiv |\varepsilon_m, n_l\hbar\omega_l, (n_{l'} - 1)\hbar\omega_{l'}\rangle. \tag{2.1.96}$$

Substituting Eqs. (2.1.93)–(2.1.96) into Eq. (2.1.26) and using Eq. (2.1.72), we obtain

$$W^{(2)}_{I \to F} = \frac{2\pi}{\hbar} \left| \sum_M \frac{V_{FM}V_{MI}}{E_M - E_I} \right|^2 \delta(E_F - E_I)$$

$$= \frac{2\pi}{\hbar^2} \left(\frac{2\pi e^2}{m^2 L^3} \right)^2 \frac{n_l n_{l'}}{\omega_l \omega_{l'}} |M^{(2)}_{fi}(\omega_l, \omega_{l'})|^2 \delta(\omega_{fi} - \omega_l - \omega_{l'}) \tag{2.1.97}$$

or

$$W^{(2)}_{i \to f} = \frac{2\pi}{\hbar^2} \left(\frac{2\pi e^2}{m^2 L^3} \right)^2 \sum_l \sum_{l'} \frac{n_l n_{l'}}{\omega_l \omega_{l'}} |M^{(2)}_{fi}(\omega_l, \omega_{l'})|^2 \delta(\omega_{fi} - \omega_l - \omega_{l'}), \tag{2.1.98}$$

where

$$M^{(2)}_{fi}(\omega_l, \omega_{l'}) = \sum_m \left(\frac{(\hat{e}_{l'} \cdot \mathbf{P}_{fm})(\hat{e}_l \cdot \mathbf{P}_{mi})}{\omega_{mi} - \omega_l} + \frac{(\hat{e}_l \cdot \mathbf{P}_{fm})(\hat{e}_{l'} \cdot \mathbf{P}_{mi})}{\omega_{mi} - \omega_{l'}} \right) \tag{2.1.99}$$

and $\omega_{fi} = \omega_l + \omega_{l'} = \omega_{fm} + \omega_{mi}$. We shall express $M^{(2)}_{fi}(\omega_l, \omega_{l'})$ in another form. Notice that by using Eq. (2.1.78) we have

$$M^{(2)}_{fi}(\omega_l, \omega_{l'}) = \frac{im}{e} \sum_m \left[\frac{\omega_{mi}(\hat{e}_{l'} \cdot \mathbf{P}_{fm})(\hat{e}_l \cdot \boldsymbol{\mu}_{mi})}{\omega_{mi} - \omega_l} + \frac{\omega_{fm}(\hat{e}_l \cdot \boldsymbol{\mu}_{fm})(\hat{e}_{l'} \cdot \mathbf{P}_{mi})}{\omega_l - \omega_{fm}} \right] \qquad (2.1.100)$$

or

$$M^{(2)}_{fi}(\omega_l, \omega_{l'}) = \frac{im}{e} \sum_m \left[(\hat{e}_{l'} \cdot \mathbf{P}_{fm})(\hat{e}_l \cdot \boldsymbol{\mu}_{mi}) - (\hat{e}_l \cdot \boldsymbol{\mu}_{fm})(\hat{e}_{l'} \cdot \mathbf{P}_{mi}) \right.$$

$$\left. + \omega_l \left(\frac{(\hat{e}_{l'} \cdot \mathbf{P}_{fm})(\hat{e}_l \cdot \boldsymbol{\mu}_{mi})}{\omega_{mi} - \omega_l} + \frac{(\hat{e}_l \cdot \boldsymbol{\mu}_{fm})(\hat{e}_{l'} \cdot \mathbf{P}_{mi})}{\omega_l - \omega_{fm}} \right) \right]. \qquad (2.1.101)$$

It can easily be shown that the first two terms on the right-hand side of Eq. (2.1.101) vanish after using the closure relation. It follows that

$$M^{(2)}_{fi}(\omega_l, \omega_{l'}) = \frac{im\omega_l}{e} \sum_m \left[\frac{(\hat{e}_{l'} \cdot \mathbf{P}_{fm})(\hat{e}_l \cdot \boldsymbol{\mu}_{mi})}{\omega_{mi} - \omega_l} + \frac{(\hat{e}_l \cdot \boldsymbol{\mu}_{fm})(\hat{e}_{l'} \cdot \mathbf{P}_{mi})}{\omega_l - \omega_{fm}} \right]. \qquad (2.1.102)$$

Repeating this process, we find

$$M^{(2)}_{fi}(\omega_l, \omega_{l'}) = \omega_l \omega_{l'} \left(\frac{im}{e} \right)^2 \sum_m \left[\frac{(\hat{e}_{l'} \cdot \boldsymbol{\mu}_{fm})(\hat{e}_l \cdot \boldsymbol{\mu}_{mi})}{\omega_{mi} - \omega_l} + \frac{(\hat{e}_l \cdot \boldsymbol{\mu}_{fm})(\hat{e}_{l'} \cdot \boldsymbol{\mu}_{mi})}{\omega_l - \omega_{fm}} \right]$$

$$= \omega_l \omega_{l'} (im)^2 S_{fi}(\omega_l, \omega_{l'}), \qquad (2.1.103)$$

where

$$S_{fi}(\omega_l, \omega_{l'}) = \sum_m \left[\frac{(\hat{e}_{l'} \cdot \mathbf{R}_{fm})(\hat{e}_l \cdot \mathbf{R}_{mi})}{\omega_{mi} - \omega_l} + \frac{(\hat{e}_l \cdot \mathbf{R}_{fm})(\hat{e}_{l'} \cdot \mathbf{R}_{mi})}{\omega_l - \omega_{fm}} \right]. \qquad (2.1.104)$$

Substituting Eq. (2.1.103) into Eq. (2.1.98) yields

$$W^{(2)}_{i \to f} = \frac{2}{\hbar^2} \left(\frac{2\pi e^2}{L^3} \right)^2 \sum_l \sum_{l'} (n_l \omega_l)(n_{l'} \omega_{l'}) |S_{fi}(\omega_l, \omega_{l'})|^2$$

$$\times \delta(\omega_{fi} - \omega_l - \omega_{l'}). \qquad (2.1.105)$$

Using Eqs. (2.1.87) and (2.1.91), Eq. (2.1.105) becomes

$$W^{(2)}_{i \to f} = \frac{2\pi}{\hbar^2} \left(\frac{2\pi e^2}{c} \right)^2 \int_0^\infty d\omega_l \omega_l I_1(\omega_l)$$

$$\times \int_0^\infty d\omega_{l'} \omega_{l'} I_2(\omega_{l'})$$

$$\times |S_{fi}(\omega_l, \omega_{l'})|^2 \delta(\omega_{fi} - \omega_l - \omega_{l'}) \qquad (2.1.106)$$

or

$$W^{(2)}_{i \to f} = \frac{2\pi}{\hbar^2} \left(\frac{2\pi e^2}{c} \right)^2 \int_0^\infty d\omega_l \omega_l \omega_{l'} \, I_1(\omega_l) I_2(\omega_{l'}) |S_{if}(\omega_l, \omega_{l'})|^2 , \quad (2.1.107)$$

where $\omega_{l'} = \omega_{fi} - \omega_l$. Depending on the spectral distributions of the incident light intensities $I_1(\omega_l)$ and $I_2(\omega_{l'})$, the integral in Eq. (2.1.107) can be simplified.

The use of two identical photon absorptions is just a particular case of $W^{(2)}_{i \to f}$ given by Eq. (2.1.107).

Next we consider the case of three-photon absorption. In this case, the initial and final states are given by

$$|I\rangle \equiv |\varepsilon_i, n_l \hbar \omega_l, n_{l'} \hbar \omega_{l'}, n_{l''} \hbar \omega_{l''} \rangle, \quad (2.1.108)$$

$$|F\rangle \equiv |\varepsilon_f, (n_l - 1)\hbar \omega_l, (n_{l'} - 1)\hbar \omega_{l'}, (n_{l''} - 1)\hbar \omega_{l''} \rangle. \quad (2.1.109)$$

Using Eq. (2.1.30), we find

$$W^{(3)}_{I \to F} = \frac{2\pi}{\hbar} \left| \sum_M \sum_K \frac{V_{FM} V_{MK} V_{KI}}{\hbar^2 \omega_{KI} \omega_{MI}} \right|^2 \delta(E_F - E_I) \quad (2.1.110)$$

or

$$W^{(3)}_{I \to F} = \frac{2\pi}{\hbar^3} \left(\frac{2e^2}{m^2 L^3} \right)^3 \frac{n_l n_{l'} n_{l''}}{\omega_l \omega_{l'} \omega_{l''}} |M^{(3)}_{fi}(\omega_l, \omega_{l'}, \omega_{l''})|^2$$
$$\times \, \delta(\omega_{fi} - \omega_l - \omega_{l'} - \omega_{l''}), \quad (2.1.111)$$

where

$$M^{(3)}_{fi}(\omega_l, \omega_{l'}, \omega_{l''}) = \sum_k \sum_m \left(\frac{(\hat{e}_{l''} \cdot \mathbf{P}_{fm})(\hat{e}_{l'} \cdot \mathbf{P}_{mk})(\hat{e}_l \cdot \mathbf{P}_{ki})}{(\omega_{ki} - \omega_l)(\omega_{mi} - \omega_l - \omega_{l'})} \right.$$

$$+ \frac{(\hat{e}_{l'} \cdot \mathbf{P}_{fm})(\hat{e}_{l''} \cdot \mathbf{P}_{mk})(\hat{e}_l \cdot \mathbf{P}_{ki})}{(\omega_{ki} - \omega_l)(\omega_{mi} - \omega_l - \omega_{l''})}$$

$$+ \frac{(\hat{e}_{l''} \cdot \mathbf{P}_{fm})(\hat{e}_l \cdot \mathbf{P}_{mk})(\hat{e}_{l'} \cdot \mathbf{P}_{ki})}{(\omega_{ki} - \omega_{l'})(\omega_{mi} - \omega_l - \omega_{l'})}$$

$$+ \frac{(\hat{e}_l \cdot \mathbf{P}_{fm})(\hat{e}_{l''} \cdot \mathbf{P}_{mk})(\hat{e}_{l'} \cdot \mathbf{P}_{ki})}{(\omega_{ki} - \omega_{l'})(\omega_{mi} - \omega_{l'} - \omega_{l''})}$$

$$+ \frac{(\hat{e}_{l'} \cdot \mathbf{P}_{fm})(\hat{e}_l \cdot \mathbf{P}_{mk})(\hat{e}_{l''} \cdot \mathbf{P}_{ki})}{(\omega_{ki} - \omega_{l''})(\omega_{mi} - \omega_{l''} - \omega_l)}$$

$$\left. + \frac{(\hat{e}_l \cdot \mathbf{P}_{fm})(\hat{e}_{l'} \cdot \mathbf{P}_{mk})(\hat{e}_{l''} \cdot \mathbf{P}_{ki})}{(\omega_{ki} - \omega_{l''})(\omega_{mi} - \omega_{l'} - \omega_{l''})} \right). \quad (2.1.112)$$

The process involved in deriving Eq. (2.1.111) from Eq. (2.1.110) is similar to that for the case of two-photon absorption and will not be given here.

From Eq. (2.1.111), we obtain

$$W^{(3)}_{i \to f} = \frac{2\pi}{\hbar^3} \left(\frac{2\pi e^2}{m^2 L^3} \right)^3 \sum_l \sum_{l'} \sum_{l''} \frac{n_l n_{l'} n_{l''}}{\omega_l \omega_{l'} \omega_{l''}} \left| M^{(3)}_{fi} (\omega_l, \omega_{l'}, \omega_{l''}) \right|^2$$

$$\times \, \delta(\omega_{fi} - \omega_l - \omega_{l'} - \omega_{l''}) \qquad (2.1.113)$$

or

$$W^{(3)}_{i \to f} = \frac{2\pi}{\hbar^3} \left(\frac{2\pi e^2}{m^2 c} \right)^3 \int_0^\infty \int_0^\infty \int_0^\infty d\omega_l \, d\omega_{l'} \, d\omega_{l''}$$

$$\times \, \frac{I_1(\omega_l) I_2(\omega_{l'}) I_3(\omega_{l''})}{\omega_l \omega_{l'} \omega_{l''}}$$

$$\times \, \left| M^{(3)}_{fi}(\omega_l, \omega_{l'}, \omega_{l''}) \right|^2 \delta(\omega_{fi} - \omega_l - \omega_{l'} - \omega_{l''}). \qquad (2.1.114)$$

Notice that

$$\omega_{fi} = \omega_l + \omega_{l'} + \omega_{l''} = \omega_{fm} + \omega_{mi} \qquad (2.1.115)$$

or

$$\omega_l - \omega_{fm} = \omega_{mi} - \omega_{l'} - \omega_{l''}, \qquad (2.1.116)$$

and so on. Note that multiphoton absorption transition probabilities can also be derived by using the semiclassical theory of radiation (McClain and Harris, 1977).

2.2 THE GREEN'S FUNCTION METHOD

The purpose of this section is to present a systematic formulation of the problem of multiphoton processes based on the Green's function (or re-solvent operator) method, which has been used successfully in a variety of problems (Mower, 1966, 1968; Goldberger and Watson, 1964). Its mathematical aspects were first explored in considerable detail by Schönberg (1951a, b). A brief discussion of it in a form useful to problems in atomic and molecular physics has been given by Messiah (1965). The most complete and self-contained exposition of the method, as applied to quantum mechanics, is that of Goldberger and Watson (1964); it is their development that is followed in this section. Actually, the way it is presented here is related somewhat to Heitler's damping theory (Heitler, 1954). It should be noted that this theoretical approach has recently been applied by Lambropoulos (1974) to treat near-resonance effects in the two-photon ionization of atoms.

2.2.1   Definition of Green's Function

We consider a quantum-mechanical system whose Hamiltonian can be written as

$$\hat{H} = \hat{H}_0 + V, \tag{2.2.1}$$

where $\hat{H}_0$ represents the unperturbed part, with the eigenstates assumed to be known, and V is a perturbation. Let $|a\rangle, |b\rangle, |c\rangle, \ldots$, be the eigenstates of $\hat{H}_0$ with respective energies $E_a^0, E_b^0, E_c^0, \ldots$. The operator for the Green's function (or the resolvent operator) for the Hamiltonian $\hat{H}$ is defined by (Goldberger and Watson, 1964)

$$\hat{G}(z) = \frac{1}{(z - \hat{H})}. \tag{2.2.2}$$

In the representation of $|a\rangle, |b\rangle, \ldots$, we have

$$\langle a'|\hat{G}(z)|a\rangle = \sum_\lambda \left\langle a' \left| \frac{1}{z - \hat{H}} \right| \lambda \right\rangle \langle \lambda|a\rangle = \sum_\lambda \frac{\langle a'|\lambda\rangle\langle\lambda|a\rangle}{z - E_\lambda}, \tag{2.2.3}$$

where $|\lambda\rangle$ and E_λ represent the eigenfunctions and eigenvalues of $\hat{H}$, respectively,

$$\hat{H}|\lambda\rangle = E_\lambda|\lambda\rangle. \tag{2.2.4}$$

To show the applications of the Green's function method, we first consider the stationary state problem. In this case we are concerned with the solution of Eq. (2.2.4). The task of finding E_λ of Eq. (2.2.4) is equivalent to that of finding the poles at $z = E_\lambda$ of the matrix elements of $\hat{G}(z)$. In other words, we have to solve the algebraic equation

$$[\langle a'|\hat{G}(z)|a\rangle]^{-1} = 0 \tag{2.2.5}$$

for the eigenvalues $z = E_\lambda$. For this purpose the diagonal matrix elements of $\hat{G}$ can be used; that is, if

$$G_{aa}(z) = \langle a|\hat{G}(z)|a\rangle, \tag{2.2.6}$$

then we solve $[G_{aa}(z)]^{-1} = 0$. To evaluate $G_{aa}(z)$, we rewrite Eq. (2.2.2) as

$$(z - \hat{H}_0)\hat{G} = 1 + V\hat{G}. \tag{2.2.7}$$

Next, a new operator $\hat{F}$ is introduced:

$$\langle a'|\hat{G}(z)|a\rangle \equiv \langle a'|\hat{F}(z)|a\rangle G_{aa}(z) \tag{2.2.8a}$$

or, for

$$\hat{G} = \hat{F}\hat{g}, \tag{2.2.8b}$$

$$\langle a'|\hat{g}|a\rangle = \delta_{a'a}G_{aa}(z) = \delta_{a'a}g_{aa}(z). \tag{2.2.9}$$

It follows that

$$\langle a|\hat{F}|a\rangle = 1. \tag{2.2.10}$$

Substituting Eq. (2.2.8b) into Eq. (2.2.7) yields

$$(z - \hat{H}_0)\hat{F}\hat{g} = 1 + V\hat{F}\hat{g} \tag{2.2.11}$$

and

$$\langle a|(z - \hat{H}_0)\hat{F}\hat{g}|a\rangle = (z - E_a^0)\langle a|\hat{F}\hat{g}|a\rangle$$
$$= (z - E_a^0)G_{aa} = 1 + \langle a|V\hat{F}\hat{g}|a\rangle$$
$$= 1 + R_a G_{aa}, \tag{2.2.12}$$

where

$$R_a = \langle a|V\hat{F}|a\rangle = R_{aa} \tag{2.2.13}$$

or

$$\hat{R} = V\hat{F}. \tag{2.2.14}$$

Here $\hat{R}$ is usually called the level-shift operator.

From Eq. (2.2.12) we can solve for G_{aa},

$$G_{aa}(z) = [z - E_a^0 - R_{aa}(z)]^{-1}. \tag{2.2.15}$$

According to Eq. (2.2.5), the eigenvalues of $\hat{H}$ are simply the solution of the algebraic equation

$$E_\lambda = E_a^0 + R_{aa}(E_\lambda). \tag{2.2.16}$$

To determine the level shift R_a, we must first calculate $\hat{F}$ from Eq. (2.2.11). Equation (2.2.11) can be rewritten as

$$(z - \hat{H}_0 - \hat{O})\hat{F}\hat{g} = 1 + (V - \hat{O})\hat{F}\hat{g} \tag{2.2.17}$$

or

$$\hat{F} = \frac{1}{z - \hat{H}_0 - \hat{O}}\frac{1}{\hat{g}} + \frac{1}{z - \hat{H}_0 - \hat{O}}(V - \hat{O})\hat{F}, \tag{2.2.18}$$

where $\hat{O}$ is an arbitrary operator. Notice that

$$\hat{F}|a\rangle = \frac{1}{z - \hat{H}_0 - \hat{O}}(z - \hat{H}_0 - R_a)|a\rangle + \frac{1}{z - \hat{H}_0 - \hat{O}}(V - \hat{O})\hat{F}|a\rangle. \tag{2.2.19}$$

To show the application of Eq. (2.2.19), we first choose

$$\hat{O} = R_a\hat{P}_a, \tag{2.2.20}$$

where $\hat{P}_a$ represents the projection operator,

$$\hat{P}_a = |a\rangle\langle a|.$$

Substituting Eq. (2.2.20) into Eq. (2.2.19) yields

$$\hat{F}|a\rangle = |a\rangle + \frac{1}{z - \hat{H}_0 - R_a\hat{P}_a}(V - R_a\hat{P}_a)\hat{F}|a\rangle. \qquad (2.2.21)$$

Using the relations

$$\langle a|(V - R_a\hat{P}_a)\hat{F}|a\rangle = 0, \qquad (2.2.22)$$

$$\hat{P}_a(1 - \hat{P}_a) = 0,$$

Eq. (2.2.21) becomes

$$\hat{F}|a\rangle = |a\rangle + \frac{1}{z - \hat{H}_0 - R_a\hat{P}_a}(1 - \hat{P}_a)(V - R_a\hat{P}_a)\hat{F}|a\rangle$$

$$= |a\rangle + \frac{1}{z - \hat{H}_0}(1 - \hat{P}_a)V\hat{F}|a\rangle. \qquad (2.2.23)$$

From Eq. (2.2.23), we can calculate the level shift operator $\hat{R}$,

$$\hat{R} = V + V\frac{1}{z - \hat{H}_0}(1 - \hat{P}_a)V\hat{F} = V + V\frac{1}{z - \hat{H}_0}(1 - \hat{P}_a)\hat{R}. \qquad (2.2.24)$$

Equation (2.2.24) provides an integral equation for $\hat{R}$. For example, by iteration we find that

$$\hat{R} = V + V\frac{1}{z - \hat{H}_0}(1 - \hat{P}_a)V + V\frac{1}{z - \hat{H}_0}(1 - \hat{P}_a)V$$

$$\times \frac{1}{z - \hat{H}_0}(1 - \hat{P}_a)V + \cdots \qquad (2.2.25)$$

and

$$R_a(z) = \langle a|V|a\rangle + \sum_{a'}^{a' \neq a}\frac{|\langle a|V|a'\rangle|^2}{z - E_{a'}^0} + \cdots. \qquad (2.2.26)$$

Substituting Eq. (2.2.26) into Eq. (2.2.16) yields

$$E_\lambda = E_a^0 + \langle a|V|a\rangle + \sum_{a'}^{a' \neq a}\frac{|\langle a|V|a'\rangle|^2}{E_\lambda - E_{a'}} + \cdots. \qquad (2.2.27)$$

This is the Wigner–Brillouin perturbation method for calculating E_λ.

2.2.2 Application to Multiphoton Processes

In Section 2.2.1, we showed how to apply the Green's function method to the time-independent problem. In this section we shall discuss the application of the Green's function method to time-dependent phenomena, especially multiphoton absorption (MPA) and ionization (MPI). Again we write the total Hamiltonian as

$$\hat{H} = \hat{H}_0 + V \tag{2.2.28}$$

and

$$\hat{H}_0 = \hat{H}_{\rm m} + \hat{H}_{\rm r}, \tag{2.2.29}$$

where $\hat{H}_{\rm m}$ is the Hamiltonian of the free molecule, $\hat{H}_{\rm r}$ the Hamiltonian of the free radiation field, and V the interaction between the molecule and the radiation field. Our central problem is now the solution of the time-dependent Schrödinger equation

$$\hat{H}\Psi = i\hbar\,(\partial\Psi/\partial t). \tag{2.2.30}$$

Applying the Laplace transform to Eq. (2.2.30) yields

$$\hat{H}\Psi(p) = i\hbar p\Psi(p) - i\hbar|a\rangle, \tag{2.2.31}$$

where the system is assumed to be in one of the eigenstates of $\hat{H}_0$ at $t = 0$, that is, $\Psi(t = 0) \equiv |a\rangle$ and $\Psi(p)$ represents the Laplace transform of $\Psi(t)$,

$$\Psi(p) = \int_0^\infty e^{-pt}\Psi(t)\,dt. \tag{2.2.32}$$

Equation (2.2.31) can be rewritten as

$$\Psi(p) = \frac{1}{p + (i\hat{H}/\hbar)}|a\rangle. \tag{2.2.33}$$

The inverse transformation of Eq. (2.2.33) gives us $\Psi(t)$ as

$$\Psi(t) = \frac{1}{2\pi i}\int_{c_1} dp\, e^{pt}\frac{1}{p + (i\hat{H}/\hbar)}|a\rangle = \hat{U}(t)|a\rangle, \tag{2.2.34}$$

where c_1 represents the usual Laplace transform contour (Goldberger and Watson, 1964). Let us now make a change in the variable of integration in Eq. (2.2.34), $p = -iz$. Equation (2.2.34) is modified to read

$$\Psi(t) = \frac{1}{2\pi i}\int_{c_2} dz\, e^{-izt}\hat{G}(z)|a\rangle, \tag{2.2.35}$$

where $\hat{G}(z)$ represents the Green's function operator defined in Section 2.2.1. The wave function $\Psi(t)$ contains a complete description of the system for $t > 0$. For example, if we are interested in the transition $a \to f$, then from

Eq. (2.2.35) we can see that it is necessary to calculate $G_{fa}(z)$. Notice that if we start with

$$\hat{H}\Psi = i\hbar\,(\partial\Psi/\partial t), \tag{2.2.36}$$

then we have

$$\Psi(t) = \frac{1}{2\pi i}\int_{c_1} dp\; e^{pt}\,\frac{1}{p + (i\hat{H}/\hbar)}|a\rangle. \tag{2.2.37}$$

Making the substitution $p = -iE/\hbar$ yields

$$\Psi(t) = \frac{1}{2\pi i}\int_c dE\; e^{-iEt/\hbar}\,\frac{1}{E - \hat{H}}|a\rangle$$

$$= \frac{1}{2\pi i}\int_c dE\; e^{-iEt/\hbar}\,\hat{G}(E)|a\rangle = \hat{U}(t)|a\rangle. \tag{2.2.38}$$

In this case, for the transition $a \to f$ we should calculate $G_{fa}(E)$ or $U_{fa}(t)$. The preceding results will be applied to MPA processes.

The total system defining the molecular states and the radiation field is assumed to be composed of an initial state $|I\rangle \equiv |\{\varepsilon_i\}; \{n_i\hbar\omega_i\}\rangle$, intermediate states $|A\rangle \equiv |\{\varepsilon_a\}; \{n_a\hbar\omega_a\}\rangle$, and a final state $|F\rangle \equiv |\{\varepsilon_f\}\rangle; \{n_f\hbar\omega_f\}\rangle$, where $\{\varepsilon_i\}$, $\{\varepsilon_a\}$, and $\{\varepsilon_f\}$ denote manifolds in the initial, intermediate, and final states of the molecule, respectively, and the $n_i\hbar\omega_i$ are the photon energies. The total Hamiltonian is given by

$$\hat{H} = \hat{H}_0 + V, \tag{2.2.39}$$

$$\hat{H}_0 = \hat{H}_m + \hat{H}_r, \tag{2.2.40}$$

and

$$V = V_{nd} + V_{mr}, \tag{2.2.41}$$

where $\hat{H}_m$ is the adiabatic molecular Hamiltonian, $\hat{H}_r$ the Hamiltonian for the free electromagnetic field, V_{nd} the off-diagonal part of the nuclear kinetic energy operator (Lin, 1966), and V_{mr} the molecule–radiation interaction.

The zeroth-order basis set satisfies

$$\hat{H}_0|I\rangle = (\hat{H}_m^0 + \hat{H}_r)|I\rangle = E_I^0|I\rangle, \tag{2.2.42}$$

where E_I^0 is the zeroth-order energy of the initial state and is given by $E_I = \varepsilon_i + \sum_i n_i\hbar\omega_i$. The transition amplitude $U_{FI}(t)$ from the initial to the final state is expressed in terms of the time-independent Green's function (Goldberger and Watson, 1964): by

$$U_{FI}(t) = \langle F|\exp\left(-\frac{it\hat{H}}{\hbar}\right)|I\rangle = \frac{1}{2\pi i}\int dE\,\exp(-iEt/\hbar)G_{FI}(E), \tag{2.2.43}$$

where G_{FI} is the matrix element of the Green's function (resolvent operator) defined by

$$(E - \hat{H})\hat{G}(E) = 1. \qquad (2.2.44)$$

The matrix elements of the Green's function are evaluated by using the Dyson equation (Fujimura and Lin, 1979a)

$$\hat{G}(E) = \hat{G}^0(E) + \hat{G}^0(E)V\hat{G}(E), \qquad (2.2.45)$$

where the zeroth-order Green's function $\hat{G}^0(E)$ is defined as

$$(E - \hat{H}_0 + i\eta)\hat{G}^0(E) = 1 \qquad (2.2.46)$$

and $\eta \to 0^+$.

Equations for the matrix elements of $\hat{G}(E)$ are written as (Fujimura and Lin, 1979a)

$$G_{FI} = \sum_A G^0_{FF}V_{FA}G_{AI}, \qquad (2.2.47)$$

$$G_{AI} = \sum_B G^0_{AA}V_{AB}G_{BI} = G^0_{AA}V_{AI}G_{II} + \sum_B^{B \neq I} G^0_{AA}V_{AB}G_{BI}, \qquad (2.2.48)$$

$$G_{II} = G^0_{II} + \sum_A G^0_{II}V_{IA}G_{AI}, \qquad (2.2.49)$$

where the summations in Eqs. (2.2.47)–(2.2.49) are over both the molecular vibronic states and the radiation field. Substituting Eqs. (2.2.47) and (2.2.49) into Eq. (2.2.48) yields

$$(E - E^0_A - \Lambda_{AA})G_{AI} = V_{AI}G^0_{II} + \sum_{A'}^{A' \neq A} \Lambda_{AA'}G_{A'I}, \qquad (2.2.50)$$

where

$$\Lambda_{AA'} = \sum_B V_{AB}G^0_{BB}V_{BA'}, \qquad (2.2.51)$$

$$\Lambda_{AA} = \sum_B |V_{AB}|^2 G^0_{BB}. \qquad (2.2.52)$$

The formal solution of G_{AI} is written as

$$G_{AI} = \frac{G^0_{II}}{E - E^0_A - \Lambda_{AA}} \left(V_{AI} + \sum_{A'}^{A' \neq A} \frac{\Lambda_{AA'}V_{A'I}}{E - E^0_{A'} - \Lambda_{A'A'}} \right.$$
$$\left. + \sum_{A'}^{A' \neq A} \sum_{A''}^{A'' \neq A'} \frac{\Lambda_{AA'}\Lambda_{A'A''}V_{A''I}}{(E - E^0_{A'} - \Lambda_{A'A'})(E - E^0_{A''} - \Lambda_{A''A''})} + \cdots \right). \qquad (2.2.53)$$

If we restrict ourselves to the lowest order in Eq. (2.2.53), we obtain

$$G_{AI} = G^0_{II}V_{AI}/(E - E^0_A - \Lambda_{AA}). \qquad (2.2.54)$$

Substituting Eq. (2.2.54) into Eq. (2.2.47) yields

$$G_{FI} = \sum_A \frac{G^0_{FF} V_{FA} V_{AI} G^0_{II}}{E - E^0_A - \Lambda_{AA}}. \tag{2.2.55}$$

Here the diagonal self-energy Λ_{AA} can be expressed as (Lambropoulos, 1974; Fujimura and Lin, 1979a)

$$\Lambda_{AA} = D_A - \tfrac{1}{2} i \Gamma_A, \tag{2.2.56}$$

where the level shift D_A and the decay width Γ_A are written as

$$D_A = P \sum_B \frac{|\langle B|V|A\rangle|^2}{E - E^0_B} \tag{2.2.57}$$

and

$$\Gamma_A = 2\pi \sum_B |\langle B|V|A\rangle|^2 \, \delta(E - E^0_B), \tag{2.2.58}$$

respectively. In Eq. (2.2.57), P denotes the principal part.

Substituting Eqs. (2.2.56) and (2.2.55) into Eq. (2.2.43), the transition amplitude $U_{FI}(t)$ can be expressed as

$$U_{FI}(t) = \sum_A V_{FA} V_{AI} \left(\frac{e^{-itE^0_I/\hbar}(1 - e^{-(it/\hbar)(E^0_F - E^0_I)})}{(E^0_I - E^0_A - \Lambda_{AA})(E^0_I - E^0_F)} \right.$$

$$\left. + \frac{e^{-(it/\hbar)(E^0_A + \Lambda_{AA})}}{(E^0_A + \Lambda_{AA} - E^0_F)(E^0_A + \Lambda_{AA} + E^0_I)} - \frac{e^{-itE^0_F/\hbar}}{(E^0_I - E^0_A - \Lambda_{AA})(E^0_F - E^0_A - \Lambda_{AA})} \right). \tag{2.2.59}$$

Using the on-the-energy-shell approximation $E^0_I \to E^0_F$ and the limit $t \to \infty$, the first term of the right-hand side in Eq. (2.2.59) makes a dominant contribution to the transition probability (Fujimura and Lin, 1979a):

$$U_{FI}(t) = \sum_A V_{FA} V_{AI} \frac{e^{-itE^0_I/\hbar}(1 - e^{-(it/\hbar)(E^0_F - E^0_I)})}{(E^0_I - E^0_A - \Lambda_{AA})(E^0_I - E^0_F)}. \tag{2.2.60}$$

As is well known, the transition probability from $|I\rangle$ to $|F\rangle$ per unit time is given by differentiating the transition probability $\sum_F |U_{FI}(t)|^2$ and letting $t \to \infty$. We find that

$$W^{(2)}_I = \sum_F \frac{d}{dt} |U_{FI}(t)|^2 = \frac{2\pi}{\hbar} \sum_F \delta(E^0_I - E^0_F) \left| \sum_A \frac{V_{FA} V_{AI}}{E^0_I - E^0_A - \Lambda_{AA}} \right|^2. \tag{2.2.61}$$

This expression is general and can be used to treat two-photon processes like Raman scattering and two-photon absorption.

Let us apply Eq. (2.2.61) to two-photon absorption. For this purpose, from Eq. (2.2.61) we have

$$W^{(2)}_{I \to F} = \frac{2\pi}{\hbar} \delta(E^0_I - E^0_F) \left| \sum_M \frac{V_{FM} V_{MI}}{E^0_I - E^0_M - \Lambda_{MM}} \right|^2. \qquad (2.2.62)$$

This should be compared with Eq. (2.1.97); the difference between these two equations is the existence of the factor Λ_{MM} in Eq. (2.2.62), which can introduce the level shift and linewidth. By substituting Eqs. (2.1.93)–(2.1.96) into Eq. (2.2.62), we obtain

$$W^{(2)}_{I \to F} = \frac{2\pi}{\hbar^2} \left(\frac{2\pi e^2}{m^2 L^3} \right)^2 \frac{n_l n_{l'}}{\omega_l \omega_{l'}} |M^{(2)}_{fi\gamma}(\omega_l, \omega_{l'})|^2 \delta(\omega_{fi} - \omega_l - \omega_{l'}), \qquad (2.2.63)$$

where

$$M^{(2)}_{fi\gamma}(\omega_l, \omega_{l'}) = \sum_m \left[\frac{(\hat{e}_{l'} \cdot \mathbf{P}_{fm})(\hat{e}_l \cdot \mathbf{P}_{mi})}{\omega_{mi} + \gamma_m - \omega_l} + \frac{(\hat{e}_l \cdot \mathbf{P}_{fm})(\hat{e}_{l'} \cdot \mathbf{P}_{mi})}{\omega_{mi} + \gamma_m + \omega_{l'}} \right] \qquad (2.2.64)$$

and $\gamma_m = \Lambda_{mm}/\hbar$. The damping constant γ_m includes not only the radiative process but also the nonradiative process [see Eqs. (2.2.56)–(2.2.58)]. A more refined treatment of the damping constant γ_m has been given by Lambropoulos (1974).

Note that in treating two-photon absorption we look for the expression of G_{FI} that involves V twice. Similarly, to treat three-photon absorption we look for the expression of G_{FI} that involves V three times by using Eqs. (2.2.47) and (2.2.48). The resulting expression for the transition probability of three-photon absorption will be similar to that given by Eq. (2.1.111), except for the appearance of the damping constants associated with the intermediate states, and will not be given here.

2.3. THE DENSITY MATRIX METHOD

In this section we shall present a density matrix formalism for calculating the transition probabilities of multiphoton processes in which the damping effect is properly taken into account. The rate expression obtained can then be used to discuss the optical phenomena in the resonance region and to study the temperature effect on the rate of multiphoton processes. Within this matrix formalism both steady-state and transient optical phenomena can be studied and both forward and reverse optical processes can be treated (Lin and Eyring, 1977a; Fujimura and Lin, 1979b; Bloembergen, 1965; Agarwal, 1974).

2.3.1   The Density Matrix

To explain the physical meaning of the density matrix, we consider a two-level system with the wave function (Sargent *et al.*, 1974)

$$\Psi(\mathbf{r}, t) = C_a(t)U_a(\mathbf{r}) + C_b(t)U_b(\mathbf{r}) \tag{2.3.1}$$

and the Hamiltonian operator

$$\hat{H} = \hat{H}_0 + V, \tag{2.3.2}$$

where $U_a(\mathbf{r})$ and $U_b(\mathbf{r})$ are the eigenfunctions of the unperturbed Hamiltonian $\hat{H}_0$ and V is the interaction energy. The amplitudes $C_a(t)$ and $C_b(t)$ can be determined from the solution of the time-dependent Schrödinger equation

$$\hat{H}\Psi = i\hbar(\partial\Psi/\partial t); \tag{2.3.3}$$

we find

$$dC_a/dt = -(iE_a/\hbar)C_a - (i/\hbar)V_{ab}C_b \tag{2.3.4}$$

$$dC_b/dt = -(iE_b/\hbar)C_b - (i/\hbar)V_{ba}C_a. \tag{2.3.5}$$

The density matrix corresponding to the wave function given by Eq. (2.3.1) is defined by the bilinear products

$$\rho_{aa} \equiv C_a C_a^*, \tag{2.3.6}$$

$$\rho_{ab} \equiv C_a C_b^*, \tag{2.3.7}$$

$$\rho_{ba} \equiv C_b C_a^* = \rho_{ab}^*, \tag{2.3.8}$$

$$\rho_{bb} = C_b C_b^*, \tag{2.3.9}$$

or in matrix notation

$$\rho = \begin{pmatrix} \rho_{aa} & \rho_{ab} \\ \rho_{ba} & \rho_{bb} \end{pmatrix}. \tag{2.3.10}$$

Notice that the expectation value of an operator $\hat{M}$ can be calculated as

$$
\begin{aligned}
\langle \hat{M} \rangle &= \langle \Psi(\mathbf{r}, t) | \hat{M} | \Psi(\mathbf{r}, t) \rangle \\
&= C_a^* C_a M_{aa} + C_a^* C_b M_{ab} + C_b^* C_a M_{ba} + C_b^* C_b M_{bb} \\
&= \rho_{aa} M_{aa} + \rho_{ba} M_{ab} + \rho_{ab} M_{ba} + \rho_{bb} M_{bb},
\end{aligned} \tag{2.3.11}
$$

which can be expressed as

$$\langle \hat{M} \rangle = T_r(\hat{\rho}\hat{M}) = T_r(\hat{M}\hat{\rho}). \tag{2.3.12}$$

That is, the expectation value of $\hat{M}$ is just the trace of the matrix product of $\hat{\rho}$ and $\hat{M}$.

We can derive the equation of motion for the density matrix from Eqs. (2.3.1)–(2.3.5). Proceeding with one matrix element at a time, we have

$$d\rho_{aa}/dt = C_a^* \frac{dC_a}{dt} + C_a \frac{dC_a^*}{dt}$$

$$= C_a^* \left(-\frac{iE_a}{\hbar} C_a - \frac{i}{\hbar} V_{ab} C_b \right) + C_a \left(\frac{iE_a}{\hbar} C_a^* + \frac{i}{\hbar} V_{ba} C_b^* \right)$$

$$= -\frac{i}{\hbar} (V_{ab}\rho_{ba} - V_{ba}\rho_{ab})$$

$$= -\frac{i}{\hbar} [\hat{V}, \hat{\rho}]_{aa}, \tag{2.3.13}$$

where $[\hat{V}, \hat{\rho}]$ represents the commutator

$$[\hat{V}, \hat{\rho}] = \hat{V}\hat{\rho} - \hat{\rho}\hat{V}. \tag{2.3.14}$$

Similarly, we find that

$$d\rho_{bb}/dt = -\frac{i}{\hbar} (V_{ba}\rho_{ab} - V_{ab}\rho_{ba}) = -\frac{i}{\hbar} [\hat{V}, \hat{\rho}]_{bb}. \tag{2.3.15}$$

This value is equal in magnitude and opposite in sign to that in Eq. (2.3.13) for ρ_{aa}, that is,

$$\frac{d\rho_{aa}}{dt} + \frac{d\rho_{bb}}{dt} = 0, \tag{2.3.16}$$

which expresses the fact that when the system's probability of being in the upper level decreases, its probability of being in the lower level increases. Mathematically, the result follows from the normalization condition

$$\rho_{aa} + \rho_{bb} = 1. \tag{2.3.17}$$

The off-diagonal matrix element ρ_{ab} can be calculated similarly as

$$\frac{d\rho_{ab}}{dt} = -\frac{i}{\hbar} (E_a - E_b)\rho_{ab} + \frac{i}{\hbar} V_{ab}(\rho_{aa} - \rho_{bb}). \tag{2.3.18}$$

Equations (2.3.13), (2.3.15), and (2.3.18) can be written in the form

$$\frac{d\hat{\rho}}{dt} = -\frac{i}{\hbar} [\hat{H}, \hat{\rho}] = -i\hat{L}\hat{\rho}, \tag{2.3.19}$$

the so-called Liouville equation, where $\hat{L}$ is the Liouville operator.

In the derivation of the Liouville equation [Eq. (2.3.19)] the pure-state system with two levels has been treated. Next, we shall consider a collection of multi-level isolated systems, that is, a mixed system with multiple levels.

The wave function of the jth system $\psi^{(j)}$ is expressed as

$$\psi^{(j)} = \sum_n C_n^{(j)}(t) U_n. \tag{2.3.20}$$

Defining P_j as the statistical distribution of the system (for example, $P_j = 1/N$ if there are N systems in the ensemble), the density matrix element ρ_{nm} has the form

$$\rho_{nm} = \sum_j P_j C_n^{(j)}(t) [C_m^{(j)}(t)]^* \tag{2.3.21}$$

This expression can be derived from the definition of the general density operator for the mixed case;

$$\hat{\rho} = \sum_j P_j |\psi^{(j)}\rangle\langle\psi^{(j)}|. \tag{2.3.22}$$

For the pure-state case, the density operator is defined as

$$\hat{\rho} = |\psi\rangle\langle\psi|.$$

It can easily be shown that the equation of motion of $\hat{\rho}(t)$ in the mixed case is given by the Liouville equation (2.3.19). From the Liouville equation for the multiple-level system, we find that

$$\frac{d\rho_{mn}}{dt} = -\frac{i}{\hbar} \sum_k (H_{nk}\rho_{kn} - \rho_{mk}H_{kn})$$

$$= -i \sum_{m'} \sum_{n'} L_{mn:m'n'}\rho_{m'n'}, \tag{2.3.23}$$

and the matrix element of the Liouville operator $\hat{L}$ is given by

$$L_{mn:m'n'} = 1/\hbar(H_{mn'}\delta_{nn'} - H_{n'n}\delta_{mm'}), \tag{2.3.24}$$

where $\delta_{nn'}$ and $\delta_{mm'}$ are the Kronecker deltas. Equation (2.3.24) is a very useful relation for practical calculation.

2.3.2   Master Equations for a System in Contact with a Heat Bath

In this section we shall derive the master equations for a system embedded in a heat bath; we shall then be able to describe the proper time-dependent behavior of both diagonal and off-diagonal portions of the density matrix of the system, and thus we can employ the master equations to study optical phenomena in which the resonance effect is important. The idea is to consider an isolated system that can be divided into a "reservoir" (or "heat bath") and a "subsystem" (or "system of interest") and to eliminate the irrelevant part of the density matrix so as to obtain the equation of motion for the reduced density matrix of the subsystem (called simply the "system").

The time dependence of the density matrix $\hat{\rho}(t)$ of an isolated system is determined by the Liouville equation of motion

$$d\hat{\rho}/dt = -(i/\hbar)(\hat{H}\hat{\rho} - \hat{\rho}\hat{H}) = -i\hat{L}\hat{\rho}, \tag{2.3.25}$$

where $\hat{L}$ again represents the Liouville operator. The total system consists of a subsystem with Hamiltonian $\hat{H}_s$ and a heat bath with Hamiltonian $\hat{H}_b$. If we let $\hat{H}_1$ represent the interaction between the two parts, then the Hamiltonian of the total system can be written as

$$\hat{H} = \hat{H}_s + \hat{H}_b + \hat{H}_1 = \hat{H}_0 + \hat{H}_1 \tag{2.3.26}$$

and the corresponding Liouville operator takes the form

$$\hat{L} = \hat{L}_s + \hat{L}_b + \hat{L}_1 = \hat{L}_0 + \hat{L}_1. \tag{2.3.27}$$

Applying the Laplace transform method to Eq. (2.3.25) yields

$$\hat{\rho}(p) = [1/(p + i\hat{L})]\hat{\rho}(0), \tag{2.3.28}$$

in which $\hat{\rho}(0)$ represents the density matrix of the total system at $t = 0$ and $\hat{\rho}(p)$ denotes the Laplace transform of $\hat{\rho}(t)$,

$$\hat{\rho}(p) = \int_0^\infty e^{-pt}\hat{\rho}(t)\, dt. \tag{2.3.29}$$

Next we introduce the transition operator $\hat{M}(p)$ defined by

$$\frac{1}{p + i\hat{L}} = \frac{1}{p + i\hat{L}_0}\left[1 + \hat{M}(p)\frac{1}{p + i\hat{L}_0}\right]. \tag{2.3.30}$$

Here $\hat{M}(p)$ can be put into the form

$$\hat{M}(p) = (-i\hat{L}_1) + (-i\hat{L}_1)\frac{1}{p + i\hat{L}}(-i\hat{L}_1). \tag{2.3.31}$$

Substituting Eq. (2.3.30) into Eq. (2.3.28) yields (Lin and Eyring, 1977b)

$$\hat{\rho}(p) = \frac{1}{p + i\hat{L}_0}\left[1 + \hat{M}(p)\frac{1}{p + i\hat{L}_0}\right]\hat{\rho}(0). \tag{2.3.32}$$

Notice that the density matrix of the system of interest at time t can be found by

$$\hat{\rho}^{(s)}(t) = \mathrm{Tr}_b[\hat{\rho}(t)] \tag{2.3.33}$$

or, equivalently,

$$\hat{\rho}^{(s)}(p) = \mathrm{Tr}_b[\hat{\rho}(p)], \tag{2.3.34}$$

where Tr_b represents the operation of carrying out a trace over the quantum states of the heat bath. To eliminate the reservoir variables, we shall assume that at $t = 0$

$$\hat{\rho}(0) = \hat{\rho}^{(s)}(0)\hat{\rho}^{(b)}(0).$$ (2.3.35)

Combining Eqs. (2.3.35) and (2.3.34) with Eq. (2.3.32), we obtain

$$\hat{\rho}^{(s)}(p) = \frac{1}{p + i\hat{L}_s}\left[1 + \langle \hat{M}(p)\rangle \frac{1}{p + i\hat{L}_s}\right]\hat{\rho}^{(s)}(0),$$ (2.3.36)

where $\langle \hat{M}(p)\rangle = \text{Tr}_b[\hat{M}(p)\hat{\rho}^{(b)}(0)]$.

Equation (2.3.36) can be conveniently expressed as

$$\hat{\rho}^{(s)}(p) = \frac{1}{p + i\hat{L}_s + \langle \hat{M}_c(p)\rangle}\hat{\rho}^{(s)}(0),$$ (2.3.37)

in which

$$\langle \hat{M}_c(p)\rangle = \frac{1}{1 + \langle \hat{M}(p)\rangle (p + i\hat{L}_s)^{-1}}(-\langle \hat{M}(p)\rangle).$$ (2.3.38)

Notice that Eq. (2.3.37) can be rewritten as

$$p\hat{\rho}^{(s)}(p) - \hat{\rho}^{(s)}(p) = -i\hat{L}_s\hat{\rho}^{(s)}(p) - \langle \hat{M}_c(p)\rangle \hat{\rho}^{(s)}(p).$$ (2.3.39)

Inverting the Laplace transform of Eq. (2.3.39), we obtain

$$\frac{d\hat{\rho}^{(s)}(t)}{dt} = -i\hat{L}_s\hat{\rho}^{(s)}(t) - \int_0^t d\tau \langle \hat{M}_c(\tau)\rangle \hat{\rho}^{(s)}(t - \tau),$$ (2.3.40)

where

$$\langle \hat{M}_c(p)\rangle = \int_0^\infty e^{-pt}\langle \hat{M}_c(t)\rangle\, dt.$$ (2.3.41)

Equations (2.3.37) and (2.3.40) are equivalent, and both are useful for finding the density matrix of the system of interest at $t > 0$. The term $\langle \hat{M}_c(t)\rangle$ is usually called the memory function or memory kernel. Equation (2.3.40) can be written as

$$\frac{d\hat{\rho}^{(s)}(t)}{dt} = -i\hat{L}_s\rho^{(1)}(t) - \int_0^t d\tau \langle \hat{M}_c(\tau)\rangle \exp\left(-\tau \frac{d}{dt}\right)\hat{\rho}^{(s)}(t).$$ (2.3.42)

Here the following relation has been used:

$$\exp\left(-\tau \frac{d}{dt}\right)\hat{\rho}^{(s)}(t) = \sum_{n=0}^\infty \frac{(-\tau)^n}{n!}\frac{d^n\hat{\rho}^{(s)}(t)}{dt^n} = \hat{\rho}^{(s)}(t - \tau).$$ (2.3.43)

Making the approximation of replacing d/dt in Eq. (2.3.42) by $-i\hat{L}_s$, we obtain

$$\frac{d\hat{\rho}^{(s)}(t)}{dt} = -i\hat{L}_s\hat{\rho}^{(s)}(t) - \int_0^t d\tau \langle \hat{M}_c(\tau)\rangle \exp(i\tau\hat{L}_s)\hat{\rho}^{(s)}(t).$$ (2.3.44)

Equation (2.3.42) is called the generalized master equation (GME), which can yield both the diagonal and off-diagonal matrix elements of the density matrix of the system. While the diagonal matrix elements of $\hat{\rho}^{(s)}(t)$ describe the time evolution of the population of the system, the off-diagonal matrix elements of $\hat{\rho}^{(s)}(t)$ provide the phase information of the system and determine the Lorentzian bandwidths and the band shifts of optical spectra.

Note that in the GME given by Eq. (2.3.44) the rate constants are not defined: They are defined only when the Markov approximation is introduced, that is,

$$\hat{\Gamma} = \int_0^\infty d\tau \, \langle \hat{M}_c(\tau) \rangle \exp(i\tau \hat{L}_s), \qquad (2.3.45)$$

where $\hat{\Gamma}$ is the damping operator. In the Markov approximation, Eq. (2.3.44) becomes

$$d\hat{\rho}^{(s)}(t)/dt = -i\hat{L}_s\hat{\rho}(t) - \hat{\Gamma}\hat{\rho}^{(s)}(t). \qquad (2.3.46)$$

The preceding treatment indicates that when the Markov approximation does not hold, to find the time-dependent behavior of the system one has to deal with the memory function rather than with the rate constant (Lin, 1980). It is important to study the validity of the commonly used assumption that $\hat{\rho}(0) = \hat{\rho}^{(s)}(0)\hat{\rho}^{(b)}(0)$.

To the second-order approximation with respect to $\hat{H}_1$, from Eq. (2.3.46) we obtain

$$\frac{d\rho_{n_s n_s}^{(s)}}{dt} = -\sum_{n_s'} \Gamma_{n_s n_s : n_s' n_s'} \rho_{n_s' n_s'}^{(s)} \qquad (2.3.47)$$

and

$$\frac{d\rho_{m_s n_s}^{(s)}}{dt} = -(i\omega_{m_s n_s} + \Gamma_{m_s n_s : m_s n_s})\rho_{m_s n_s}^{(s)}. \qquad (2.3.48)$$

The calculation of the matrix elements $\Gamma_{n_s n_s : n_s' n_s'}$ and $\Gamma_{m_s n_s : m_s n_s}$ is described in the following. From Eq. (2.3.45) we can see that to calculate the matrix elements of $\hat{\Gamma}$ we need to calculate the matrix elements of $\langle \hat{M}_c(\tau) \rangle$ or $\langle \hat{M}_c(p) \rangle$ first. Notice that in the second-order approximation we have

$$\langle \hat{M}_c(p) \rangle^{(2)} = -\langle \hat{M}(p) \rangle^{(2)} = \left\langle \hat{L}_1 \frac{1}{p + i\hat{L}_0} \hat{L}_1 \right\rangle \qquad (2.3.49)$$

or, for example,

$$\langle \hat{M}_c(p) \rangle_{n_s n_s : n_s' n_s'}^{(2)} = \sum_{n_b} \sum_{n_b'} \rho_{n_b' n_b}^{(b)}(0) \left(\hat{L}_1 \frac{1}{p + i\hat{L}_0} \hat{L}_1 \right)_{NN:N'N'}, \qquad (2.3.50)$$

where $N \equiv n_s n_b$ and $N' \equiv n_s' n_b'$. Here it has been assumed that $\langle \hat{L}_1 \rangle = 0$, which is reasonable provided that there is no external field present.

Using Eq. (2.3.24), we obtain

$$\left(\hat{L}_1 \frac{1}{p + i\hat{L}_0} \hat{L}_1\right)_{NN:N'N'} = \frac{\delta_{NN'}}{\hbar^2} \sum_M |H'_{MN}|^2 \left(\frac{1}{p + i\omega_{MN}} + \frac{1}{p + i\omega_{NM}}\right)$$

$$- \frac{|H'_{NN'}|^2}{\hbar^2(p + i\omega_{N'N})} - \frac{|H'_{NN'}|^2}{\hbar^2(p + i\omega_{NN'})}. \quad (2.3.51)$$

Thus for the case $n_s \neq n'_s$, we have

$$\langle \hat{M}_c(\tau)\rangle^{(2)}_{n_s n_s : n'_s n'_s} = -\frac{1}{\hbar^2} \sum_{n_b} \sum_{n'_b} \rho^{(b)}_{n'_b n'_b}(0) |H'_{n_s n_b, n'_s n'_b}|^2$$

$$\times \left(\frac{1}{p + i\omega_{n_s n_b, n'_s n'_b}} + \frac{1}{p + i\omega_{n'_s n'_b, n_s n_b}}\right) \quad (2.3.52)$$

or

$$\langle \hat{M}_c(\tau)\rangle^{(2)}_{n_s n_s : n'_s n'_s} = -\frac{2}{\hbar^2} \sum_{n_b} \sum_{n'_b} \rho^{(b)}_{n'_b n'_b}(0) |H'_{n_s n_b, n'_s n'_b}|^2 \cos \omega_{n_s n_b, n'_s n'_b} \tau. \quad (2.3.53)$$

Here for convenience we set $\hat{H}' = \hat{H}_1$. Substituting Eq. (2.3.53) into Eq. (2.3.45) yields

$$\Gamma_{n_s n_s : n'_s n'_s} = -\frac{2\pi}{\hbar^2} \sum_{n_b} \sum_{n'_b} \rho^{(b)}_{n'_b n'_b}(0) |H'_{n_s n_b, n'_s n'_b}|^2 \, \delta(\omega_{n_s n_b, n'_s n'_b})$$

$$= -k_{n'_s \to n_s}. \quad (2.3.54)$$

Similarly, $\Gamma_{n_s n_s : n_s n_s}$ can be calculated as

$$\langle \hat{M}_c(p)\rangle^{(2)}_{n_s n_s : n_s n_s}$$

$$= \frac{1}{\hbar^2} \sum_{n_b} \sum_{m_s} \sum_{m_b} \rho^{(b)}_{n_b n_b}(0) |H'_{m_s m_b, n_s n_b}|^2$$

$$\times \left(\frac{1}{p + i\omega_{m_s m_b, n_s n_b}} + \frac{1}{p + i\omega_{n_s n_b, m_s m_b}}\right) - \frac{1}{\hbar^2} \sum_{n_b} \sum_{n'_b} \rho^{(b)}_{n'_b n'_b}(0)$$

$$\times |H'_{n_s n_b, n_s n'_b}|^2 \left(\frac{1}{p + i\omega_{n_s n'_b, n_s n_b}} + \frac{1}{p + i\omega_{n_s n_b, n_s n'_b}}\right)$$

$$= \frac{1}{\hbar^2} \sum_{n_b} \sum_{m_s}^{m_s \neq n_s} {}' \sum_{m_b} \rho^{(b)}_{n_b n_b}(0) |H'_{m_s m_b, n_s n_b}|^2$$

$$\times \left(\frac{1}{p + i\omega_{m_s m_b, n_s n_b}} + \frac{1}{p + i\omega_{n_s n_b, m_s m_b}}\right), \quad (2.3.55)$$

$$\langle \hat{M}_c(\tau) \rangle^{(2)}_{n_s n_s : n_s n_s}$$

$$= \frac{2}{\hbar^2} \overset{m_s \neq n_s}{\underset{m_s}{\sum}}{}' \sum_{n_b} \sum_{m_b} \rho^{(b)}_{n_b n_b}(0) |H'_{m_s m_b, n_s n_b}|^2 \cos \tau \omega_{m_s m_b, n_s n_b}, \qquad (2.3.56)$$

and

$$\Gamma_{n_s n_s : n_s n_s} = \int_0^\infty d\tau \langle \hat{M}_c(\tau) \rangle^{(2)}_{n_s n_s : n_s n_s}$$

$$= \frac{2\pi}{\hbar^2} \overset{m_s \neq n_s}{\underset{m_s}{\sum}}{}' \sum_{n_b} \sum_{m_b} \rho^{(b)}_{n_b n_b}(0) |H'_{m_s m_b, n_s n_b}|^2 \delta(\omega_{m_s m_b, n_s n_b})$$

$$= - \overset{m_s \neq n_s}{\underset{m_s}{\sum}}{}' \Gamma_{m_s m_s : n_s n_s} = - \overset{n'_s \neq n_s}{\underset{n'_s}{\sum}}{}' \Gamma_{n'_s n'_s : n_s n_s}$$

$$= \overset{n'_s \neq n_s}{\underset{n'_s}{\sum}}{}' k_{n_s \to n'_s}. \qquad (2.3.57)$$

Using Eqs. (2.3.54) and (2.3.57), the master equation for $\rho^{(s)}_{n_s n_s}$ given by Eq. (2.3.47) can be written as

$$\frac{d\rho^{(s)}_{n_s n_s}}{dt} = \sum_{n'_s}{}' (k_{n'_s \to n_s} \rho^{(s)}_{n_s n'_s} - k_{n_s \to n'_s} \rho^{(s)}_{n_s n_s}). \qquad (2.3.58)$$

Next we consider the calculation of $\Gamma_{m_s n_s : m_s n_s}$. For this purpose, we calculate

$$\langle \hat{M}_c(p) \rangle^{(2)}_{m_s n_s : m_s n_s} = \sum_{n_b} \sum_{n'_b} \rho^{(b)}_{n_b n_b}(0) \left(\hat{L}_1 \frac{1}{p + i\hat{L}_0} \hat{L}_1 \right)_{MN : M'N'}, \qquad (2.3.59)$$

where $M \equiv m_s n_b$, $N \equiv n_s n_b$, $M' \equiv m_s n'_b$, and $N' \equiv n_s n'_b$. Notice that

$$\left(\hat{L}_1 \frac{1}{p + i\hat{L}_0} \hat{L}_1 \right)_{MN : M'N'} = \frac{1}{\hbar^2} \sum_{M''} \left(\frac{H'_{MM''} H'_{M''M'}}{p + i\omega_{M''N'}} \delta_{NN'} + \frac{H'_{N'M''} H'_{M''N}}{p + i\omega_{MM''}} \delta_{MM'} \right)$$

$$- \frac{1}{\hbar^2} H'_{N'N} H'_{MM'} \left(\frac{1}{p + i\omega_{M'N}} + \frac{1}{p + i\omega_{M'N}} \right)$$

$$(2.3.60)$$

and

$$\langle \hat{M}_c(p) \rangle^{(2)}_{m_s n_s : m_s n_s}$$

$$= \frac{1}{\hbar^2} \sum_{n_b} \sum_{m'_s} \sum_{m_b} \rho^{(b)}_{n_b n_b}(0) \left(\frac{|H'_{m_s n_b, m'_s m_b}|^2}{p + i\omega_{m'_s m_b, m_s n_b}} + \frac{|H'_{n_s n_b, m'_s m_b}|^2}{p + i\omega_{n_s n_b, m'_s m_b}} \right)$$

$$- \frac{2}{\hbar^2} \sum_{n_b} \sum_{n_b} \rho^{(b)}_{n_b n_b}(0) H'_{n_s n_b, n_s n_b} \frac{H'_{m_s n_b, m_s n_b}}{p + i\omega_{m_s n_b, m_s n_b}}. \qquad (2.3.61)$$

Substituting Eq. (2.3.61) into Eq. (2.3.45) yields

$$\Gamma_{m_sn_s:m_sn_s} = \frac{1}{\hbar^2} \sum_{m_s'} \sum_{m_b} \sum_{n_b} \rho^{(b)}_{n_bn_b}(0)(|H'_{m_sn_b,m_s'm_b'}|^2 \eta(\omega_{m_s'm_b',m_sn_b})$$

$$+ |H'_{n_sn_b,m_s'm_b'}|^2 \eta(\omega_{n_sn_b,m_s'm_b'})$$

$$- \frac{2}{\hbar^2} \sum_{n_b} \sum_{n_b'} \rho^{(b)}_{n_bn_b'}(0) H'_{n_sn_b',n_sn_b} H'_{m_sn_b,m_sn_b'} \eta(\omega_{n_b'n_b}), \quad (2.3.62)$$

where, for example,

$$\eta(\omega_{m_s'm_b',n_sn_b}) = \lim_{t \to \infty} \int_0^t d\tau \exp(-i\tau\omega_{m_s'm_b',n_sn_b}). \quad (2.3.63)$$

Notice that $\eta(x)$ can be written as

$$\eta(x) = \pi\delta(x) - iP(1/x), \quad (2.3.64)$$

where $P(1/x)$ means that the principal value of $1/x$ is to be taken. Using Eq. (2.3.64), Eq. (2.3.62) becomes

$$\Gamma_{m_sn_s:m_sn_s} = \Gamma_{m_sn_s:m_sn_s}(r) + i\Gamma_{m_sn_s:m_sn_s}(i) \quad (2.3.65)$$

where

$$\Gamma_{m_sn_s:m_sn_s}(r) = \tfrac{1}{2}(\Gamma_{m_sm_s:m_sm_s} + \Gamma_{n_sn_s:n_sn_s}) + \Gamma^{(d)}_{m_sn_s:m_sn_s}, \quad (2.3.66)$$

$$\Gamma^{(d)}_{m_sn_s:m_sn_s} = \frac{\pi}{\hbar^2} \sum_{n_b} \sum_{n_b'} \rho^{(b)}_{n_bn_b}(0)(H'_{n_sn_b,n_sn_b'} - H'_{m_sn_b',m_sn_b})^2 \delta(\omega_{n_b',n_b}), \quad (2.3.67)$$

and

$$\Gamma_{m_sn_s:m_sn_s}(i) = \Delta\omega_{m_sn_s}$$

$$= -\frac{1}{\hbar^2} P \sum_{n_b} \sum_{m_s'} \sum_{m_b'} \rho^{(b)}_{n_bn_b}(0)\left(\frac{|H'_{m_sn_b,m_s'm_b'}|^2}{\omega_{m_s'm_b',m_sn_b}} + \frac{|H'_{n_sn_b,m_s'm_b'}|^2}{\omega_{n_sn_b,m_s'm_b'}}\right)$$

$$+ \frac{2}{\hbar^2} P \sum_{n_b} \sum_{n_b'} \rho^{(b)}_{n_bn_b}(0) \frac{H'_{n_sn_b,n_sn_b'} H'_{m_sn_b',m_sn_b}}{\omega_{n_b,n_b'}} \quad (2.3.68)$$

The term $\Gamma_{m_sn_s:m_sn_s}(r)$ is usually called the dephasing rate constant; it consists of the inelastic part $\tfrac{1}{2}(\Gamma_{m_sm_s:m_sm_s} + \Gamma_{n_sn_s:n_sn_s})$ and the elastic part $\Gamma^{(d)}_{m_sn_s:m_sn_s}$, which is called the pure dephasing. The term $\Gamma_{m_sn_s:m_sn_s}(i)$ or $(\Delta\omega_{m_sn_s})$ represents the energy level shift of the system due to its interaction with the heat bath.   Equation (2.3.48) can then be written as

$$\frac{d\rho^{(s)}_{m_sn_s}}{dt} = -i(\omega_{m_sn_s} + \Delta\omega_{m_sn_s})\rho^{(s)}_{m_sn_s} - \Gamma_{m_sn_s:m_sn_s}(r)\rho^{(s)}_{m_sn_s}. \quad (2.3.69)$$

The phase of the system $\rho^{(s)}_{m_s n_s}$ is known to determine the spectral position through $\omega_{m_s n_s} + \Delta\omega_{m_s n_s}$ and the Lorentzian spectral bandwidth through the dephasing constant $\Gamma_{m_s n_s : m_s n_s}(r)$. From the preceding discussion, we can see that the bandwidth is determined not only by the total decay rates $\Gamma_{m_s m_s : m_s m_s}$ and $\Gamma_{n_s n_s : n_s n_s}$ but also by the pure dephasing $\Gamma^{(d)}_{m_s n_s : m_s n_s}$. Due to the interaction between the system and the heat bath, there will be a spectral shift $\Delta\omega_{m_s n_s}$; which in the case of a condensed phase will vary with temperature.

2.3.3 Application of the Density Matrix Method to Optical Spectroscopy

In this section, we shall present a density matrix formalism for calculating the rates of multiphoton processes in which damping effects originating from nonadiabatic interaction in the molecular system and/or from the molecule–heat bath interaction are properly taken into account so that the rate expression obtained can be used to discuss the optical phenomena in the resonance region and to study the temperature effect on the rates of multiphoton processes. We first derive a master equation for multiphoton processes by using the projection operator method (Agarwal, 1974; Fujimura and Lin, 1981). The expressions for the one-, two-, and three-photon transition probabilities are then given by using the derived master equation.

We consider a total system consisting of the molecular system, heat bath, and radiation field. The Liouville equation that describes the motion of the density matrix for the total system is given by

$$\partial\hat{\rho}/\partial t = -i\hat{L}\hat{\rho}, \tag{2.3.70}$$

where the Liouville operator $\hat{L}$ is defined as

$$\hat{L} = [\hat{H}, \cdot]/\hbar, \tag{2.3.71}$$

in which the Hamiltonian $\hat{H}$ of the total system has the form

$$\hat{H} = \hat{H}_0 + \hat{H}', \tag{2.3.72}$$

with $\hat{H}_0 = \hat{H}_s + \hat{H}_b + \hat{H}_r$ and $\hat{H}' = \hat{H}'_{sr} + \hat{H}'_{sb}$. Here the Hamiltonians with the subscripts s, b, and r denote those of the molecular system, the heat bath, and the radiation field, respectively. The Liouville operators will be denoted by $\hat{L}^0_s$, $\hat{L}^0_b$, and $\hat{L}^0_r$. Here $\hat{H}'_{sr}$ and $\hat{H}'_{sb}$ represent the interaction Hamiltonian between the system and the radiation field and that between the system and the heat bath, respectively. The interaction Liouville operators will be denoted by $\hat{L}'_{sr}$ and $\hat{L}'_{sb}$.

Utilizing the Laplace transform [see Eqs. (2.3.28) and (2.3.29)], the expression for the density matrix of the total system [Eq. (2.3.70)] can be transformed as

$$(p + i\hat{L})\hat{\rho}(p) = \hat{\rho}(0), \tag{2.3.74}$$

where $\hat{\rho}(0)$ represents $\hat{\rho}(t)$ at $t = 0$ and is assumed to be

$$\hat{\rho}(0) = \hat{\rho}^{(s)}(0)\hat{\rho}^{(b)}(0)\hat{\rho}^{(r)}(0),$$

in which the density operators with superscripts s, b, and r denote those for the molecular system, the heat bath, and the radiation field in equilibrium, respectively.

Applying the projection operator defined by $\hat{A} = \rho^{(b)}\,\mathrm{Tr}_b$ to both sides of Eq. (2.3.74) from the left-hand side, we obtain

$$\hat{A}(p + i\hat{L}_0 + i\hat{L}')\hat{A}\hat{\rho} + \hat{A}(p + i\hat{L}_0 + i\hat{L}')(1 - \hat{A})\hat{\rho} = \hat{A}\hat{\rho}. \tag{2.3.75}$$

Similarly, applying the complement of the operator $\hat{A}$, $(1 - \hat{A})$, to Eq. (2.3.74), we obtain

$$(1 - \hat{A})(p + i\hat{L}_0 + i\hat{L}')\hat{A}\hat{\rho}(p) + (1 - \hat{A})(p + i\hat{L}_0 + i\hat{L}')(1 - \hat{A})\hat{\rho}(p) = 0 \tag{2.3.76}$$

or

$$(1 - \hat{A})\hat{\rho}(p) = -[(1 - \hat{A})(p + i\hat{L}_0 + i\hat{L}')(1 - \hat{A})]^{-1}(1 - \hat{A})(i\hat{L}')\hat{A}\hat{\rho}(p), \tag{2.3.77}$$

where $\hat{A}\hat{L}' = 0$ has been assumed. Substituting Eq. (2.3.77) into Eq. (2.3.75) yields

$$\hat{A}(p + i\hat{L}_0 + i\hat{L}')\hat{A}\hat{\rho}(p) + \hat{A}\hat{L}'[(1 - \hat{A})(p + i\hat{L}_0 + i\hat{L}')(1 - \hat{A})]^{-1}\hat{L}'\hat{A}\hat{\rho}(p)$$
$$= \hat{A}\hat{\rho}(0), \tag{2.3.78}$$

which can be rewritten as

$$[p + i\hat{L}_s + i\hat{L}_r + \langle\Sigma(p)\rangle_b]\hat{\rho}^{(sr)}(p) = \hat{\rho}^{(sr)}(0), \tag{2.3.79}$$

where $\hat{\rho}^{(sr)} = \mathrm{Tr}_b\,\hat{\rho}$ and $\langle\Sigma(p)\rangle_b = \mathrm{Tr}_b\,\Sigma(p)\hat{\rho}^{(b)}(0)$. Here $\Sigma(p)$ is defined as

$$\Sigma(p) = i\hat{L}'_{sb} + i\hat{L}'_{sb}(1 - \hat{A})(p + i\hat{L})^{-1}(1 - \hat{A})\hat{L}'_{sb} \tag{2.3.80}$$

or

$$\Sigma(p) = i\hat{L}'_{sb} - i\hat{L}'_{sb}(1 - \hat{A})(p + i\hat{L}^0_s + i\hat{L}'_{sr})^{-1}(1 - \hat{A})\Sigma(p). \tag{2.3.81}$$

By applying the projection operator defined by $\hat{\alpha} = \hat{\rho}^{(r)}(0)\,\mathrm{Tr}_R$ to Eq. (2.3.79) as we did previously, we can trace out the radiation field variables to obtain the expression for the equation of the reduced density matrix for the molecular system,

$$[p + i\hat{L}_s^0 + \{\hat{M}(p)\}]\hat{\rho}^{(s)}(p) = \hat{\rho}^{(s)}(0), \qquad (2.3.82)$$

where $\hat{\rho}^{(s)}(p) = \text{Tr}_R \, \hat{\rho}^{(sr)}(p)$ and

$$(1 - \hat{\alpha})\hat{\rho}^{(sr)} = -\{(1 - \hat{\alpha})[p + i\hat{L}_s^0 + i\hat{L}_r^0 + i\hat{L}_{sr}' + \langle\Sigma(p)\rangle_b](1 - \hat{\alpha})\}^{-1}$$
$$\times \, [i\hat{L}_{sr}' + \langle\Sigma(p)\rangle_b]\hat{\alpha}\hat{\rho}^{(sr)}, \qquad (2.3.83)$$

where $\{\cdots\} = \text{Tr}_R \cdots \hat{\rho}^{(r)}(0)$ and

$$\hat{M}(p) = i\hat{L}_{sr}' + \langle\Sigma(p)\rangle_b - [i\hat{L}_{sr}' + \langle\Sigma(p)\rangle_b](1 - \hat{\alpha})$$
$$\times \, [p + i\hat{L}_s^0 + i\hat{L}_r^0 + i\hat{L}_{sr}' + \langle\Sigma(p)\rangle_b]^{-1}(1 - \hat{\alpha})$$
$$\times \, [i\hat{L}_{sr}' + \langle\Sigma(p)\rangle_b]. \qquad (2.3.84)$$

Applying the inverse Laplace transformation to Eq. (2.3.82), we obtain the equation for the reduced density matrix $\hat{\rho}^{(s)}(t)$ as

$$\frac{d\hat{\rho}^{(s)}(t)}{dt} = -i\hat{L}_s\hat{\rho}^{(s)}(t) - \int_0^t d\tau \, \{\hat{M}(\tau)\}\hat{\rho}^{(s)}(t - \tau), \qquad (2.3.85)$$

where

$$\{\hat{M}(\tau)\} = \frac{1}{2\pi i} \int_{-i\infty + c}^{i\infty + c} dp \, \{\hat{M}(p)\} \exp(p\tau). \qquad (2.3.86)$$

In the Markov approximation, that is, where the molecule–radiation field interaction time is instantaneously short compared with the time of the change of the molecular system density matrix, Eq. (2.3.85) becomes

$$\frac{d\hat{\rho}^{(s)}(t)}{dt} = (-i\hat{L}_s - \hat{k})\hat{\rho}^{(s)}(t), \qquad (2.3.87)$$

where

$$\hat{k} = \lim_{t \to \infty} \int_0^\infty d\tau \, \{\hat{M}(\tau)\} \exp(i\hat{L}_s\tau). \qquad (2.3.88)$$

Equation (2.3.87) is the master equation for the multiphoton processes of the molecular system in the presence of the heat bath. The transition operator $\hat{M}(p)$ can be expanded as

$$\{\hat{M}(p)\} = \{\hat{M}^{(0)}(p)\} + \{\hat{M}^{(1)}(p)\} + \{\hat{M}^{(2)}(p)\} + \{\hat{M}^{(3)}(p)\} + \cdots, \quad (2.3.89)$$

where the transition operators with superscripts related to the photon number involved in the processes are defined as

$$\{\hat{M}^{(0)}(p)\} = \{\langle\Sigma(p)\rangle_b\} = \{\langle\hat{L}_{sb}'\hat{G}(p)\hat{L}_{sb}'\rangle_b\}, \qquad (2.3.90)$$

which represents the transition operator for radiationless processes in the presence of the radiation field,

$$\{\hat{M}^{(1)}(p)\} = \{\hat{L}'_{sr}\hat{G}(p)\hat{L}'_{sr}\}, \tag{2.3.91}$$

$$\{\hat{M}^{(2)}(p)\} = -\{\hat{L}'_{sr}\hat{G}(p)\hat{L}'_{sr}\hat{G}(p)\hat{L}'_{sr}\hat{G}(p)\hat{L}'_{sr}\}, \tag{2.3.92}$$

and

$$\{\hat{M}^{(3)}(p)\} = \{\hat{L}'_{sr}\hat{G}(p)\hat{L}'_{sr}\hat{G}(p)\hat{L}'_{sr}\hat{G}(p)\hat{L}'_{sr}\hat{G}(p)\hat{L}'_{sr}\} \tag{2.3.93}$$

for the one-, two-, and three-photon processes, respectively. Here

$$\hat{G}(p) = [p + i\hat{L}^0_s + i\hat{L}^0_r + \langle\Sigma(p)\rangle_b]^{-1}. \tag{2.3.94}$$

The transition operator for the n-photon processes can be generated from the operator for the $(n-1)$-photon processes (Fujimura and Lin, 1983)

$$\hat{C}^{(n-1)}(p) = \hat{G}(p)\hat{M}^{(n-1)}(p)\hat{G}(p) \tag{2.3.95}$$

as

$$\hat{M}^{(n)}(p) = -\hat{L}'_{sr}\hat{C}^{(n-1)}(p)\hat{L}'_{sr} \qquad (n \geq 2). \tag{2.3.96}$$

The matrix element of the transition operator is given by

$$\hat{M}^{(n)}(p)_{NN:AA}$$

$$= \frac{1}{\hbar^2}\sum_L\sum_M[\{H'_{NM}H'_{LA}\hat{C}^{(n-1)}(p)_{MN:LA} + H'_{MN}H'_{AL}\hat{C}^{(n-1)}(p)_{NM:AL}\}$$

$$- \{H'_{NM}H'_{AL}\hat{C}^{(n-1)}(p)_{MN:AL} + H'_{MN}H'_{LA}\hat{C}^{(n-1)}(p)_{NM:LA}\}], \tag{2.3.97}$$

where $\hat{H}' = \hat{H}'_{sr}$, $N = nn_r$, $M = mm_r$, and so on. Here the subscript s for molecular system has been omitted.

We note that taking the trace over the radiation field variables in Eq. (2.3.97) can be written as

$$\{\hat{M}^{(n)}(p)\}_{nn:aa} = \sum_{n_r}\sum_{a_r}\rho^{(r)}_{a_r a_r}(0)[\hat{M}^{(n)}(p)]_{nn_r,nn_r:aa_r,aa_r}$$

$$= \sum_{n_r}\sum_{a_r}\rho^{(r)}_{a_r a_r}(0)[\hat{M}^{(n)}(p)]_{NN:AA}. \tag{2.3.98}$$

In deriving Eq. (2.3.97), the following expression for the matrix element of the interaction Liouville operator $\hat{L}'_{sr}$ has been utilized:

$$(\hat{L}'_{sr})_{NN:LM} = (1/\hbar)[(\hat{H}'_{sr})_{NL}\delta_{NM} - (\hat{H}'_{sr})_{MN}\delta_{NL}]. \tag{2.3.99}$$

We are now in a position to evaluate the one-, two-, and three-photon transition probabilities. Let us first consider the one-photon processes. The one-photon transition probabilities are defined as

$$W^{(1)} = \sum_a \sum_n \lim_{t \to \infty} \frac{d\rho_{nn}^{(s)}(t)}{dt} = \sum_a \sum_n \lim_{t \to \infty} \int_0^t d\tau \, \{\hat{M}^{(1)}(\tau)\}_{nn:aa} \hat{\rho}_{aa}^{(s)}(0),$$

(2.3.100)

where

$$\{\hat{M}^{(1)}(p)\}_{nn:aa} = \{\hat{L}'_{sr} G(p) \hat{L}'_{sr}\} \simeq \{\hat{L}'_{sr} \hat{G}^{(0)}(p) \hat{L}'_{sr}\}_{nn:aa}$$

$$= \sum_{n_r} \sum_{a_r} \rho_{a_r a_r}^{(r)}(0) [\hat{L}'_{sr} \hat{G}^{(0)}(p) \hat{L}'_{sr}]_{nn_r nn_r : aa_r aa_r}$$

$$= \sum_{n_r} \sum_{a_r} \rho_{a_r a_r}^{(r)}(0) [\hat{M}^{(1)}(p)]_{NN:AA}$$

(2.3.101)

Here the effect of the radiation field–molecule interaction in $\hat{G}(p)$ has been neglected;

$$\hat{G}^{(0)}(p) = [p + i\hat{L}_s^0 + i\hat{L}_r^0 + \langle \Sigma^{(0)}(p) \rangle_b]^{-1},$$

(2.3.102)

with

$$\langle \Sigma^{(0)}(p) \rangle_b = \langle \hat{L}'_{sb}(p + i\hat{L}_s^0 + i\hat{L}_b^0)^{-1} \hat{L}'_{sb} \rangle.$$

(2.3.103)

The matrix element of $\hat{M}^{(1)}(p)$ [Eq. (2.3.98)] can be expressed as

$$M^{(1)}(p)_{NN:AA} = -\frac{1}{\hbar^2} \sum_{LM} [\{H'_{NM} H'_{LA} G^{(0)}(p)_{MA} + H'_{MN} H'_{AL} G^{(0)}(p)_{AM}\} \delta_{NA} \delta_{ML}$$

$$- \{H'_{NM} H'_{AL} G^{(0)}(p)_{AN} + H'_{MN} H'_{LA} G^{(0)}(p)_{NA}\} \delta_{MA} \delta_{NL}].$$

(2.3.104)

Since $A \neq N$ for the absorption and emission processes, the preceding equation can be reduced to

$$\hat{M}^{(1)}(p)_{NN:AA} = \frac{2}{\hbar^2} |H'_{NA}|^2 \operatorname{Re}[G^{(0)}(p)_{NA}]$$

(2.3.105)

where

$$G^{(0)}(p)_{NA} = G^{(0)}(p)_{NA:NA} = (p + i\omega_{nn_r, aa_r} + \Gamma_{na})^{-1}.$$

(2.3.106)

Here $\Gamma_{na} = \langle \Sigma^{(0)}(0) \rangle_{na:na}$ is the dephasing rate constant relevant to the n and a molecular states and has already been derived in Section 2.3.2 [see Eq. (2.3.65)].

From Eq. (2.3.86) with Eq. (2.3.103), we find

$$\{M^{(1)}(\tau)\}_{nn:aa} = \frac{1}{\hbar^2} \sum_{n_r} \sum_{a_r} \rho^{(r)}_{a_r a_r}(0) \, \mathrm{Re} \left| H'_{nn_r:aa_r} \right|^2$$

$$\times \exp[-(i\omega_{nn_r,aa_r} + i\Delta\omega_{na} + \Gamma_{na})]. \qquad (2.3.107)$$

In the dipole approximation, Eq. (2.3.107) becomes

$$\{\hat{M}^{(1)}(\tau)\}_{nn:aa}$$

$$= \frac{e^2}{\hbar^2 m^2} \sum_{l} \sum_{l'} \left(\frac{2\pi\hbar}{\omega_l L^3} \right)^{1/2} \left(\frac{2\pi\hbar}{\omega'_l L^3} \right)^{1/2} \mathrm{Re} \sum_{n_r} \sum_{a_r} \rho^{(r)}_{a_r a_r}(0)$$

$$\times \langle aa_r| \hat{e}_l \cdot \mathbf{P}(\hat{a}_l + \hat{a}_l^\dagger)|nn_r \rangle \langle nn_r|(\hat{a}_{l'}^\dagger + \hat{a}_{l'})\hat{e}_{l'} \cdot \mathbf{P}|aa_r \rangle$$

$$\times \exp[-\tau(i\omega_{nn_r,aa_r} + i\Delta\omega_{na} + \Gamma_{na})]. \qquad (2.3.108)$$

Here the quantized radiation field description has been adopted, and the following molecule–radiation field interaction in the dipole approximation has been used:

$$\hat{H}'_r = -\sum_l (\hat{e}_l \cdot \mathbf{P}) \frac{e}{m} \left(\frac{2\pi\hbar}{\omega_l L^3} \right)^{1/2} (\hat{a}_l + \hat{a}_l^\dagger), \qquad (2.3.109)$$

where the subscript l specifies the photon mode and the polarization, and $\hat{a}_l^\dagger$ and $\hat{a}_l$ are the boson operators and they satisfy the commutation relation $[\hat{a}_l, \hat{a}_l^\dagger] = 1$.

For all radiation fields, except for the case of coherent radiation, Eq. (2.3.108) can be expressed as

$$\{M^{(1)}(\tau)\}_{nn:aa}$$

$$= \frac{e^2}{\hbar^2 m^2} \sum_l \frac{2\pi\hbar}{\omega_l L^3} \mathrm{Re} \sum_{n_r} \sum_{a_r} \rho^{(r)}_{a_r a_r}(0) |\hat{e}_l \cdot \mathbf{P}_{na}|^2 [a_l \delta_{n_l, a_l - 1}$$

$$+ (a_l + 1)\delta_{n_l, a_l + 1}] \exp[-\tau(i\omega_{nn_r,aa_r} + i\Delta\omega_{na} + \Gamma_{na})]$$

$$= \frac{2\pi e^2}{m^2 \hbar L^3} \mathrm{Re} \sum_l \frac{|\hat{e}_l \cdot \mathbf{P}_{na}|^2}{\omega_l} \sum_{a_l} \rho^{(r)}_{a_l a_l}(0) \{a_l \exp[-i\tau(\omega_{na} - \omega_l)]$$

$$+ (a_l + 1) \exp[-i\tau(\omega_{na} + \omega_l)]\} \exp[-\tau(i\Delta\omega_{na} + \Gamma_{na})] \quad (2.3.110)$$

or

$$\{M^{(1)}(\tau)\}_{nn:aa}$$

$$= \frac{4\pi}{\hbar L^3} \mathrm{Re} \sum_l \frac{\omega_{na}^2}{\omega_l} |\hat{e}_l \cdot \mathbf{R}_{na}|^2 \sum_{a_l} \rho^{(r)}_{a_l a_l}(0) \{a_l \exp[-i\tau(\omega_{na} - \omega_l)]$$

$$+ (a_l + 1) \exp[-i\tau(\omega_{na} + \omega_l)]\} \exp[-\tau(i\Delta\omega_{na} + \Gamma_{na})] \quad (2.3.111)$$

The first and second terms of Eqs. (2.3.110) and (2.3.111) correspond to the absorption and emission processes, respectively.

Let us first consider one-photon absorption. Substituting the first term in Eq (2.3.111) into Eq. (2.3.100) yields

$$W_{abs}^{(1)} = \frac{4\pi}{\hbar L^3} \sum_n \sum_a \rho_{aa}^{(s)}(0) \sum_l \omega_{na} |\hat{e}_l \cdot \mathbf{R}_{na}|^2 \sum_{a_l} \rho_{a_l a_l}^{(r)}(0)$$

$$\times \, a_l \frac{\Gamma_{na}}{(\bar{\omega}_{na} - \omega_l)^2 + \Gamma_{na}^2}, \tag{2.3.112}$$

where $\bar{\omega}_{na} = \omega_{na} + \Delta\omega_{na}$ and $\omega_{na}^2/\omega_l \approx \omega_{na}$ have been used.

Depending on the statistical nature of the radiation field, the various types of density operator of the photon $\hat{\rho}^{(r)}(0)$ can be considered. For absorption of monochromatic radiation field, in which the density operator is given by

$$\hat{\rho}^{(r)}(0) = |a_r\rangle\langle a_r|, \tag{2.3.113}$$

Eq. (2.3.112) becomes

$$W_{abs}^{(1)} = \frac{4\pi}{\hbar} \frac{a_r}{L^3} \sum_a \sum_n \omega_{na} \, \rho_{aa}^{(s)}(0) |\hat{e}_r \cdot \mathbf{R}_{na}|^2 \frac{\Gamma_{na}}{(\bar{\omega}_{na} - \omega_r)^2 + \Gamma_{na}^2}. \tag{2.3.114}$$

This expression shows that the absorption spectrum is characterized by the Lorentzian band shape whose width Γ_{na} originates from nonadiabatic perturbations for large molecules and/or the molecule–heat bath interaction. Any width and level shift originating from the molecule–radiation interaction has not been taken into account because $\hat{G}^{(0)}(p)$ of Eq. (2.3.102) has been used instead of $\hat{G}(p)$. This width and level shift plays an important role in the case in which nonradiative interactions and the molecule–heat bath interaction can be neglected.

The density operator for the thermal excitation of photons in a single mode at temperature T is given by

$$\hat{\rho}^{(r)}(0) = [1 - \exp(-\beta\hbar\omega_r)] \sum_{a_r} \exp(-\beta a_r \hbar\omega_r) |a_r\rangle\langle a_r|, \tag{2.3.115}$$

and its diagonal matrix element is expressed as

$$\rho_{a_r a_r}^{(r)}(0) = [1 - \exp(-\beta\hbar\omega_r)] \exp(-\beta a_r \hbar\omega_r), \tag{2.3.116}$$

where $\beta = 1/k_B T$ and k_B denotes Boltzmann's constant.

$$W_{abs}^{(1)} = \frac{4\pi}{\hbar} \frac{\bar{a}_r}{L^3} \sum_a \sum_n \omega_{na} \, \rho_{aa}^{(s)}(0) |\hat{e}_r \cdot \mathbf{R}_{na}|^2 \frac{\Gamma_{na}}{(\bar{\omega}_{na} - \omega_r)^2 + \Gamma_{na}}, \tag{2.3.117}$$

where $\bar{a}_r$ denotes the mean photon number, defined as

$$\bar{a}_r = [\exp(\beta\hbar\omega_r) - 1]^{-1}. \tag{2.3.118}$$

In deriving Eq. (2.3.117), the following relation has been used:

$$\sum_{a_r} a_r \exp(-\beta a_r \hbar \omega_r) = -\frac{\partial}{\partial \beta \hbar \omega_r} \sum_{a_r} \exp(-\beta a_r \hbar \omega_r)$$

$$= -\frac{\partial}{\partial \beta \hbar \omega_r} [1 - \exp(-\beta \hbar \omega_r)]^{-1}. \quad (2.3.119)$$

The one-photon absorption transition probability has the same structure as that in the case of monochromatic photon absorption.

Another important density operator is that for coherent states. Coherent states can be generated in a laser operating above its threshold (Loudon, 1973). The density operator at $t = 0$ for the radiation field in a coherent state has the form (Appendix I, Section B)

$$\hat{\rho}^{(r)}(0) = |\alpha_r\rangle\langle\alpha_r|. \quad (2.3.120)$$

From the definition of the coherent state, the eigenket is an eigenfunction of the annihilation operator $\hat{a}_r$:

$$\hat{a}_r|\alpha_r\rangle = \alpha_r|\alpha_r\rangle, \quad (2.3.121)$$

where α_r is complex and is written in terms of the amplitude $|\alpha_r|$ and the phase ϕ as (Loudon, 1973)

$$\alpha_r = |\alpha_r| \exp(i\phi). \quad (2.3.122)$$

In this case, because of nonorthogonality between the coherent states $\langle\alpha_r|\beta_r\rangle \neq 0$, the matrix elements relevant to the photon field in Eq. (2.3.108) can be written as

$$\sum_{\beta_k}\sum_{\alpha_k}\langle\alpha_k|\rho^{(r)}_{\alpha_r\alpha_r}(0)(\hat{a}_r + \hat{a}_r^\dagger)|\beta_k\rangle\langle\beta_k|(\hat{a}_r^\dagger + \hat{a}_r)|\alpha_k\rangle \exp[\tau(i\omega_{\alpha_k} - l\omega_{\beta_k})]$$

$$= \langle\alpha_r|(\hat{a}_r + \hat{a}_r^\dagger)[\exp(-i\omega_r t)\hat{a}_r^\dagger + \exp(i\omega_r t)\hat{a}_r]|\alpha_r\rangle$$

$$= \langle\alpha_r|\hat{a}_r^\dagger\hat{a}_r|\alpha_r\rangle \exp(i\omega_r t) + \langle\alpha_r|\hat{a}_r\hat{a}_r^\dagger|\alpha_r\rangle \exp(-i\omega_r t)$$

$$+ \langle\alpha_r|(\hat{a}_r^\dagger)^2|\alpha_r\rangle \exp(-i\omega_r t) + \langle\alpha_r|(\hat{a}_r)^2|\alpha_r\rangle \exp(i\omega_r t)$$

$$= |\alpha_r|^2 \exp(i\omega_r t) + (1 + |\alpha_r|^2) \exp(-i\omega_r t) + |\alpha_r|^2 \exp[i(\omega_r t + 2\phi)]$$

$$+ |\alpha_r|^2 \exp[-i(\omega_r t + 2\phi)]. \quad (2.3.123)$$

The terms with the phase factor ϕ make no significant contribution to one-photon processes for $t \to \infty$. The first and second terms of Eq. (2.3.123) express the one-photon absorption and emission of radiation in the coherent state, respectively. The one-photon absorption probability is then given by

$$W^{(1)}_{abs} = \frac{4\pi|\alpha_r|^2}{\hbar L^3} \sum_a \sum_n \rho^{(s)}_{aa}(0)\omega_{na}|\hat{\mathbf{e}}_r \cdot \mathbf{R}_{na}|^2 \frac{\Gamma_{na}}{(\bar{\omega}_{na} - \omega_r)^2 + \Gamma_{na}^2}. \quad (2.3.124)$$

We should note that $W_{abs}^{(1)}$, Eq. (2.3.114), and Eq. (2.3.117) represent the transition probabilities (or rate constants in reciprocal seconds) of the one-photon absorption. The absorption cross section (or absorption coefficient per molecule) $\alpha_{abs}^{(1)}$ can be obtained from $W_{abs}^{(1)}$ by using the relation of intensity $I = \hbar\omega_r a_r c/L^3$, $\hbar\omega_r \bar{a}_r c/L^3$, or $\hbar\omega_r |\alpha_r|^2 c/L^3$ (in units of erg s^{-1} cm^{-2}), where c is the velocity of light, that is,

$$W_{abs}^{(1)} = (\alpha_{abs}^{(1)}/\hbar\omega_r)I. \qquad (2.3.125)$$

Here

$$\alpha_{abs}^{(1)} = \frac{4\pi}{hc}\sum_a \sum_n \omega_{na}\rho_{aa}^{(s)}(0)|\hat{e}_r \cdot \mathbf{\mu}_{na}|^2 \frac{\Gamma_{na}}{(\bar{\omega}_{na} - \omega_r)^2 + \Gamma_{na}^2} \qquad (2.3.126)$$

in units of square centimeters.

For randomly oriented systems, the spatial average of $\alpha_{abs}^{(1)}$ given by the preceding equation has to be carried out; in this case, $|\hat{e}_r \cdot \mathbf{\mu}_{na}|^2$ is to be replaced by $\frac{1}{3}|\mathbf{\mu}_{na}|^2$.

Next we consider one-photon emission. From Eq. (2.3.111), the transition probability $W_{emi}^{(1)}$ is written as

$$W_{emi}^{(1)} = \frac{4\pi}{\hbar L^3}\sum_a \sum_n \omega_{an}\rho_{aa}^{(s)}(0)\sum_l \sum_{a_l}\rho_{a_l a_l}^{(r)}(0)(1 + a_l)|\hat{e}_l \cdot \mathbf{R}_{na}|^2$$

$$\times \frac{\Gamma_{na}}{(\bar{\omega}_{an} - \omega_l)^2 + \Gamma_{na}^2}, \qquad (2.3.127)$$

where $\bar{\omega}_{an} = \omega_{an} + \Delta\omega_{an}$.

For spontaneous emission, that is, with no radiation at $t = 0$, we set $a_l = 0$ and, because $\sum_{a_l}\rho_{a_l a_l}^{(r)}(0) = 1$, Eq. (2.3.127) becomes

$$W_{spon}^{(1)} = \frac{4\pi}{\hbar L^3}\sum_a \sum_n \omega_{an}\rho_{aa}^{(s)}(0)\sum_l |\hat{e}_l \cdot \mathbf{R}_{na}|^2 \frac{\Gamma_{na}}{(\bar{\omega}_{an} - \omega_l)^2 + \Gamma_{na}^2}. \qquad (2.3.128)$$

Here we should note that summing over the mode of radiation l, or integration over ω_1 [see Eq. (2.1.87)], leads to divergence of the total spontaneous emission probability, because a Lorentzian band shape with constant width has been assumed. This divergence originates from using the Markov approximation: In this approximation the decay width has been assumed to be independent of the photon frequency. Far from the maximum band intensity of the spectrum, this approximation in general breaks down, since the molecule–radiation interaction time becomes of the same order as that of the molecule–heat bath interaction. Equation (2.3.128) therefore should be restricted to the ranges of the band maximum of the spectrum.

For induced emission, the transition probability can be expressed as

$$W_{\text{ind}}^{(1)} = \frac{4\pi}{\hbar L^3} \sum_a \sum_n \omega_{an} \rho_{aa}^{(s)}(0) \sum_l \sum_{a_l} \rho_{a_l a_l}^{(r)}(0) a_l |\hat{e}_l \cdot \mathbf{R}_{na}|^2$$

$$\times \frac{\Gamma_{na}}{(\bar{\omega}_{an} - \omega_l)^2 + \Gamma_{na}^2}. \qquad (2.3.129)$$

For monochromatic radiation in which the bandwidth is narrower than Γ_{na}, Eq. (2.3.129) can be reduced to

$$W_{\text{ind}}^{(1)} = \frac{4\pi}{\hbar L^3} \sum_a \sum_n \omega_{an} \rho_{aa}^{(s)}(0) a_r |\hat{e}_r \cdot \mathbf{R}_{na}|^2 \frac{\Gamma_{na}}{(\bar{\omega}_{an} - \omega_r)^2 + \Gamma_{na}^2}, \qquad (2.3.130)$$

which has the same structure as that of the one-photon absorption transition probability, Eq. (2.3.114). As we have just seen, for radiation fields character-ized by thermal excitation of photons and by coherent states, $W_{\text{ind}}^{(1)}$ can be derived and it has the same structure as that of Eq. (2.3.130).

Next let us consider two-photon processes. One of the merits of apply-ing the density matrix method to multiphoton processes is that one can take into account the effects of dephasings that originate from the molecule–heat bath interaction. The derived transition probabilities of multiphoton transi-tions involve not only those of simultaneous processes but also those of sequential ones that cannot be obtained by using the conventional time-dependent perturbation discussed in Section 2.1. We can see how the transi-tion probabilities of the different types of processes can be derived by using the density matrix method.

From Eq. (2.3.88), the two-photon transition probability $W^{(2)}$ is given by

$$W^{(2)} = \lim_{t \to \infty} \sum_a \sum_n \int_0^t d\tau \frac{1}{2\pi i} \int_{-i\infty+c}^{i\infty+c} dp \exp(p\tau)\{\hat{M}^{(2)}(p)\}_{nn:aa} \rho_{aa}^{(s)}(0), \qquad (2.3.131)$$

where the matrix element of $\hat{M}^{(2)}(p)$ has the form

$$\{M^{(2)}(p)\}_{nn:aa} = \sum_{n_r} \sum_{a_r} \rho_{a_r a_r}^{(r)}(0) [M^{(2)}(p)]_{NN:AA}. \qquad (2.3.132)$$

With Eq. (2.3.97), this matrix element can be expressed as

$$[M^{(2)}(p)]_{NN:AA} = \frac{1}{\hbar^2} \sum_M \sum_L [H'_{NM} H'_{LA} G^{(0)}(p)_{MN} G^{(0)}(p)_{LA} (\hat{M}^{(1)})_{MN:LA}$$

$$+ H'_{MN} H'_{AL} G^{(0)}(p)_{NM} G^{(0)}(p)_{AL} (\hat{M}^{(1)})_{NM:AL}$$

$$- H'_{NM} H'_{AL} G^{(0)}(p)_{MN} G^{(0)}(p)_{AL} (\hat{M}^{(1)})_{MN:AL}$$

$$- H'_{MN} H'_{LA} G^{(0)}(p)_{NM} G^{(0)}(p)_{LA} (\hat{M}^{(1)})_{NM:LA}] \qquad (2.3.133)$$

or

$$M^{(2)}(p)_{NN:AA} = \frac{2}{\hbar^2} \operatorname{Re} \sum_M \sum_L [H'_{NM} H'_{LA} G^{(0)}(p)_{MN} G^{(0)}(p)_{LA} (\hat{M}^{(1)})_{MN:LA}$$

$$- H'_{NM} H'_{AL} G^{(0)}(p)_{MN} G^{(0)}(p)_{AL} (\hat{M}^{(1)})_{MN:AL}], \qquad (2.3.134)$$

where the symmetry properties $M^{(1)}_{IJ:KL} = M^{(1)*}_{JI:LK}$ and $G^{(0)}(p)_{IJ} = G^{(0)*}(p)_{JI}$ have been used. Noting that for $H'_{AN} \neq 0$, we obtain

$$(M^{(1)})_{MN:LA} = \frac{1}{\hbar^2} \sum_I H'_{IN} H'_{AI} G^{(0)}(p)_{MI} \delta_{ML} \qquad (2.3.135)$$

and

$$(M^{(1)})_{MN:AL} = \frac{1}{\hbar^2} \sum_I [\delta_{NL} H'_{MI} H'_{IA} G^{(0)}(p)_{IN} + \delta_{MA} H'_{IN} H'_{LI} G^{(0)}(p)_{AI}]$$

$$- \frac{1}{\hbar^2} H'_{MA} H'_{LN} [G^{(0)}(p)_{ML} + G^{(0)}(p)_{AN}], \qquad (2.3.136)$$

Eq. (2.3.134) becomes

$$M^{(2)}(p)_{NN:AA} = -\frac{2}{\hbar^2} \operatorname{Re} \sum_M \sum_L H'_{AL} H'_{LN} H'_{NM} H'_{MA} \{ G^{(0)}(p)_{MN} G^{(0)}(p)_{AL} G^{(0)}(p)_{AN}$$

$$+ G^{(0)}(p)_{MN} G^{(0)}(p)_{ML} [G^{(0)}(p)_{MA} + G^{(0)}(p)_{AL}] \}. \qquad (2.3.137)$$

This expression has been derived by Lin and Eyring (1977a). It is possible to decompose Eq. (2.3.137) into three terms, that is, simultaneous, sequential, and mixed terms (Fujimura and Lin, 1981):

$$M^{(2)}(p)_{NN:AA} = M^{(2)}_{\text{sim}}(p)_{NN:AA} + M^{(2)}_{\text{seq}}(p)_{NN:AA} + M^{(2)}_{\text{mix}}(p)_{NN:AA}. \qquad (2.3.138)$$

This can be accomplished by using the relation

$$G^{(0)}(p)_{MN} G^{(0)}(p)_{AL} G^{(0)}(p)_{AN} = G^{(0)}(p)_{MA} G^{(0)}(p)_{AL} G^{(0)}(p)_{AN}$$

$$+ [G^{(0)}(p)_{MN} - G^{(0)}(p)_{MA}]$$

$$\times G^{(0)}(p)_{AL} G^{(0)}(p)_{AN}$$

$$= G^{(0)}(p)_{MA} G^{(0)}(p)_{AL} G^{(0)}(p)_{AN}$$

$$+ [-1 + (\Gamma_{MA} - \Gamma_{MN} + \Gamma_{AN}) G^{(0)}(p)_{AN}]$$

$$\times G^{(0)}(p)_{MN} G^{(0)}(p)_{MA} G^{(0)}(p)_{AL}. \qquad (2.3.139)$$

Each term in Eq. (2.3.138) is expressed as

$$
M_{\text{sim}}^{(2)}(p)_{NN:AA} = \frac{2}{\hbar^4} \, \text{Re} \, G^{(0)}(p)_{AN} \left| \sum_{M} H'_{NM} G^{(0)}(p)_{MA} H'_{MA} \right|^2,
\tag{2.3.140}
$$

$$
M_{\text{seq}}^{(2)}(p)_{NN:AA} = \frac{2}{\hbar^4} \, \text{Re} \sum_{M} \sum_{L} (\Gamma_{MA} - \Gamma_{ML} + \Gamma_{AL}) H'_{AL} H'_{LN} H'_{NM} H'_{MA} G^{(0)}(p)_{MN}
$$
$$
\times \, G^{(0)}(p)_{AL} G^{(0)}(p)_{ML} G^{(0)}(p)_{MA},
\tag{2.3.141}
$$

and

$$
M_{\text{mix}}^{(2)}(p)_{NN:AA} = \frac{2}{\hbar^4} \, \text{Re} \sum_{M} \sum_{L} (\Gamma_{MA} - \Gamma_{MN} + \Gamma_{AN}) H'_{AL} H'_{LN} H'_{NM} H'_{MA}
$$
$$
\times \, G^{(0)}(p)_{AN} G^{(0)}(p)_{MN} G^{(0)}(p)_{MA} G^{(0)}(p)_{AL}.
\tag{2.3.142}
$$

Substituting Eq. (2.3.137) into Eq. (2.3.131) yields

$$
W^{(2)} = \frac{2}{\hbar^4} \, \text{Re} \sum_{a} \sum_{n} \sum_{a_r} \sum_{n_r} \sum_{M} \sum_{K} \rho_{aa}^{(s)}(0) \rho_{a_r a_r}^{(r)}(0) \frac{H'_{AL} H'_{LN} H'_{NM} H'_{MA}}{i\omega_{MA} + \Gamma_{MA}}
$$
$$
\times \left[\frac{1}{(i\omega_{AL} + \Gamma_{AL})(i\omega_{AN} + \Gamma_{AN})} \right.
$$
$$
\left. + \frac{1}{(i\omega_{ML} + \Gamma_{ML})} \left(\frac{1}{(i\omega_{MA} + \Gamma_{MA})} + \frac{1}{(i\omega_{AL} + \Gamma_{AL})} \right) \right],
\tag{2.3.143}
$$

where bar over the frequency ω has been omitted for simplicity. In terms of the simultaneous, sequential, and mixed processes, the two-photon transition probability becomes

$$
W^{(2)} = W_{\text{sim}}^{(2)} + W_{\text{seq}}^{(2)} + W_{\text{mix}}^{(2)},
\tag{2.3.144}
$$

where the transition probability of the simultaneous process is given as

$$
W_{\text{sim}}^{(2)} = \frac{2}{\hbar^4} \sum_{a} \sum_{n} \sum_{a_r} \sum_{n_r} \rho_{aa}^{(s)}(0) \rho_{a_r a_r}^{(r)}(0) \frac{\Gamma_{NA}}{\omega_{NA}^2 + \Gamma_{NA}^2} \left| \sum_{M} \frac{H'_{NM} H'_{MA}}{i\omega_{MA} + \Gamma_{MA}} \right|^2,
\tag{2.3.145}
$$

where the subscripts in capital letters specify both the molecular state and radiation fields.

For the sequential process, we have

$$
W_{\text{seq}}^{(2)} = \frac{2}{\hbar^4} \, \text{Re} \sum_{a} \sum_{n} \sum_{a_r} \sum_{n_r} \rho_{aa}^{(s)}(0) \rho_{a_r a_r}^{(r)}(0) \sum_{M} \sum_{L} (\Gamma_{MA} - \Gamma_{ML} + \Gamma_{AL})
$$
$$
\times \frac{H'_{AL} H'_{LN} H'_{NM} H'_{MA}}{(i\omega_{MN} + \Gamma_{MN})(i\omega_{AL} + \Gamma_{AL})(i\omega_{MA} + \Gamma_{MA})(i\omega_{ML} + \Gamma_{ML})},
\tag{2.3.146}
$$

and for the mixed process

$$W_{\text{mix}}^{(2)} = \frac{2}{\hbar^4} \operatorname{Re} \sum_a \sum_n \sum_{a_r} \sum_{n_r} \rho_{aa}^{(s)}(0) \rho_{a_r a_r}^{(r)}(0) \sum_M \sum_L (\Gamma_{MA} - \Gamma_{MN} + \Gamma_{AN})$$

$$\times \frac{H'_{AL} H'_{LN} H'_{NM} H'_{MA}}{(i\omega_{AN} + \Gamma_{AN})(i\omega_{MN} + \Gamma_{MN})(i\omega_{MA} + \Gamma_{MA})(i\omega_{AL} + \Gamma_{AL})}. \quad (2.3.147)$$

Using Eq. (2.3.66), the factors involving the damping constants in $W_{\text{seq}}^{(2)}$ and $W_{\text{mix}}^{(2)}$ can be expressed as

$$\Gamma_{MA} - \Gamma_{ML} + \Gamma_{AL} = \Gamma_{AA} + \Gamma_{MA}^{(d)} - \Gamma_{ML}^{(d)} + \Gamma_{AL}^{(d)}$$

and

$$\Gamma_{MA} - \Gamma_{MN} + \Gamma_{AN} = \Gamma_{AA} + \Gamma_{MA}^{(d)} - \Gamma_{MN}^{(d)} + \Gamma_{AN}^{(d)}$$

for $W_{\text{seq}}^{(2)}$ and $W_{\text{mix}}^{(2)}$, respectively.

For ordinary molecular systems in which the population decay in the initial state is negligibly small, $\Gamma_{AA} = 0$, the previous factors can be considered to be expressed in terms of the pure dephasing constants relevant to the states involved in the transition. In other words, if the effects of the pure dephasings can be made negligibly small compared with the population decay constants of the intermediate and/or final states, the simultaneous process makes a dominant contribution to the two-photon process.

Equation (2.3.146) can be rewritten as

$$W_{\text{seq}}^{(2)} = \frac{2}{\hbar^4} \sum_a \sum_n \sum_{a_r} \sum_{n_r} \rho_{aa}^{(s)}(0) \rho_{a_r a_r}^{(r)}(0) \left[\sum_M \frac{(2\Gamma_{MA} - \Gamma_{MM})\Gamma_{MN}}{\Gamma_{MM}} \right.$$

$$\times \frac{|H'_{AM}|^2 |H'_{MN}|^2}{(\omega_{AM}^2 + \Gamma_{AM}^2)(\omega_{MN}^2 + \Gamma_{MN}^2)}$$

$$\left. + \sum_M \sum_{\substack{L \\ (M \neq L)}} \frac{(\Gamma_{MA} - \Gamma_{ML} + \Gamma_{AL}) H'_{AL} H'_{LN} H'_{NM} H'_{MA}}{(i\omega_{MN} + \Gamma_{MN})(i\omega_{AL} + \Gamma_{AL})(i\omega_{MA} + \Gamma_{MA})(i\omega_{ML} + \Gamma_{ML})} \right].$$

$$(2.3.148)$$

The first term represents the sequential two-photon processes, $a \to m \to n$ transitions. This term originates mainly from the pure dephasing between the ground and intermediate states. The role of pure dephasing in resonant light scattering has been discussed by Jones and Zewail (1978) and Fujimura et al. (1981).

The derived expressions for $W^{(2)}$ can be applied to any two-photon process. Since the damping effect that originates from the molecule–heat bath interaction is included, the expressions for $W^{(2)}$ can be applied to resonant two-photon processes. We shall apply Eq. (2.3.145) for $W^{(2)}$ to two-photon absorption (TPA). It has been shown (Fujimura and Lin, 1981) that $W^{(2)}$ for TPA can be separated into simultaneous TPA, sequential TPA, and their mixed terms, as seen previously. For simultaneous TPA, substituting Eqs. (2.2.93)–(2.1.96) into Eq. (2.3.145) and using Eq. (2.1.72), we find that

$$
W_{\text{sim}}^{(2)} = \frac{2}{\hbar^2}\left(\frac{2\pi e^2}{m^2 L^3}\right)^2 \sum_i \sum_f \sum_l \sum_{l'} \rho_{ii}^{(s)}(0)\rho_{ll}^{(r)}(0)\rho_{l'l'}^{(r)}(0)\frac{n_l n_{l'}'}{\omega_l \omega_{l'}}
$$

$$
\times \left| \sum_m \left[\frac{(\hat{e}_{l'} \cdot \mathbf{P}_{fm})(\hat{e}_l \cdot \mathbf{P}_{mi})}{i(\omega_{mi} - \omega_l) + \Gamma_{mi}} + \frac{(\hat{e}_l \cdot \mathbf{P}_{fm})(\hat{e}_{l'} \cdot \mathbf{P}_{mi})}{i(\omega_{mi} - \omega_{l'}) + \Gamma_{mi}} \right] \right|^2
$$

$$
\times \frac{\Gamma_{fi}}{(\omega_{fi} - \omega_l - \omega_{l'})^2 + \Gamma_{fi}^2}, \tag{2.3.149}
$$

where i, m, and f denote the initial, intermediate, and final vibronic states of molecules, respectively. This expression should be compared with Eq. (2.1.98). Equation (2.3.149) for $W_{\text{abs}}^{(2)}$ can be useful for studying the temperature and resonance effects on TPA and for analyzing the line shapes of the TPA spectra. For example, the preceding expression has been applied in discussing the temperature effect on resonant Raman scattering (Fujimura and Lin, 1979b).

Finally, we consider the three-photon processes. The three-photon transition probability $W^{(3)}$ is given by

$$
W^{(3)} = \sum_a \sum_n \lim_{t\to\infty} \int_0^t d\tau \frac{1}{2\pi i}\int_{-i\infty+c}^{i\infty+c} dp\,\exp(p\tau)\{\hat{M}^{(3)}(p)\}_{nn:aa}\rho_{aa}^{(s)}(0), \tag{2.3.150}
$$

where

$$
\{M^{(3)}(p)\}_{nn:aa} = \sum_{n_r}\sum_{a_r}\rho_{a_r a_r}^{(r)}(0)[\hat{M}^{(3)}(p)]_{NN:AA}, \tag{2.3.151}
$$

where $N = nn_r$ and $A = aa_r$. To calculate $W^{(3)}$, we have to evaluate $[\hat{M}^{(3)}(p)]_{NN:AA}$ first. The evaluation of $[\hat{M}^{(3)}(p)]_{NN:AA}$ using Eq. (2.3.97) is straightforward but somewhat tedious. For the case in which $A \neq N$ and $H'_{AN} = 0$, we find (Fujimura and Lin, 1983)

$$
[\hat{M}^{(3)}(p)]_{NN:AA} = -\frac{1}{\hbar^2}\sum_L\sum_M [H'_{NM}H'_{LA}C^{(2)}(p)_{MN:LA}
$$

$$
+ H'_{MN}H'_{AL}C^{(2)}(p)_{NM:AL} - H'_{NM}H'_{AL}C^{(2)}(p)_{MN:AL}
$$

$$
- H'_{MN}H'_{LA}C^{(2)}(p)_{MN:LA}], \tag{2.3.152}
$$

which can be reduced to the form

$$
[\hat{M}^{(3)}(p)]_{NN:AA} = [\hat{M}^{(3)}(p)]_{NN:AA}^{(\alpha)} + [\hat{M}^{(3)}(p)]_{NN:AA}^{(\beta)}. \tag{2.3.153}
$$

Here

$$[M^{(3)}(p)]_{NN:AA}^{(\alpha)}$$

$$= \frac{2}{\hbar^6} \operatorname{Re} \sum_I \sum_K \sum_L \sum_M H'_{AI} H'_{IK} H'_{KN} H'_{NM} H'_{ML} H'_{LA} \{ G^{(0)}_{MN} G^{(0)}_{LA}$$

$$\times [G^{(0)}_{LK} G^{(0)}_{LN} G^{(0)}_{LI} + G^{(0)}_{MI} G^{(0)}_{MK} G^{(0)}_{MA} + (G^{(0)}_{LK} + G^{(0)}_{MI}) G^{(0)}_{MK} G^{(0)}_{LI}]$$

$$+ G^{(0)}_{MN} G^{(0)}_{AI} [G^{(0)}_{LN} G^{(0)}_{LI} G^{(0)}_{LK} + G^{(0)}_{MK} G^{(0)}_{AK} G^{(0)}_{LK}$$

$$+ G^{(0)}_{LN} G^{(0)}_{AK} (G^{(0)}_{AN} + G^{(0)}_{LK}) + G^{(0)}_{MK} G^{(0)}_{LI} (G^{(0)}_{LK} + G^{(0)}_{MI})] \} \qquad (2.3.154)$$

and

$$[M^{(3)}(p)]_{NN:AA}^{(\beta)}$$

$$= -\frac{2}{\hbar^6} \operatorname{Re} \sum_I \sum_K \sum_L \sum_M H'_{AI} H'_{IN} H'_{NM} H'_{MK} H'_{KL} H'_{LA}$$

$$\times \{ G^{(0)}_{MN} G^{(0)}_{LA} [G^{(0)}_{KI} G^{(0)}_{KN} G^{(0)}_{KA} + G^{(0)}_{KI} G^{(0)}_{MI} G^{(0)}_{LI}$$

$$+ (G^{(0)}_{KI} + G^{(0)}_{MA}) G^{(0)}_{MI} G^{(0)}_{KA} + (G^{(0)}_{LN} + G^{(0)}_{KI}) G^{(0)}_{KN} G^{(0)}_{LI}]$$

$$+ G^{(0)}_{MN} G^{(0)}_{AI} [G^{(0)}_{KN} G^{(0)}_{AN} G^{(0)}_{LN} + G^{(0)}_{KN} G^{(0)}_{LI} (G^{(0)}_{LN} + G^{(0)}_{KI}) + G^{(0)}_{MI} G^{(0)}_{LI} G^{(0)}_{KI}] \}$$

$$- \frac{2}{\hbar^6} \operatorname{Re} \sum_I \sum_K \sum_L \sum_M H'^*_{AI} H'^*_{IN} H'^*_{NM} H'^*_{MK} H'^*_{KL} H'^*_{LA}$$

$$\times \{ G^{(0)*}_{NI} G^{(0)*}_{AI} G^{(0)*}_{KI} G^{(0)*}_{MI} G^{(0)*}_{LI} + G^{(0)*}_{NI} G^{(0)*}_{LA} [G^{(0)*}_{NA} G^{(0)*}_{KA} G^{(0)*}_{MA}$$

$$+ G^{(0)*}_{MI} G^{(0)*}_{LI} G^{(0)*}_{KI} + G^{(0)*}_{MI} G^{(0)*}_{KA} (G^{(0)*}_{MA} + G^{(0)*}_{KI})] \}, \qquad (2.3.155)$$

where $G^{(0)}_{MN} = G^{(0)}(p)_{MN}$, and so on. Equations (2.3.154) and (2.3.155) represent three-photon processes and the interference between two- and four-photon processes, respectively. Neglecting the interference effect, we have

$$[\hat{M}^{(3)}(p)]_{NN:AA} \simeq [\hat{M}^{(3)}(p)]_{NN:AA}^{(\alpha)}. \qquad (2.3.156)$$

Substituting Eq. (2.3.156) into Eq. (2.3.150) yields

$$W^{(3)} = \frac{2}{\hbar^6} \operatorname{Re} \sum_a \sum_n \sum_{a_r} \sum_{n_r} \rho^{(s)}_{aa}(0) \rho^{(r)}_{a_r a_r}(0) \sum_I \sum_K \sum_L \sum_M H'_{AI} H'_{MI} H'_{NM} H'_{ML} H'_{LA} H'_{KN}$$

$$\times \left\{ \frac{1}{(i\omega_{AN} + \Gamma_{AN})(i\omega_{AI} + \Gamma_{AI})(i\omega_{AK} + \Gamma_{AK})(i\omega_{MN} + \Gamma_{MN})(i\omega_{LN} + \Gamma_{LN})} \right.$$

$$+ \frac{1}{i\omega_{MN} + \Gamma_{MN}} \left[\left(\frac{1}{i\omega_{LA} + \Gamma_{LA}} + \frac{1}{i\omega_{AI} + \Gamma_{AI}} \right) \right.$$

$$\times \left. \left(\frac{1}{i\omega_{LK} + \Gamma_{LK}} + \frac{1}{i\omega_{MI} + \Gamma_{MI}} \right) \frac{1}{(i\omega_{MK} + \Gamma_{MK})(i\omega_{LI} + \Gamma_{LI})} \right.$$

$$+ \left(\frac{1}{(i\omega_{LA} + \Gamma_{LA})} \frac{1}{(i\omega_{MI} + \Gamma_{MI})(i\omega_{MA} + \Gamma_{MA})} \right.$$

$$+ \frac{1}{(i\omega_{AI} + \Gamma_{AI})(i\omega_{AK} + \Gamma_{AK})(i\omega_{LK} + \Gamma_{LK})} \frac{1}{i\omega_{MK} + \Gamma_{MK}}$$

$$+ \frac{1}{(i\omega_{LN} + \Gamma_{LN})(i\omega_{LK} + \Gamma_{LK})} \left(\frac{1}{i\omega_{LA} + \Gamma_{LA}} + \frac{1}{i\omega_{AI} + \Gamma_{AI}} \right) \frac{1}{i\omega_{LI} + \Gamma_{LI}}$$

$$\left. + \frac{1}{(i\omega_{AI} + \Gamma_{AI})(i\omega_{LN} + \Gamma_{LN})(i\omega_{AK} + \Gamma_{AK})(i\omega_{LK} + \Gamma_{LK})} \right] \right\}.$$

$$(2.3.157)$$

Equation (2.3.157) can be used to obtain the transition probabilities for various three-photon processes, such as three-photon absorption/ionization, two-photon excitation fluorescence, and hyper-Raman scattering, including resonance phenomena.

Let us consider three-photon absorption processes. For nonresonant three-photon absorption, the first term of Eq. (2.3.157) makes a dominant contribution to the transition probability:

$$W^{(3)} = \frac{2}{\hbar^6} \sum_a \sum_n \sum_{a_r} \sum_{n_r} \rho_{aa}^{(s)}(0) \rho_{a_r a_r}^{(r)}(0) \frac{\Gamma_{AN}}{\omega_{AN}^2 + \Gamma_{AN}^2} \left| \sum_I \sum_K \frac{H'_{AI} H'_{IK} H'_{KN}}{\omega_{AI} \omega_{AK}} \right|^2.$$

$$(2.3.158)$$

This equation corresponds to Eq. (2.1.32), which can be derived using ordinary time-dependent perturbation theory if we let $\Gamma_{an} \to 0$.

For three-photon absorption processes in which a resonance condition is satisfied during photon absorption, two types of transition processes can be considered: two- and one-photon resonant processes are symbolically described by (2 + 1) and (1 + 2), respectively. To obtain the transition probability appropriate for the (2 + 1) process we set $K = M$ in Eq. (2.3.157). After performing some algebra, we find that the transition probability for the (2 + 1) process is expressed approximately as

$$W^{(2+1)} \simeq W_a^{(2+1)} + W_b^{(2+1)} + W_c^{(2+1)}, \qquad (2.3.159)$$

where

$$W_a^{(2+1)} = \frac{2}{\hbar^6} \sum_a \sum_n \sum_{a_r} \sum_{n_r} \rho_{aa}^{(s)}(0) \rho_{a_r a_r}^{(r)}(0) \frac{\Gamma_{AN}}{\omega_{AN}^2 + \Gamma_{AN}^2}$$

$$\times \sum_K \left| \frac{H'_{NK}}{i\omega_{AK} + \Gamma_{AK}} \right|^2 \left| \sum_I \frac{H'_{AI} H'_{IK}}{i\omega_{AI} + \Gamma_{AI}} \right|^2, \qquad (2.3.160)$$

$$W_b^{(2+1)} = \frac{2}{\hbar^6} \sum_a \sum_n \sum_{a_r} \sum_{n_r} \rho_{aa}^{(s)}(0)\rho_{a_r a_r}^{(r)}(0) \sum_K \frac{(2\Gamma_{KA}^{(d)} + \Gamma_{AA})}{\Gamma_{KK}\Gamma_{KA}}$$

$$\times \, \text{Re} \, \frac{|H'_{NK}|^2}{(i\omega_{KN} + \Gamma_{KN})} \, \text{Re} \, \frac{1}{(i\omega_{AK} + \Gamma_{AK})} \left| \sum_I \frac{H'_{AI}H'_{IK}}{i\omega_{AI} + \Gamma_{AI}} \right|^2,$$

$$(2.3.161)$$

$$W_c^{(2+1)} = \frac{2}{\hbar^6} \, \text{Re} \sum_a \sum_n \sum_{a_r} \sum_{n_r} \rho_{aa}^{(s)}(0)\rho_{a_r a_r}^{(r)}(0) \sum_K |H'_{NK}|^2$$

$$\times \, \frac{(\Gamma_{NA} + \Gamma_{KA} - \Gamma_{NK})}{(i\omega_{AN} + \Gamma_{AN})(i\omega_{KN} + \Gamma_{KN})(\omega_{AK}^2 + \Gamma_{AK}^2)} \left| \sum_I \frac{H_{AI}H_{IK}}{i\omega_{AI} + \Gamma_{AI}} \right|^2,$$

$$(2.3.162)$$

and the other terms with nonresonant character have been neglected. This equation should be compared with Eq. (2.3.144) for two-photon absorption: Eq. (2.3.160) has a simultaneous character. On the other hand, the process expressed in terms of Eq. (2.3.161) represents a sequential one and originates from the pure dephasing between the initial and resonant states and from the initial population decay, as was discussed in the two-photon processes. The last term [Eq. (2.3.162)] represents the mixing effect of the processes expressed by Eqs. (2.3.160) and (2.3.161). In the case in which the condition $\Gamma_{NK} \approx \Gamma_{KA} > \Gamma_{AN}$ is satisfied, the mixing term can safely be omitted relative to the other two terms. In this case, Eq. (2.3.159) becomes

$$W^{(2+1)} \simeq W_a^{(2+1)} + W_b^{(2+1)}. \qquad (2.3.163)$$

For the $(1 + 2)$ process, a similar expression for the transition probability can be derived, and the dephasing effect for the process can be discussed.

In this chapter we have presented the perturbation method, the Green's function method, and the density matrix method for treating multi-photon processes. Other methods like the coherent state method and the susceptibility method are also commonly used; they are briefly discussed in Appendix I, Sections B and C, and as well as in Section 2.4.

2.4   THE SUSCEPTIBILITY METHOD

The susceptibility method has been widely used to explain nonlinear optical phenomena such as harmonic generation, sum- and difference-frequency generation, stimulated scattering, and multiphoton absorption (see, for example, Shen, 1976). In this section, a classical treatment of the susceptibility method is first described on the basis of Maxwell's equations for a macroscopic ensemble; then the semiclassical theory is treated in terms of the density matrix method (Bloembergen, 1965; Yariv, 1975; Sargent *et al.*, 1974).

2.4.1   Classical Treatment

The interaction of an optical wave with atoms or molecules in the absence of magnetization and current due to free charges is described by Maxwell's equations in cgs units as

$$\mathbf{V} \times \mathbf{E} = -\frac{1}{c}\frac{\partial \mathbf{H}}{\partial t}, \tag{2.4.1}$$

$$\mathbf{V} \times \mathbf{H} = \frac{1}{c}\frac{\partial \mathbf{D}}{\partial t}, \tag{2.4.2}$$

$$\mathbf{V} \cdot \mathbf{D} = 0, \tag{2.4.3}$$

and

$$\mathbf{V} \cdot \mathbf{H} = 0, \tag{2.4.4}$$

where the electric displacement field $\mathbf{D}$ is expressed in terms of the electric field $\mathbf{E}$ and the electric polarization $\mathbf{P}$ as

$$\mathbf{D} = \mathbf{E} + 4\pi\mathbf{P}. \tag{2.4.5}$$

Substituting Eq. (2.4.5) into Eq. (2.4.2) and eliminating the magnetic field $\mathbf{H}$ in Eq. (2.4.1), we obtain the classical electric field equation with a source term originating from the polarization,

$$\mathbf{V} \times \mathbf{V} \times \mathbf{E} + \frac{1}{c^2}\frac{\partial^2 \mathbf{E}}{\partial t^2} = -\frac{4\pi}{c^2}\frac{\partial^2 \mathbf{P}}{\partial t^2}. \tag{2.4.6}$$

Similarly, we obtain

$$\mathbf{V} \cdot \mathbf{E} = -4\pi\mathbf{V} \cdot \mathbf{P}. \tag{2.4.7}$$

Note that the electric polarization $\mathbf{P}(\mathbf{r}, t)$ depends on the electric field $\mathbf{E}(\mathbf{r}, t)$: In a weak field $\mathbf{P}$ is linear in $\mathbf{E}$, but in general $\mathbf{P}$ is a nonlinear function of $\mathbf{E}$.

Assuming that $\mathbf{E}$ and $\mathbf{P}$ can be expressed in terms of sums of functions with frequencies of the applied field, we expand $\mathbf{E}$ and $\mathbf{P}$ into Fourier components as

$$\mathbf{E}(\mathbf{r}, t) = \frac{1}{2}\sum_{i=1}^{k}[\mathbf{E}(\mathbf{r}, i)\exp(i\omega_i t) + \text{c.c.}] \tag{2.4.8}$$

and

$$P(\mathbf{r}, t) = \frac{1}{2} \sum_{i=1} [P(\mathbf{r}, i) \exp(i\omega_i t) + \text{c.c.}], \qquad (2.4.9)$$

where $E(\mathbf{r}, i)$ and $P(\mathbf{r}, i)$ represent the amplitudes of the electric field and the polarization for the ith component, respectively, and c.c. denotes the complex conjugates. The polarization amplitude $P(\mathbf{r}, i)$ can be expressed as a power series of the applied electric field in the pure electric dipole case as

$$P(\mathbf{r}, i) = P^{(1)}(\mathbf{r}, i) + P^{(2)}(\mathbf{r}, i) + P^{(3)}(\mathbf{r}, i) + \ldots, \qquad (2.4.10)$$

where

$$P^{(n)}(\mathbf{r}, i) = \chi^{(n)}(-\omega_i; \omega_1, \omega_2, \ldots, \omega_n) E(\mathbf{r}, 1) E(\mathbf{r}, 2) \ldots E(\mathbf{r}, n) \qquad (2.4.11)$$

and $\chi^{(n)}$ is the susceptibility tensor of rank $n + 1$. The αth component of $P^{(n)}(\mathbf{r}, i)$ is written as

$$P_\alpha^{(n)}(\mathbf{r}, i) = \sum_\beta \sum_\gamma \cdots \sum_\nu \chi_{\beta\gamma\ldots\nu}^{(n)}(-\omega_i; \omega_1, \omega_2, \ldots, \omega_n)$$
$$\times E_\beta(\mathbf{r}, 1) E_\gamma(\mathbf{r}, 2) \ldots E_\nu(\mathbf{r}, n). \qquad (2.4.12)$$

Substitution of Eqs. (2.4.8) and (2.4.9) into Eq. (2.4.6) yields

$$\nabla \times \nabla \times E(\mathbf{r}, i) - (\omega_i/c)^2 E(\mathbf{r}, i) = 4\pi(\omega_i/c)^2 P(\mathbf{r}, i). \qquad (2.4.13)$$

Let us consider an isotropic system such as a gas or liquid. Neglecting the nonlinear polarization terms and setting $P(\mathbf{r}, i) \doteq \chi^{(1)} E(\mathbf{r}, i)$, Eqs. (2.4.13) and (2.4.7) are expressed as

$$\nabla \times \nabla \times E(\mathbf{r}, i) - (\omega_i/c)^2 \varepsilon_i E(\mathbf{r}, i) = 0 \qquad (2.4.14)$$

and

$$\nabla \cdot \varepsilon_i E(\mathbf{r}, i) = 0, \qquad (2.4.15)$$

where ε_i is the dielectric constant and is given by

$$\varepsilon_i = 1 + 4\pi\chi^{(1)}(\omega_i; \omega_i). \qquad (2.4.16)$$

Utilizing the vector equality

$$\nabla \times \nabla \times E = \nabla(\nabla \cdot E) - \nabla^2 E, \qquad (2.4.17)$$

the electric field amplitude is given by

$$E(\mathbf{r}, i) = E(i) \exp(i\mathbf{k}_i \cdot \mathbf{r}), \qquad (2.4.18)$$

with

$$E(i) \cdot \mathbf{k}_i = 0, \qquad (2.4.19)$$

where $k_i = \sqrt{\varepsilon_i}(\omega_i/c) = n_i\omega_i/c$ and n_i is the index of refraction of the macroscopic system. Substituting Eq. (2.4.18) into Eq. (2.4.8) yields

$$\mathbf{E}(\mathbf{r}, t) = \frac{1}{2}\sum_i[\mathbf{E}(i)\exp(i\mathbf{k}_i \cdot \mathbf{r} - i\omega_i t) + \text{c.c.}]. \qquad (2.4.20)$$

This expresses plane waves with a polarization direction orthogonal to the propagation direction **k**.

The time-averaged field intensity of the ith mode is given by

$$I_i = (cn_i/8)|\mathbf{E}(i)|^2. \qquad (2.4.21)$$

For isotropic media, the lowest nonlinear susceptibility $\chi^{(2)}$ vanishes in the pure electric-dipole approximation (see Section 2.4.2). The nonlinearity for these media is usually described in terms of $\chi^{(3)}$, neglecting the higher order susceptibilities; $\chi^{(3)}$ is normally too small to observe nonlinear optical effects for gases because of the low density of the molecules (Shen, 1976). The signal is proportional to the square of the number of molecules N^2. However, resonance enhancement of $\chi^{(3)}$ enables us to observe the nonlinear optical effects.

Inclusion of the third-order nonlinear susceptibility into Eqs. (2.4.14) and (2.4.15) yields

$$\mathbf{\nabla} \times \mathbf{\nabla} \times \mathbf{E}(\mathbf{r}, i) - (\omega_i/c)^2\varepsilon_i\mathbf{E}(\mathbf{r}, i) = 4\pi(\omega_i/c)^2\mathbf{P}^{(3)}(\mathbf{r}, i) \qquad (2.4.22)$$

and

$$\mathbf{\nabla} \cdot [\varepsilon_i\mathbf{E}(\mathbf{r}, i) + 4\pi\mathbf{P}^{(3)}(\mathbf{r}, i)] = 0. \qquad (2.4.23)$$

There exist various kinds of nonlinear optical processes involving $\chi^{(3)}$, such as third-harmonic generation, sum-frequency generation (SFG), and coherent anti-Stokes Raman scattering (CARS), as shown in Fig. 2.1. From

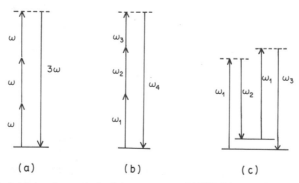

(a)   (b)   (c)

Fig. 2.1   Typical third-order optical mixing processes. (a) Third-harmonic generation, (b) sum-frequency generation (SFG), and (c) coherent anti-Stokes Raman scattering (CARS).

Eq. (2.4.12), the component of the third-order electric polarization $P_\alpha^{(3)}(\mathbf{r}, i)$ is expressed as

$$P_\alpha^{(3)}(\mathbf{r}, i) = \sum_\beta \sum_\alpha \sum_\delta \chi_{\alpha\beta\gamma\delta}^{(3)}(-\omega; \omega_1, \omega_2, \omega_3) E_\beta(\mathbf{r}, 1)$$

$$\times E_\gamma(\mathbf{r}, 2) E_\delta(\mathbf{r}, 3). \qquad (2.4.24)$$

In $\chi^{(3)}$, the frequencies that are converted in the nonlinear process are denoted by positive signs and the frequencies that are generated are denoted by negative signs; for example, $\chi^{(3)}(-\omega_4; \omega_1, \omega_2, \omega_3)$ for sum-frequency generation and $\chi^{(3)}(-\omega_3; \omega_1, \omega_1, -\omega_2)$ for CARS.

In Eq. (2.4.24) there are 27 terms for each component of $P^{(3)}$. By using the symmetry restriction, these terms can be reduced (see, for example, Nibler and Knighten, 1979): There are six equivalent terms, since $\chi^{(3)}$ is invariant to the permutations of the pairs $(\beta\omega_1)$, $(\gamma\omega_2)$, and $(\delta\omega_3)$. In the case in which n identical frequencies are applied $n!$ of the permutations are indistinguishable:

$$P_\alpha^{(3)}(\mathbf{r}, i) = \frac{6}{n!} \sum_\beta \sum_\alpha \sum_\delta \chi_{\alpha\beta\gamma\delta}^{(3)}(-\omega_i; \omega_1, \omega_2, \omega_3) E_\beta(\mathbf{r}, 1)$$

$$\times E_\gamma(\mathbf{r}, 2) E_\delta(\mathbf{r}, 3). \qquad (2.4.25)$$

For an isotropic medium, if we consider its symmetry properties, there are 21 nonzero elements, of which three terms are independent of each other. These are classified as (Nibler and Knighten, 1979)

$$\chi_{1111}^{(3)} = \chi_{xxxx}^{(3)} = \chi_{yyyy}^{(3)} = \chi_{zzzz}^{(3)},$$

$$\chi_{1122}^{(3)} = \chi_{xxyy}^{(3)} = \chi_{yyxx}^{(3)} = \chi_{xxzz}^{(3)} = \chi_{zzxx}^{(3)} = \chi_{yyzz}^{(3)} = \chi_{zzyy}^{(3)},$$

$$\chi_{1212}^{(3)} = \chi_{xyxy}^{(3)} = \chi_{yxyx}^{(3)} = \chi_{xzxz}^{(3)} = \chi_{zxzx}^{(3)} = \chi_{yzyz}^{(3)} = \chi_{zyzy}^{(3)},$$

$$\chi_{1221}^{(3)} = \chi_{xyyx}^{(3)} = \chi_{yxxy}^{(3)} = \chi_{xzzx}^{(3)} = \chi_{zxxz}^{(3)} = \chi_{yzzy}^{(3)} = \chi_{zyyz}^{(3)},$$

with

$$\chi_{1111}^{(3)} = \chi_{1122}^{(3)} + \chi_{1212}^{(3)} + \chi_{1221}^{(3)}. \qquad (2.4.26)$$

For sum-frequency generation (SFG) (Fig. 2.1b), $P^{(3)}$ is expressed as

$$P_\alpha^{(3)}(\mathbf{r}, 4) = 6\chi_{1122}^{(3)}(-\omega_4; \omega_1, \omega_2, \omega_3) \sum_\beta E_\alpha(\mathbf{r}, 1) E_\beta(\mathbf{r}, 2)$$

$$\times E_\beta(\mathbf{r}, 3) + 6\chi_{1212}^{(3)}(-\omega_4; \omega_1, \omega_2, \omega_3) \sum_\beta E_\beta(\mathbf{r}, 1)$$

$$\times E_\alpha(\mathbf{r}, 2) E_\beta(\mathbf{r}, 3) + 6\chi_{1221}^{(3)}(-\omega_4; \omega_1, \omega_2, \omega_3)$$

$$\times \sum_\beta E_\beta(\mathbf{r}, 1) E_\beta(\mathbf{r}, 2) E_\alpha(\mathbf{r}, 3) \qquad (2.4.27)$$

or

$$\mathbf{P}^{(3)}(\mathbf{r}, 4) = 6[\chi^{(3)}_{1122}(-\omega_4; \omega_1, \omega_2, \omega_3)\mathbf{E}(\mathbf{r}, 1)\mathbf{E}(\mathbf{r}, 2) \cdot \mathbf{E}(\mathbf{r}, 3)$$
$$+ \chi^{(3)}_{1212}(-\omega_4; \omega_1, \omega_2, \omega_3)\mathbf{E}(\mathbf{r}, 2)\mathbf{E}(\mathbf{r}, 1) \cdot \mathbf{E}(\mathbf{r}, 3)$$
$$+ \chi^{(3)}_{1221}(-\omega_4; \omega_1, \omega_2, \omega_3)\mathbf{E}(\mathbf{r}, 3)\mathbf{E}(\mathbf{r}, 1) \cdot \mathbf{E}(\mathbf{r}, 2)]. \quad (2.4.28)$$

In the case in which the incident electric fields given by Eq. (2.4.18) are applied, the preceding expression reduces to

$$\mathbf{P}^{(3)}(\mathbf{r}, 4) = 6[\chi^{(3)}_{1122}(-\omega_4; \omega_1, \omega_2, \omega_3)\mathbf{e}_1(\mathbf{e}_2 \cdot \mathbf{e}_3)$$
$$+ \chi^{(3)}_{1212}(-\omega_4; \omega_1, \omega_2, \omega_3)\mathbf{e}_2(\mathbf{e}_1 \cdot \mathbf{e}_3)$$
$$+ \chi^{(3)}_{1221}(-\omega_4; \omega_1, \omega_2, \omega_3)\mathbf{e}_3(\mathbf{e}_1 \cdot \mathbf{e}_2)]$$
$$\times E_1 E_2 E_3 \exp[i(\mathbf{k}_1 + \mathbf{k}_2 + \mathbf{k}_3) \cdot \mathbf{r}]$$
$$= \chi^{(3)}_{SFG}\mathbf{e}_{SFG} E_1 E_2 E_3 \exp[i(\mathbf{k}_1 + \mathbf{k}_2 + \mathbf{k}_3) \cdot \mathbf{r}], \quad (2.4.29)$$

where

$$\chi^{(3)}_{SFG}\mathbf{e}_{SFG} = 6[\chi^{(3)}_{1122}\mathbf{e}_1(\mathbf{e}_2 \cdot \mathbf{e}_3) + \chi^{(3)}_{1212}\mathbf{e}_2(\mathbf{e}_1 \cdot \mathbf{e}_3) + \chi^{(3)}_{1221}\mathbf{e}_3(\mathbf{e}_1 \cdot \mathbf{e}_2)], \quad (2.4.30)$$

and $\mathbf{e}_1$, $\mathbf{e}_2$, and $\mathbf{e}_3$ denote the unit vectors of the polarization direction. For CARS, if we note that

$$\chi^{(3)}_{1122}(-\omega_3; \omega_1, \omega_1, -\omega_2) = \chi^{(3)}_{1212}(-\omega_3; \omega_1, \omega_1, -\omega_2),$$

we obtain $\mathbf{P}^{(3)}(\mathbf{r}, 3)$ as

$$\mathbf{P}^{(3)}(\mathbf{r}, 3) = [6\chi^{(3)}_{1122}(-\omega_3; \omega_1, \omega_1, -\omega_2)\mathbf{e}_1(\mathbf{e}_1 \cdot \mathbf{e}_2)$$
$$+ 3\chi^{(3)}_{1221}(-\omega_3; \omega_1, \omega_1, -\omega_2)\mathbf{e}_2]E_1^2 E_2^* \exp[i(2\mathbf{k}_1 - \mathbf{k}_2) \cdot \mathbf{r}]$$
$$= \chi_{CARS}\mathbf{e}_{CARS} E_1^2 E_2^* \exp[i(2\mathbf{k}_1 - \mathbf{k}_2) \cdot \mathbf{r}], \quad (2.4.31)$$

where

$$\chi_{CARS}\mathbf{e}_{CARS} = 6\chi^{(3)}_{1122}(-\omega_3; \omega_1, \omega_1, -\omega_2)\mathbf{e}_1(\mathbf{e}_1 \cdot \mathbf{e}_2)$$
$$+ 3\chi^{(3)}_{1221}(-\omega_3; \omega_1, \omega_1, -\omega_2)\mathbf{e}_2. \quad (2.4.32)$$

We are now in a position to evaluate the sum-frequency mixing intensity and the CARS intensity. From Eqs. (2.4.22), (2.4.23), and (2.4.29), the equations for the sum frequency are expressed as

$$\nabla \times \nabla \times \mathbf{E}(\mathbf{r}, 4) - (\omega_4/c)^2 \varepsilon_4 \mathbf{E}(\mathbf{r}, 4)$$
$$= 4\pi(\omega_4/c)^2 \chi_{SFG}\mathbf{e}_{SFG} E_1 E_2 E_3 \exp[i(\mathbf{k}_1 + \mathbf{k}_2 + \mathbf{k}_3) \cdot \mathbf{r}] \quad (2.4.33)$$

and

$$\varepsilon_4 \nabla \cdot \mathbf{E}(\mathbf{r}, 4) = -4\pi i(\omega_4/c)^2 \chi_{SFG}\mathbf{e}_{SFG} \cdot (\mathbf{k}_1 + \mathbf{k}_2 + \mathbf{k}_3)E_1 E_2 E_3$$
$$\times \exp[i(\mathbf{k}_1 + \mathbf{k}_2 + \mathbf{k}_3) \cdot \mathbf{r}]. \quad (2.4.34)$$

Using the vector equality, Eq. (2.4.17), and $\mathbf{e}_{SFG} \cdot (\mathbf{k}_1 + \mathbf{k}_2 + \mathbf{k}_3) = 0$, in which $\mathbf{k}_1 + \mathbf{k}_2 + \mathbf{k}_3$ is in the z direction, Eq. (2.4.33) has the form

$$-\nabla^2 \mathbf{E}(z, 4) - (\omega_4/c)^2 \varepsilon_4 \mathbf{E}(z, 4)$$
$$= 4\pi(\omega_4/c)^2 \chi_{SFG} \mathbf{e}_{SFG} E_1 E_2 E_3 \exp[i(k_1 + k_2 + k_3)z]. \quad (2.4.35)$$

We assume that the solution is expressed as

$$\mathbf{E}(\mathbf{r}, 4) = \mathbf{e}_{SFG} E_4^0 \exp(ik_4 z), \quad (2.4.36)$$

where $k_4 = n_4 \omega_4/c$. The amplitude E_4^0 in Eq. (2.4.36) satisfies the equation

$$\mathbf{e}_{SFG} \frac{\partial^2 E_4^0}{\partial z^2} + 2i\mathbf{e}_{SFG} k_4 \frac{\partial E_4^0}{\partial z}$$

$$= -4\pi \left(\frac{\omega_4}{c}\right)^2 \chi_{SFG} \mathbf{e}_{SFG} E_1 E_2 E_3 \exp[i(k_1 + k_2 + k_3 - k_4)z]. \quad (2.4.37)$$

Assuming that E_4^0 is a slowly varying function of z, $\partial^2 E_4^0/\partial z^2 \approx 0$, the solution for E_4^0 at $z = l$ in Eq. (2.4.37) is given by

$$E_4^0 \simeq \frac{2\pi}{k_4} \left(\frac{\omega_4}{c}\right)^2 \chi_{SFG} E_1 E_2 E_3 \frac{\exp(i\,\Delta kl) - 1}{\Delta k}, \quad (2.4.38)$$

where the boundary condition $E_4^0 = 0$ at $z = 0$ has been used and the phase match parameter Δk is expressed as $\Delta k = k_1 + k_2 + k_3 - k_4$. The time-averaged intensity, Eq. (2.4.21), is then given by

$$I_4 = \frac{n_4 c}{8} |\mathbf{E}(\mathbf{r}, 4)|^2$$

$$= \frac{256\pi^4 \omega_4^2}{c^4 n_1 n_2 n_3 n_4} |\chi_{SFG}|^2 I_2 I_2 I_3 l^2 \left[\frac{\sin(\Delta kl/2)}{\Delta kl/2}\right]^2. \quad (2.4.39)$$

The intensity generated at the phase match $\Delta k = 0$ over a distance l is

$$I_4 = \frac{256\pi^4 \omega_4^2}{c^4 n_1 n_2 n_3 n_4} |\chi_{SFG}|^2 I_1 I_2 I_3 l^2. \quad (2.4.40)$$

For CARS, by using a method similar to that just described, the field amplitude is given by

$$E_3^0 = \frac{2\pi\omega_3}{n_3 c} \chi_{CARS} E_1^2 E_2^* \frac{\exp(i\,\Delta kl) - 1}{\Delta k}, \quad (2.4.41)$$

and the intensity is

$$I_3 = \frac{256\pi^4 \omega_3^2}{c^4 n_1^2 n_2 n_3} |\chi_{CARS}|^2 I_1^2 I_1 l^2 \left[\frac{\sin(\Delta kl/2)}{\Delta kl/2}\right]^2. \quad (2.4.42)$$

From the preceding derivation, we notice that the output signal is proportional to the product of the intensities of the input laser beams. Therefore the intensity can be very high if the input laser intensities are high and if the resonance condition is satisfied. The CARS method is in fact widely applied to low-concentration gas-phase diagnostics (see, for example, Harvey, 1981).

In deriving the expression for the intensities, zeroth-order solutions of the field amplitudes have been assumed. To obtain the improved solution, we have to solve the coupled nonlinear wave equations for the field amplitudes (Bloembergen, 1965).

2.4.2 Semiclassical Treatment

In the classical treatment of the susceptibility method described in Section 2.4.1 the nonlinear susceptibility $\chi^{(3)}$ was taken as an unknown parameter. In order to find the structure of $\chi^{(3)}$ and the dependence of the applied optical frequencies and to investigate the resonant enhancement of the intensity, we need to develop a quantum-mechanical treatment taking into account the internal (rovibronic) structure of the molecule and the dephasing (damping) effect. In this section, we shall describe the semiclassical treatment of the derivation of the susceptibility. The semiclassical approach has been carried out by Armstrong et al. (1962), based on time-dependent perturbation theory, which has been treated in Section 2.1; these authors have discussed the frequency dependence of the nonlinear susceptibility. We shall now apply the density matrix formalism developed in Section 2.3 to derive an expression for the susceptibility of an ensemble of molecules embedded in a heat bath and interacting with a time-harmonic electromagnetic field (Bloembergen, 1964; Yariv, 1975; Sargent et al., 1974). An application of the susceptibility method to intensity behavior in molecular two- and three-photon excitation spectra has been reported by Meredith (1981).

The macroscopic polarization $\mathbf{P}(t)$ is defined in terms of the expectation value of the dipole moment $\boldsymbol{\mu} = e\mathbf{R}$ as

$$\mathbf{P}(t) = N\langle \boldsymbol{\mu} \rangle = N \, \mathrm{Tr}[\hat{\rho}(t)\boldsymbol{\mu}], \qquad (2.4.43)$$

where N represents the number of molecules per unit volume and $\hat{\rho}(t)$ is the density matrix of the molecules. The semiclassical treatment based on the density matrix method consists of the following procedure.

(i) The density matrix $\hat{\rho}(t)$, which satisfies the equation

$$d\hat{\rho}(t)/dt = -i\hat{L}^0\hat{\rho}(t) - i\hat{L}'\hat{\rho}(t) - \Gamma\hat{\rho}(t), \qquad (2.4.44)$$

is solved by using the perturbation expansion. In Eq. (2.4.44), $\hat{L}^0 = (1/\hbar)[\hat{H}^0, \cdot]$ and $\hat{L}' = (1/\hbar)[\hat{H}'(t), \cdot]$, where $\hat{H}'(t)$ is the interaction Hamiltonian and Γ is the dephasing constant. The interaction Hamiltonian is assumed to be of the dipole type,

$$\hat{H}'(t) = -\boldsymbol{\mu} \cdot \mathbf{E}(t), \tag{2.4.45}$$

where the electric field $\mathbf{E}(t)$ is given by

$$\mathbf{E}(t) = \frac{1}{2} \sum_i [\mathbf{E}(i) \exp(it\omega_i) + \text{c.c.}] \tag{2.4.46}$$

Here we omitted the terms involving the propagation vector of the electromagnetic wave $\mathbf{k}_i$. In solving Eq. (2.4.44), a rapid relaxation of the off-diagonal density matrix element to its steady state is assumed relative to that of the diagonal one.

(ii) An expression for the linear or nonlinear susceptibility is obtained by comparing the component of the time-independent term of the resulting equation [Eq. (2.4.43)] with the expression

$$\mathbf{P} = \chi^{(1)} \cdot \mathbf{E} + \chi^{(2)} : \mathbf{EE} + \chi^{(3)} : \mathbf{EEE} + \cdots. \tag{2.4.47}$$

The off-diagonal density matrix element $\hat{\rho}_{an}^{(1)}(t)$, which is the solution to the first-order radiation–molecule interaction, is needed to evaluate the linear susceptibility $\chi^{(1)}$. The solutions of the off-diagonal density matrix elements to the second and the third orders are needed for $\chi^{(2)}$ and $\chi^{(3)}$.

Let us now consider the linear susceptibility. From Eq. (2.4.44), the first-order off-diagonal density matrix element $\rho_{an}^{(1)}$ is expressed as

$$d\rho_{an}^{(1)}/dt = -\omega_{an}\rho_{an}^{(1)} - (iH_{an}'/\hbar)\Delta\rho_{an}^{(0)} - \Gamma_{an}\rho_{an}^{(1)}, \tag{2.4.48}$$

where $\omega_{an} = (H_{aa}^0 - H_{nn}^0)/\hbar$ and $\Delta\rho_{an}^{(0)} = \rho_{nn}^{(0)} - \rho_{aa}^{(0)}$. By introducing $\sigma_{an}^{(1)}(t)$, the slowly varying function of t defined by

$$\rho_{an}^{(1)}(t) = \sigma_{an}^{(1)}(t) \exp(it\omega_r), \tag{2.4.49}$$

Eq. (2.4.48) can be written as

$$\frac{d\sigma_{an}^{(1)}}{dt} = -[i(\omega_{an} + \omega_r) + \Gamma_{an}]\sigma_{an}^{(1)} + \frac{i\boldsymbol{\mu}_{an} \cdot \mathbf{E}(r)}{2\hbar}$$

$$\times [1 + \exp(-2i\omega_r t)]\Delta\rho_{an}^{(0)}. \tag{2.4.50}$$

The solution of this equation with $d\sigma_{an}^{(1)}/dt = 0$ in the adiabatic approximation [neglecting the term involving $\exp(-2i\omega_r t)$ (i.e., in the rotating wave approximation)] is given by

$$\sigma_{an}^{(1)} = \frac{i}{2\hbar} \boldsymbol{\mu}_{an} \cdot \mathbf{E}(\omega_r) \frac{\Delta\rho_{an}^{(0)}}{i(\omega_r + \omega_{an}) + \Gamma_{an}}. \tag{2.4.51}$$

Substituting Eq. (2.4.49) with Eq. (2.4.51) into the expression $\mathbf{P}^{(1)}(t) = N \sum_a \sum_n \rho_{an}^{(1)}\boldsymbol{\mu}_{na}$ and comparing the resulting expression with Eq. (2.4.47), we obtain an expression for the linear susceptibility as

$$\chi^{(1)} = -\frac{i}{\hbar} \sum_a \sum_n \Delta N^{(0)} \frac{\mu_{an}\mu_{na}}{i(\omega_r + \omega_{an}) + \Gamma_{an}}, \qquad (2.4.52)$$

where $\Delta N^{(0)}$ is the population difference in the absence of an electromagnetic field, that is, $\Delta N^{(0)} = N(\rho_{aa}^{(0)} - \rho_{nn}^{(0)})$. For the absorption process of visible or UV laser light from states a to n, $\Delta N^{(0)} \doteq N_a = N\rho_{aa}^{(0)}$.

The susceptibility can be divided into a real part $\chi^{(1)'}$ and an imaginary part $\chi^{(1)''}$:

$$\chi^{(1)} = \chi^{(1)'} - i\chi^{(1)''}. \qquad (2.4.53)$$

Here $\chi^{(1)'}$, which is related to dispersion, is given by

$$\chi^{(1)'} = -\frac{1}{\hbar} \sum_a \sum_n N_a \frac{\mu_{an}\mu_{na}(\omega_r + \omega_{an})}{(\omega_r + \omega_{an})^2 + \Gamma_{an}^2} \qquad (2.4.54)$$

and $\chi^{(2)''}$ is given by

$$\chi^{(1)''} = \frac{1}{\hbar} \sum_a \sum_n N_a \frac{\mu_{an}\mu_{na}\Gamma_{an}}{(\omega_r + \omega_{an})^2 + \Gamma_{an}^2}. \qquad (2.4.55)$$

We can see that Eq. (2.4.55) is proportional to the absorption cross section discussed in Section 2.3.

Next we consider the second-order nonlinear susceptibility $\chi^{(2)}$. The second-order off-diagonal density matrix element $\rho_{an}^{(2)}(t)$ satisfies the equation

$$\frac{d\rho_{an}^{(2)}}{dt} = -i\omega_{an}\rho_{an}^{(2)} - \frac{i}{\hbar} \sum_m (H'_{am}\rho_{mn}^{(1)} - \rho_{am}^{(1)}H'_{mn}) - \Gamma_{an}\rho_{an}^{(2)}. \qquad (2.4.56)$$

Letting

$$\rho_{an}^{(2)} = \sigma_{an}^{(2)}(t)\exp[it(\omega_1 + \omega_2)] \equiv \rho_{an}^{(2)}(\omega_1, \omega_2), \qquad (2.4.57)$$

substituting Eqs. (2.4.57) and (2.4.49) with Eq. (2.4.51) into Eq. (2.4.56), and invoking the adiabatic and rotating wave approximations, we obtain an expression for $\sigma_{an}^{(2)}$:

$$\sigma_{an}^{(2)} = \left(\frac{i}{2\hbar}\right)^2 \frac{1}{i(\omega_1 + \omega_2 + \omega_{an}) + \Gamma_{an}}$$

$$\times \sum_m \left\{ \left[\frac{\boldsymbol{\mu}_{am} \cdot \mathbf{E}(2)\boldsymbol{\mu}_{mn} \cdot \mathbf{E}(1)}{i(\omega_1 + \omega_{mn}) + \Gamma_{mn}} + \frac{\boldsymbol{\mu}_{am} \cdot \mathbf{E}(1)\boldsymbol{\mu}_{mn} \cdot \mathbf{E}(2)}{i(\omega_2 + \omega_{mn}) + \Gamma_{mn}} \right] \Delta\rho_{mn}^{(0)} \right.$$

$$\left. - \left[\frac{\boldsymbol{\mu}_{am} \cdot \mathbf{E}(2)\boldsymbol{\mu}_{mn} \cdot \mathbf{E}(1)}{i(\omega_2 + \omega_{am}) + \Gamma_{am}} + \frac{\boldsymbol{\mu}_{am} \cdot \mathbf{E}(1)\boldsymbol{\mu}_{mn} \cdot \mathbf{E}(2)}{i(\omega_1 + \omega_{am}) + \Gamma_{am}} \right] \Delta\rho_{am}^{(0)} \right\}. \qquad (2.4.58)$$

From Eq. (2.4.58) and the definition of the second-order nonlinear susceptibility, $\chi^{(2)}$ is given as

$$
\chi^{(2)} = -\frac{N}{\hbar^2} \sum_a \sum_n \sum_m \frac{\mu_{am}\mu_{mn}\mu_{na}}{i(\omega_1 + \omega_2 + \omega_{an}) + \Gamma_{an}}
$$
$$
\times \left[\left(\frac{1}{i(\omega_1 + \omega_{mn}) + \Gamma_{mn}} + \frac{1}{i(\omega_2 + \omega_{mn}) + \Gamma_{mn}} \right) \Delta\rho_{mn}^{(0)} \right.
$$
$$
\left. - \left(\frac{1}{i(\omega_2 + \omega_{am}) + \Gamma_{am}} + \frac{1}{i(\omega_1 + \omega_{am}) + \Gamma_{am}} \right) \Delta\rho_{am}^{(0)} \right]. \quad (2.4.59)
$$

We can see from this expression that $\chi^{(2)}$ vanishes if the molecular system
has inversion symmetry.

For the system in which $\rho_{ii}^{(0)}(0) = 0$ for $i \neq a$ and the resonance condition
$\omega_1 + \omega_{am} \simeq 0$ is satisfied, Eq. (2.4.59) can be rewritten as

$$
\chi^{(2)} = -\frac{1}{\hbar^2} \sum_a \sum_n \sum_m N_a^{(0)} \frac{\mu_{am}\mu_{mn}\mu_{na}}{[i(\omega_1 + \omega_2 + \omega_{an}) + \Gamma_{an}][i(\omega_1 + \omega_{am}) + \Gamma_{am}]}
$$
$$
+ \chi_{nr}^{(2)}, \quad (2.4.60)
$$

where $\chi_{nr}^{(2)}$ represents the second-order nonlinear susceptibility originating
from the nonresonant contribution.

In a similar manner, we can derive an expression for $\chi^{(3)}$. The third-order
off-diagonal density matrix element $\rho_{an}^{(3)}$ is given by

$$
\rho_{an}^{(3)} = \frac{i}{2\hbar} \frac{\exp[it(\omega_1 + \omega_2 + \omega_3)]}{i(\omega_1 + \omega_2 + \omega_3 + \omega_{an}) + \Gamma_{an}} \sum_k [\mu_{ak} \cdot \mathbf{E}(1)\rho_{kn}^{(2)}(\omega_2, \omega_3)
$$
$$
+ \mu_{ak} \cdot \mathbf{E}(2)\rho_{kn}^{(2)}(\omega_1, \omega_3) + \mu_{ak} \cdot \mathbf{E}(3)\rho_{kn}^{(2)}(\omega_1, \omega_2)
$$
$$
- \rho_{ak}^{(2)}(\omega_2, \omega_3)\mu_{kn} \cdot \mathbf{E}(1) - \rho_{ak}^{(2)}(\omega_1, \omega_3)\mu_{kn} \cdot \mathbf{E}(2)
$$
$$
- \rho_{ak}^{(2)}(\omega_1, \omega_2)\mu_{kn} \cdot \mathbf{E}(3)]. \quad (2.4.61)
$$

For the system in which $\rho_{ii}^{(0)} = 0$ for $i \neq a$ and the resonance conditions
$\omega_1 + \omega_{am} \simeq 0$ and $\omega_1 + \omega_2 + \omega_{ak} \simeq 0$ are satisfied, $\chi^{(3)}$ can be expressed as

$$
\chi^{(3)} = \left(\frac{i}{\hbar^3} \right) \sum_a \sum_n \sum_m \sum_k \frac{N_a\mu_{am}\mu_{mk}\mu_{kn}\mu_{na}}{[i(\omega_1 + \omega_2 + \omega_3 + \omega_{an}) + \Gamma_{an}]}
$$
$$
\times \frac{1}{[i(\omega_1 + \omega_2 + \omega_{ak}) + \Gamma_{ak}][i(\omega_1 + \omega_{am}) + \Gamma_{am}]} + \chi_{nr}^{(3)} \quad (2.4.62)
$$

where $\chi_{nr}^{(3)}$ represents the nonresonant contribution.

In the preceding derivation of the susceptibility, where we used the perturbation of the density matrix, saturation effects and effects of the population decay constant were omitted. In Appendix I, Section C, a more explicit derivation of the linear susceptibility, including those effects, is given.

In this section we have treated the semiclassical approach to the derivation of the susceptibility. The quantum-mechanical treatment based on the density matrix method can be accomplished starting from Eqs. (2.3.79) and (2.3.83).

2.5 MULTIPHOTON IONIZATION

The transition probabilities derived in the previous sections can be applied to both multiphoton absorption (MPA) and ionization (MPI). To apply them to MPI, it should be noted that the final state is in the continuum. To be explicit, let us treat the two-photon ionization case. For this purpose, we may use Eq. (2.1.106),

$$W^{(2)}_{i \to f} = \frac{2\pi}{\hbar^2} \left(\frac{2\pi e^2}{c} \right)^2 \int_0^\infty d\omega_l \, \omega_l I_1(\omega_l) \int_0^\infty d\omega_{l'} \, \omega_{l'} I_2(\omega_{l'})$$

$$\times \, |M^{(2)}_{fi}(\omega_l, \omega_{l'})|^2 \delta(\omega_{fi} - \omega_l - \omega_{l'}). \tag{2.5.1}$$

Since the final state $|f\rangle$ is in the continuum, if the wave vector of the photoelectron is $\mathbf{K}$, then the state $|f\rangle$ will be characterized by the wave vector $\mathbf{K}$, that is, $|f\rangle = |\mathbf{K}\rangle$.

Notice that the energy of the photoelectron with wave vector $\mathbf{K}$ is given by

$$E_K = P_K^2/2m = K^2\hbar^2/2m = \hbar\omega_K, \tag{2.5.2}$$

whereas the density of final states $\rho(\omega_K)$ is given by

$$\rho(\omega_K) \, d\omega_K \, d\Omega_{\mathbf{K}} = (mK/8\pi^3\hbar) \, d\omega_K \, d\Omega_{\mathbf{K}}, \tag{2.5.3}$$

where $\Omega_{\mathbf{K}}$ is the direction of $\mathbf{K}$. Equation (2.5.3) can be derived as follows. As shown in Appendix 2.1, the number of states in the range $(\mathbf{K}, \mathbf{K} + d\mathbf{K})$ can be expressed as

$$dN = (L/2\pi)^3 K^2 \, dK \, d\Omega_{\mathbf{K}} = \rho(\omega_K) \, d\omega_K \, d\Omega_{\mathbf{K}}. \tag{2.5.4}$$

That is,

$$\rho(\omega_K) = \left(\frac{L}{2\pi} \right)^3 K^2 \frac{dK}{d\omega_K} = \left(\frac{L}{2\pi} \right)^3 \frac{mK}{\hbar}, \tag{2.5.5}$$

where L^3 is the volume of the container. Here Eq. (2.5.2) has been used to calculate $dK/d\omega_K$. The factor L^3 in Eq. (2.5.5) is often absorbed in the normalization factor of the wave function of the photoelectron; in this case $\rho(\omega_K)$ is given by

$$\rho(\omega_K) = mK/8\pi^3\hbar. \tag{2.5.6}$$

If the ionization energy of the molecule from the initial state $|i\rangle$ is $\hbar\omega_I$, then

$$\omega_f = \omega_I + \omega_K, \tag{2.5.7}$$

which implies that, for example,

$$\omega_{fm} = \omega_I + \omega_K - \omega_m = \omega_K + \omega_{Im}, \tag{2.5.8}$$

where $\hbar\omega_{Im}$ gives the energy of ionization from the intermediate excited state $|m\rangle$.

The MPI transition probability is then obtained from Eq. (2.5.1) by integrating over all final energies of the photoelectron, and the result, which describes the photoelectron angular distribution, is given by

$$W^{(2)}_{i\to k}(\Omega_{\mathbf{K}}) = \frac{2\pi}{\hbar^2}\left(\frac{2\pi e^2}{c}\right)^2 \int_0^\infty d\omega_l\, \omega_l I_1(\omega_l) \int_0^\infty d\omega_{l'}$$
$$\times \omega_{l'} I_2(\omega_{l'}) \rho(\omega_K) |M^{(2)}_{Ki}(\omega_l, \omega_{l'})|^2, \tag{2.5.9}$$

which depends on the photoelectron direction of propagation. Notice that

$$\omega_{fi} = \omega_{Ii} + \omega_K = \omega_l + \omega_{l'}. \tag{2.5.10}$$

It is evident that the total transition probability per unit time for two-photon ionization is calculated from Eq. (2.5.9) by integrating over all photoelectron directions.

In this chapter we have presented the perturbation, Green's function, density matrix, and susceptibility methods used to treat multiphoton processes. The best method for investigating a particular multiphoton process depends on the molecular system, the laser excitation condition, the molecule–radiation interaction strength, and so forth.

REFERENCES

Agarwal, G. S. (1974). "Quantum Optics." Springer-Verlag, Berlin and New York.
Armstrong, J., Bloembergen, N., Ducuing, J., and Pershan, P. S. (1962), *Phys. Rev.* **127**, 1918.
Bloembergen, N. (1965). "Nonlinear Optics." Benjamin, Reading, Massachusetts.
Eyring, H., Walter, J., and Kimball, G. E. (1944). "Quantum Chemistry." Wiley (Interscience), New York.

Fong, F. K. (1975). "Theory of Molecular Relaxation Processes: Applications in Chemistry and Biology," Wiley, New York.

Fujimura, Y., and Lin, S. H. (1979a). *J. Chem. Phys.* **70**, 247.

Fujimura, Y., and Lin, S. H. (1979b). *J. Chem. Phys.* **71**, 3733.

Fujimura, Y., and Lin, S. H. (1981). *J. Chem. Phys.* **74**, 3726.

Fujimura, Y., and Lin, S. H. (1983). *J. Chem. Phys.* **78**, 6468.

Fujimura, Y., Kono, H., Nakajima, T., and Lin, S. H. (1981). *J. Chem. Phys.* **75**, 99.

Goldberger, M. L., and Watson, K. M. (1964). "Collision Theory." Wiley, New York.

Harvey, A. B. (1981). "Chemical Applications of Nonlinear Raman Spectroscopy." Academic Press, New York.

Heitler, W. (1954). "The Quantum Theory of Radiation." Oxford Univ. Press, London and New York.

Jones, K.E., and Zewail, A. H. (1978). *Springer Ser. Chem. Phys.* **3**, 196.

Lambropoulos, P. (1974). *Phys. Rev. A* **9**, 1992.

Lee, P. S., and Lee, Y. C. (1973). *Phys. Rev. A* **8**, 1727.

Lee, P. S., Lee, Y. C., and Chang, C. T. (1973). *Phys. Rev. A* **8**, 1722.

Levine, R. D. (1969). "Quantum Mechanics of Molecular Rate Processes." Oxford Univ. Press, London and New York.

Lin, S. H. (1966). *J. Chem. Phys.* **44**, 3759.

Lin, S. H. (1967). *In* "Physical Chemistry" (H. Eyring, ed.) Vol. 2, Chapter 3. Academic Press, New York.

Lin, S. H. (1972). *J. Chem. Phys.* **56**, 2645.

Lin, S. H. (1973). *J. Chem. Phys.* **58**, 5760.

Lin, S. H., ed. (1980). "Radiationless Transitions, p. 363." Academic Press, New York.

Lin, S. H., and Eyring, H. (1977a). *Proc. Natl. Acad. Sci. U.S.A.* **74**, 3105.

Lin, S. H., and Eyring, H. (1977b). *Proc. Natl. Acad. Sci. U.S.A.* **74**, 3623.

Louisell, W. H. (1973). "Quantum Statistical Properties of Radiation." Wiley (Interscience), New York.

Loudon, R. (1973). "The Quantum Theory of Light." Oxford Univ. Press (Clarendon), London and New York.

McClain, W. M., and Harris R. A. (1977). *Excited States* **3**, 1.

Meredith, G. R. (1981). *J. Chem. Phys.* **75**, 4317.

Messiah, A. (1965). "Mécanique Quantique." Dunod, Paris.

Mower, L. (1966). *Phys. Rev.* **142**, 799.

Mower, L. (1968). *Phys. Rev.* **165**, 145.

Nibler, J. W., and Knighten, G. V. (1979). *Top. Curr. Phys.* **11**, 253.

Sargent, M., III, Scully, M. O., and Lamb, W. E., Jr. (1974). "Laser Physics." Addison-Wesley, Reading, Massachusetts.

Schiff, L. I. (1955). "Quantum Mechanics." McGraw-Hill, New York.

Schönberg, M. (1951a). *Nuovo Cimento* **8**, 651.

Schönberg, M. (1951b). *Nuovo Cimento* **8**, 817.

Shen, Y. R. (1976). *Rev. Mod. Phys.* **48**, 1.

Yariv, A. (1975). "Quantum Electronics." Wiley, New York.

CHAPTER
3

Experimental
Methods

This chapter begins with a short review of the early experimental results of two-photon spectroscopy. We shall discuss the principal experimental methods and show what kind of spectroscopic information can be obtained by means of them. The technique of detecting two- or three-photon absorption by multiphoton ionization of molecules is described in detail. In an additional section, we shall briefly consider some results regarding a new type of optical mass spectroscopy that combines spectroscopic information derived from multiphoton absorption with mass spectrometic information.

3.1 EARLY EXPERIMENTAL RESULTS OF TWO-PHOTON SPECTROSCOPY

3.1.1 First Experimental Observation of Two-Photon Absorption in the Visible Region of the Spectrum

In the first article on two-photon absorption, which was published in 1929, Maria Goeppert-Mayer briefly discussed the feasibility of experimentally detecting two-photon processes. She wrote that because of the quadratic intensity dependence, it would be difficult to observe two-photon absorption. This quadratic intensity dependence is the reason why no progress was made in two-photon absorption experiments over a period of 30 years. The first successful experiment was performed by Kaiser and Garrett (1961). They used a high-power monochromatic ruby laser with small divergence, so that high light intensities, unobtainable with conventional light sources, were obtained within the focused beam. The red ruby light at $\lambda = 6943$ Å was focused into a Eu^{2+}-doped CaF_2 crystal, and the emission of blue light from the two-photon excited band was observed.

3.1.2 Early Experiments with Molecules in Liquid and Solid Phases

The first experiments on molecular systems were restricted to materials of high particle density. Tunable light sources are necessary for measurement of two-photon spectra of molecular systems. In the early 1960s, however, no tunable laser sources in the visible region of the spectrum were available, and spectra could only be obtained by a combination of a solid state laser with a fixed wavelength and a conventional light source with tunable output.

Two-photon absorption in a molecular system was observed for the first time by Peticolas et al. (1963). They used just one ruby laser and therefore obtained only a single point at $\lambda = 6943$ Å in the two-photon spectrum of crystalline anthracene. Fröhlich and Mahr (1966) used a combination of a ruby laser and a conventional Xe flash lamp and were able to measure point by point the first two-photon spectrum of the same material (with low resolution). They were not able to decide from their experimental results whether they were observing two-photon transitions into excited "gerade" electronic states or a vibrationally induced two-photon transition into already known "ungerade" electronic states. This fundamental question will be discussed in more detail in Chapters 5 and 6. It was with this experimental setup of a ruby laser and a Xe flash lamp that the first experiments in a solution of organic molecules were performed. Investigations of α-chlornaphthalene (Eisenthal et al., 1968; Monson and McClain, 1970) revealed a strong absorption band between 38,000 and 42,000 cm^{-1}. The polarization behavior corresponded to a totally symmetric two-photon absorption tensor (see Chapter 4) and it was concluded that a $^1A_g^- \leftarrow {}^1A_g^-$ transition had been observed (Monson and McClain, 1972).

The resolution of two-photon spectra obtained by means of this technique was limited to about 100 cm^{-1}. The reasons for this low resolution were the low intensity level of the Xe flash lamp and the intensity loss in the monochromator that selected the light wavelength out of the broad emission band of the flash lamp. Furthermore, the spectral range was restricted by the low energy of the ruby laser photons (1.79 eV). The energy of the second photon from the flash lamp had to be not higher than about 5 eV, since at higher photon energies the onset of one-photon absorption, which is stronger by several orders of magnitude than the second-order two-photon absorption process, would take place. This was why the dye laser essentially improved detection sensitivity and spectral resolution and turned two-photon absorption into a practical kind of spectroscopy. The first experiments were performed with a ruby laser–pumped dye laser source (Bergman and Jortner, 1972). The frequency width of this dye laser was as large as 5 Å, and because of the low repetition rate of the ruby laser, the two-photon spectra were measured point by point across a spectrum width of several hundred angstroms. Various organic materials were investigated in crystals as well as in solution. New measurements on the prototype material anthracene then yielded spectral results differing from those of the old experiments (Fröhlich and Mahr, 1966). Finally, a comparison of two-photon spectra in solution and in crystals (Webman and Jortner, 1969) yielded the new result that indeed two-photon absorption is vibrationally induced in the region of the $^1B_{2u}$ state. Similar experiments were performed with crystalline naphthalene by Bergman and Jortner (1974), with higher resolution for naphthalene in solution (Mikami and Ito, 1975). Drucker and McClain (1974a) systematically investigated organic substances such as biphenyl and 0–0-bridged biphenyls. These molecules are of special interest in exciton theory (Monson and McClain, 1972; Drucker and McClain, 1974b).

Detailed review articles on these early two-photon experiments were written by Worlock (1972) and Bredikhin and Galanin (1973), the latter dealing mainly with Russian contributions to the field.

3.1.3   Two-Photon Spectroscopy in Molecular Crystals and Matrices at Low Temperatures

The vibrationally resolved two-photon spectroscopy of organic molecules was initiated by measurements in crystals at low temperatures. In 1973 Hochstrasser and co-workers published the two-photon spectrum of a biphenyl crystal at 2 K. The spectrum was obtained by continuously scanning the wavelength of the dye laser. They used a dye laser transversely pumped by a N_2 laser according to the equipment setup published by Hänsch in 1972. Shortly afterward, measurements on benzene crystals (Hochstrasser et al., 1974a), naphthalene crystals (Hochstrasser et al., 1974b), and naphthalene in a durene matrix (Hochstrasser et al., 1974b) were published. The spectra revealed a sharp well-resolved structure due to the vibronic activity in the range of the $S_1 \leftarrow S_0$ transition. Even though the polarization behavior of the various bands is easily obtained in the solid state, an unambiguous assignment of the vibronic bands was not possible in these measurements. Problems arose because the molecule is distorted when it is located in the crystal or matrix and when point groups are changed (Genkin and Kitai, 1975).

3.1.4 Two-Photon Spectroscopy of Gas-Phase Molecules

Two-photon spectra of gas-phase molecules were only obtained as late as 1974. This is due to the low particle density of low-pressure gases. The low particle density allows the molecules to remain relatively unperturbed by their surroundings, and true vibrational frequencies can be measured. Rotational spectroscopy in the gas phase yields detailed information about the rotational constants of the molecules and hence about their molecular structure.

The feasibility of two-photon spectroscopy of gas-phase molecules was demonstrated for the first time by Hochstrasser *et al.* (1974c). Shortly afterward, the first assignment of a spectrum based on hot-band analysis was made by Wunsch *et al.* (1975). Their result was confirmed by the polarization measurements of Bray *et al.* (1975) and Friedrich and McClain (1975) and finally by observation of the spectrally resolved resonance fluorescence by Knight and Parmenter (1976).

Naphthalene was the second polyatomic molecule of which a two-photon spectrum was obtained in the gas phase (Boesl *et al.*, 1976). The analysis of the two-photon spectrum yielded hitherto unknown vibrational frequencies in the excited S_1 state. Furthermore, additional information not accessible in one-photon spectroscopy was obtained from the hot band spectrum concerning vibrational frequencies in the electronic ground state.

Nitric oxide was the first small diatomic molecule to be studied by two-photon spectroscopy (Bray *et al.*, 1974). In this spectrum the rotational transitions are easily resolved, and therefore a comparison with the theoretical results for rotational line strength and polarization behaviors (Bray and Hochstrasser, 1976) is possible. In the homonuclear molecule I_2 (Rousseau and Williams, 1974; Danyluk and King, 1976), two-photon spectroscopy yielded additional information about five gerade electronic states already known from electric discharge fluorescence measurements (Wieland *et al.*, 1972).

3.2 EXPERIMENTAL TECHNIQUES I

Since two-photon absorption is a second-order process, it is rather weak at the moderate light intensities available from tunable dye lasers. This demands a sensitive technique for detecting only a few two-photon absorption events in the sample. In principle, higher light intensities could be used, but then higher order processes become more probable and the measured spectrum may be a superposition of two-, three-, four-, or more than four-photon spectra. Even ionization and fragmentation of molecules is possible at high light intensities, and in this case the multiphoton spectrum of a fragment may be superposed. For this reason in most cases one should use highly sensitive detection techniques and moderate light intensities rather than high light intensities and nonsensitive detection techniques. First a series of sensitive detection techniques will be discussed.

3.2.1 Measurement of Two-Photon Absorption

The measurement of absorption requires measuring the light power first in front of the sample and then after it has passed through the sample. The detection limit is strongly dependent on the time-integrating technique used in the experiment, among other things. Usually, one cannot detect differences in light intensity smaller than 0.1%. Qualitatively, it is clear that true absorption measurements in two-photon spectroscopy are possible only for samples of high density, for example, liquids and solids; it is not a feasible method for gas-phase spectroscopy, since it is impossible to keep the laser light focused over a long absorption path. The first absorption measurements in two-photon spectroscopy were published by Hopfield and co-workers in 1963. They used the combination of a ruby laser and a continuum flash lamp. Two-photon absorption was monitored on an oscilloscope as a short dip in the transmitted flash lamp light intensity which coincided with the laser pulse. This experiment required accurately overlapping both light beams over a long distance in order to obtain a high level of absorption. A special setup was devised by Hopfield et al. (1963), in which a crystal acted as a light guide for both light beams.

Swofford and McClain (1975) used a combination of a high-power pulsed dye laser and a fixed-frequency continuous wave (cw) Kr^+ ion laser and were able to detect absorption differences as small as 0.1%. Even though in this way a spectrum is hard to measure, they were able to obtain very accurate absolute values of the two-photon absorption cross section for diphenyl-1,3-butadiene at a special wavelength. In addition to the spectrum, the absolute two-photon cross section yields further evidence of whether two-photon absorption is purely electronic or vibrationally induced.

Staginnus et al. (1968) automatically processed absorption measurement data. In their setup accuracy was increased by integrating over several laser pulses.

A very sensitive method for the detection of weak absorption is the intra-cavity absorption technique (Peterson et al., 1971; Schröder et al., 1975; Hänsch et al., 1972). Here the weakly absorbing sample is placed in the cavity of a dye laser. As a consequence, the dye laser emission is quenched at those wavelengths absorbed by the material. In this way an absorption as small as 10^{-4} can be detected in a time interval of 20 ns. This technique is appropriate for detecting small amounts of materials with sharply structured spectra rather than for measuring a complete spectrum. So far only one application has been made to two-photon spectroscopy: two-photon absorption in an anthracene solution (Kleinschmidt *et al.*, 1974).

3.2.2 Detection of Fluorescence after Two-Photon Excitation of Molecules

The most convenient and sensitive method for measurement of two-photon spectra is the detection of the fluorescence of the excited molecules. As a consequence of two-photon absorption, most molecules emit a photon at about twice the energy of the absorbed photons. These emitted photons are detected with high sensitivity. A typical setup is shown in Fig. 3.1. The light from a tunable dye laser is focused on a fluorescence cell that contains the molecular gas or solution. Simultaneous absorption of two (visible) photons takes place in the small focus and (UV) photons are emitted with a quantum yield that is typical for the molecule under investigation. The emitted UV photons are observed with a high-gain photomultiplier. Special filters are used to separate out the intense exciting visible light. This can easily be accomplished, since in two-photon spectroscopy there usually is a large frequency shift between the exciting visible light and the emitted UV light. This greatly improves the signal-to-noise ratio relative to one-photon excitation, which is a side benefit of this detection technique. In gases and in solutions of low concentration no emitted photons are reabsorbed; however, this might be a problem in pure liquids and in crystals with a large one-photon absorption cross section. The signal of the photomultiplier is then

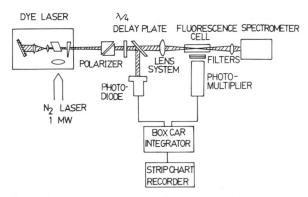

Fig. 3.1 Experimental setup for recording two-photon fluorescence excitation spectra of gas-phase molecules.

fed into a boxcar integrator and integrated over a short time interval (10 ns to a few picoseconds, depending on the lifetime of the fluorescence). The integrated signal is then recorded on a strip chart recorder. Finally, by continuously scanning the wavelength of the dye laser one can obtain the two-photon excitation spectrum of the molecule under investigation. For very weak signals and for time-resolved fluorescence measurements the boxcar integrator may be replaced by a transient digitizer and a data-processing system.

The solid angle for the observation of fluorescence is about 6×10^{-1} sr in this setup. The detection limit is reached when the signal-to-noise ratio in the measured spectrum is better than one-to-one. For a conventional photomultiplier we have found that it is possible to record two-photon spectra when 200 photons are emitted from the focus for a single exciting laser pulse.

For molecules with a fluorescence quantum yield of unity, this means that some 200 two-photon absorption events should have taken place during the laser pulse in order for a two-photon excitation spectrum to be measured. Unfortunately, the fluorescence quantum yield of most molecules is less than unity, which is the fundamental disadvantage of the fluorescence detection method. In this case detection by resonance-enhanced multiphoton ionization might be useful (see Section 3.3). It is also possible to detect phosphorescence if the molecules undergo a fast intersystem crossing process into the triplet system. This has been demonstrated for a pyrazine crystal at low temperatures by Esherick *et al.* (1975). For the detection of vibronic states with high excess energies above S_1, it might be useful to have a high-pressure buffer gas produce a fast collisional deactivation of the excited levels down to the vibrationless ground state. At high pressures this collisional deactivation might compete with the internal radiationless process, and fluorescence from the thermalized S_1 state with large fluorescence quantum yields can be observed. By this method vibronic states with excess energies as much as 6000 cm^{-1} above the vibrationless electronic S_1 state have been observed (Wunsch *et al.*, 1981).

3.3   EXPERIMENTAL TECHNIQUES II: DETECTION OF TWO- OR THREE-PHOTON ABSORPTION BY MULTIPHOTON IONIZATION OF MOLECULES

3.3.1   General Remarks

After a molecule has been excited in an intense light field by two- or three-photon absorption, it is very likely to absorb further photons, which finally results in its ionization. Another sensitive detection technique for

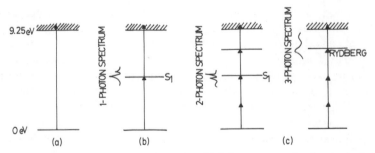

Fig. 3.2 Several possible resonantly enhanced multiphoton processes leading to the ionization of a polyatomic molecule: (a) one-photon, (b) two-photon, and (c) four-photon ionization.

two- or three-photon absorption processes in molecular gases is based on the subsequent ionization of the molecule after two- or three-photon excitation.

Several possible processes for ionization of a polyatomic molecule such as benzene are shown in Fig. 3.2. For direct one-photon ionization, photon energies greater than 9.25 eV would be necessary, for example, vacuum UV light of wavelength shorter than 1340 Å. Higher order processes, such as two-, three-, and even four-photon ionization, are possible if real molecular states are in resonance with the energy of one, two, or three photons. This strongly enhances the probability of the multiphoton ionization process. The rate-determining step is usually the first step, a two-or three-photon absorption via virtual intermediate states. At the high intensities necessary for the observation of two- or three-photon absorption, the subsequent one-photon steps into the ionization continuum are automatically saturated and every molecule that has been excited by a two- or three-photon absorption is ionized. The underlying rate equation model for multiphoton ionization is discussed in Chapter 4 in a quantitative manner.

The interesting feature of multiphoton ionization from a spectroscopic point of view is the resonance enhancement by resonant intermediate states. Since the ionization efficiency is strongly enhanced when the photon energy is in resonance with real intermediate states, scanning the wavelength of the laser leads to a modulation of the ion current, which reflects the spectrum of the intermediate states. Thus it is possible to measure the intermediate-state (two- or three-photon) spectrum by measuring the ion currents. Some examples will be given at the end of this chapter. This method was demonstrated for the first time by Lineberger and Patterson (1972), and Johnson (1975, 1976) measured two-photon spectra of polyatomic molecules by three- and four-photon ionization.

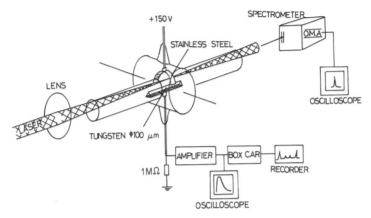

Fig. 3.3 Ionization cell and setup for recording multiphoton intermediate-state spectra of molecules by multiphoton ionization.

3.3.2 The Ionization Cell

In Fig. 3.3 the ionization cell is shown in detail. The laser is focused into the cell containing the molecular gas at a typical pressure of a few Torr. It is very common to use a device with a thin wire that is axially positioned in a cylindrical metal plate biased with a positive voltage of about 100 V: The potential drives the free electrons produced by the multiphoton ionization process to the positive electrode. If enough voltage is applied between the electrodes and if the particle density is sufficient, charge amplification by collisions can take place, increasing the detectability of the electrons. If the gas pressure is very low, a buffer gas is needed to achieve charge multiplication in the ionization cell. The voltage produced by the current at a 1-MΩ impedance is then amplified in a preamplifier by one order of magnitude, fed into a boxcar integrator, and then integrated with a gate width of some 10 μs. The modulation of the current as a function of wavelength is then recorded on a strip chart recorder and reflects the intermediate state resonance. Without a charge amplification in the ionization cell the empirically determined detection limit is about 1000 ions produced within one laser pulse.

3.3.3 Typical Experimental Results Obtained
with the Ionization Cell

To illustrate the feasibility of the ionization detection method, Fig. 3.4 shows the results for benzene when the ion current was measured as a function of the laser wavelength between 36,500 and 39,000 cm^{-1}. The result is

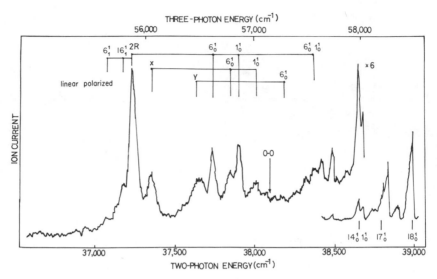

Fig. 3.4 Three- and two-photon spectrum (right side) of benzene (C_6H_6) as measured by multiphoton ionization.

typical for the ionization method and can be directly compared with the excitation spectrum in the same wavelength range in Fig. 3.5, which was obtained by observing the fluorescence as described in Section 3.2.2. Even though both spectra cover the same wavelength range, they are completely different.

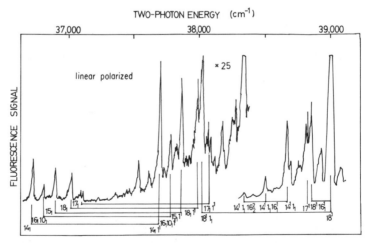

Fig. 3.5 The two-photon spectrum of benzene (C_6H_6) measured by detection of fluorescence in the spectral range corresponding to that of Fig. 3.4.

In the excitation spectrum, the two-photon spectrum of benzene is shown with high sensitivity. The hot band region is located on the red side of the spectrum with the inducing vibrations excited in the ground state (Wunsch *et al.*, 1977). On the blue side some of the corresponding cold bands (17_0^1, 18_0^1) are shown. These bands are roughly two orders of magnitude more intense than the hot bands. Only the blue side of the multiphoton ionization spectrum with the cold bands resembles the fluorescence excitation spectrum. Between 38,500 and 36,500 cm^{-1} a completely different spectrum appears; it was found to be the three-photon spectrum of benzene in the region of the E_{1u} state of the molecule. The relatively sharp, strong peaks correspond to Rydberg transitions (2R), and the broad, weak peaks are supposed to be due to valence states (probably the vibrational structure of the $^1E_{1u}$ state). The $^1E_{1u}$ state itself cannot be measured with an energy of 55,400 cm^{-1}, since the three-photon spectrum is cut off at 55,898 cm^{-1}. This corresponds to the threshold photon energy necessary for three-photon ionization of the molecule. Hence the spectrum between 55,900 and 57,500 cm^{-1} is governed by three-photon resonances and the underlying absorption process is shown in the right-hand side of Fig. 3.2. The virtual intermediate state for this three-photon absorption might be the $|B_{2u}, 14^1\rangle$ state with vibronic A_{1g} symmetry and with a distance from resonance of only 2000 cm^{-1}.

The following conclusions can be drawn from the two differing spectra obtained with alternative detection methods:

(i)   The ionization technique is more sensitive to states with high excitation energies that barely fluoresce and are near the ionization continuum.

(ii)   The ionization technique is especially sensitive for the detection of Rydberg states whose relaxation rates are supposed to be not as fast as relaxation from valence states.

(iii)   By fluorescence detection one preferentially obtains pure two-photon spectra, since fluorescence from higher three-photon states that might be simultaneously excited by the absorption of a third photon is not very probable.

(iv)   In many cases the ionization technique may produce overlapping spectra from two- or three-photon intermediate states and therefore assignment is not always unambiguous.

3.3.4   Ion Detection in a Mass Spectrometer

Ion detection in an ionization cell is the simplest method and a very sensitive one as well. However, this technique provides no information about the type of ions produced in the multiphoton ionization process. As will be discussed in more detail in Chapter 7, there is often a strong fragmentation

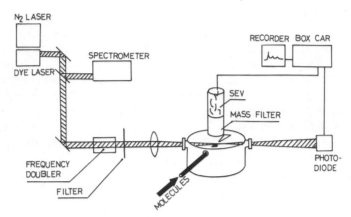

Fig. 3.6 Experimental setup for mass-selective detection of ions produced in a resonantly enhanced multiphoton ionization process. [From Boesl *et al.* (1981).]

of ions in the multiphoton process (Zandee and Bernstein, 1979; Boesl *et al.*, 1980).

In order to shed light on the ionization process, it is useful to detect the ions in a mass spectrometer, which allows determination of the mass of the ions. The scheme of the setup for mass-selective ion detection is shown in Fig. 3.6. Tunable laser light or frequency-doubled light is focused into an effusive molecular beam close to the aperture of the nozzle. The ions produced by the multiphoton ionization process are withdrawn through an ion lens system into the mass filter, analyzed for mass, and finally recorded with an ion multiplier.

Our first experiments were performed with a commercial quadrupole mass filter (UTI 100 C) (Boesl *et al.*, 1978). A beam of benzene molecules was produced with an effusive nozzle (0.2 mm ϕ) with a flow rate of 3×10^{-3} Torr $cm^3 \ s^{-1}$. The molecular density n_0 in the ionization region close to the nozzle aperture was $10^{13} \ cm^{-3}$. The light was focused on the axis of the molecular beam in front of the nozzle, resulting in an illuminated volume of about $2 \times 10^{-5} \ cm^3$ which contains about 2×10^8 benzene molecules at the moment of irradiation. A tunable dye laser was used, the light of which was frequency doubled in a lithium formate crystal, yielding a peak UV power of 300 W at 2500 Å. Benzene molecules were ionized via two-photon ionization when the laser wavelength was tuned to a real intermediate state in the S_1 system. The energy level scheme of this two-photon ionization is shown in Fig. 3.7, with the mass spectrum for a single parent-ion peak obtained in this experiment at low light intensities. At higher intensities ($> 10^7 \ W/cm^2$) a strong fragmentation is observed (see Fig. 3.8). Alternatively, it is also possible to use visible light. In this case the laser wavelength has to be tuned

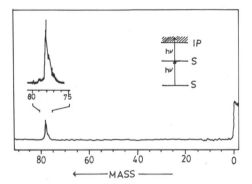

Fig. 3.7 Mass spectrum of benzene (C_6H_6) ionized by two-step photoionization with UV laser light at 2590.1 Å. At low light intensities ($<10^7$ W cm^{-2}) only parent molecular ions appear. (Measured with a quadrupole mass spectrometer.) [From Boesl *et al.* (1978).]

to a real intermediate two-photon state of the molecule (Boesl *et al.*, 1980), as shown in Fig. 3.9. At the high intensity of the laser light in this experiment multiphoton ionization produces a rich fragmentation of the ions that is detected by the mass spectrometer. This experiment clearly revealed that in a high intensity two-photon experiment molecules are not only excited to the two-photon state but are also ionized, and a rich pattern of fragment ions is formed. Therefore in a two-photon or, even more important, a three-photon experiment with highly intense laser light one must check whether these fragments may influence the measured two- or three-photon spectrum. This is certainly important for three- and four-photon

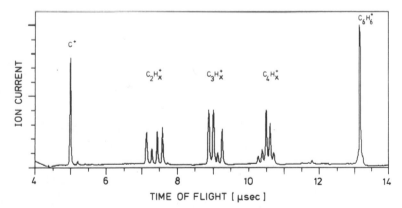

Fig. 3.8 Fragmentation pattern in the mass spectrum of benzene (C_6H_6) obtained by two-step photoionization with UV laser light at 2590.1 Å. At high light intensities ($>10^7$ W cm^{-2}) smaller molecular ions are observed. (Measured with the time-of-flight mass spectrometer shown in Fig. 3.10.) [From Boesl *et al.* (1978).]

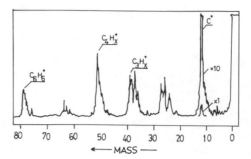

Fig. 3.9 Fragmentation pattern in the mass spectrum of benzene (C_6H_6) obtained by four-photon ionization with intense visible laser light at 5043.8 Å. [From Boesl *et al.* (1980).]

ionization experiments because of the high laser intensity necessary for the ionization process.

The ions produced in a multiphoton ionization process have characteristic features: They are produced within a time interval of some nanoseconds during the laser pulse in the small volume of about 10^{-5} cm^3 provided by the focus of the laser light. Therefore, mass analysis by time-of-flight detection is the method of choice for this type of ionization. The scheme of the laser mass spectrometer for time-of-flight analysis is shown in Fig. 3.10.

The nozzle ($+300$ V) that produces the effusive molecular beam is placed between two cylindrical electrodes separated by 12 mm. The repeller elec-

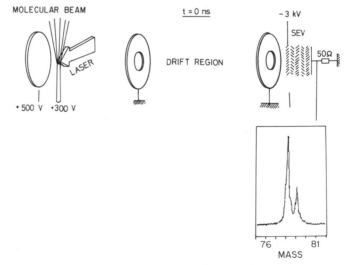

Fig. 3.10 Scheme of a time-of-flight laser mass spectrometer for multiphoton ionization. The mass resolution achieved with this simple setup is demonstrated in the insert for the mass spectrum of benzene isotopes (78,79 amu).

trode is at a voltage of $+500$ V and the second electrode is at 0 V. The ions produced in front of the nozzle are accelerated on their way to the electrode, pass through its concentric aperture, and then enter the drift tube. After traveling a distance of 282 mm, they pass through another electrode at ground (0) potential and finally hit the cathode of the ion multiplier (venetian blind multiplier), separated in time according to their mass. The construction scheme is so simple because there is a negligibly small potential gradient across the small ionization region (100 mm ϕ) and the ion source is already pulsed due to the short duration (~ 5 ns) of the laser pulse.

The time-resolved ion signal is then digitized in a transient digitizer (Tektronix 7512) and processed in a batch transfer to a computer. Finally, the time-of-flight spectrum is plotted, without averaging, as obtained from a single laser shot. Even though the concept of this time-of-flight mass spectrometer is very simple, its mass resolution is $M/\Delta M \sim 250$ at 78 amu. This is demonstrated in the small insert of Fig. 3.10 where a small portion of the time-of-flight mass spectrum of benzene in natural isotopic abundance is shown. Using the proper wavelength we obtained an enhancement of a factor 6 of $^{13}C^{12}C_5H_6$ (79 amu) in the ion mass spectrum. The two mass peaks due to "light" and "heavy" benzene are well separated in the mass spectrum resulting in a mass resolution of 250. The mass resolution was increased to $M/\Delta M \sim 4000$ by using a reflecting field setup that compensates for differences in the kinetic energy of the ions (Boesl et al., 1982).

3.4 THE PHOTOACOUSTIC DETECTION METHOD

The photoacoustic detection method (Rosencwaig, 1978) is principally different from the most convenient methods discussed in Sections 3.2.2 and 3.3. Fluorescence detection, and to some degree ionization detection, decreases in sensitivity when a fast competing intra- or intermolecular relaxation takes place from the multiphoton excited level. In this case there is a dissipation of the energy selectively released in the excited level into thermal energy. After pulsed excitation there is a rapid conversion of absorbed energy into pressure fluctuations, which then can be detected by a microphone. This means that photoacoustic spectroscopy is based on the detection of those effects that are loss channels in fluorescence and ionization detection. Apparently, photoacoustic detection techniques should be suitable principally for observing two-photon spectra in weakly fluorescing materials.

There have been few successful attempts to measure multiphoton spectra of molecules by photoacoustic methods. The two-photon spectrum of liquid benzene has been published, including the main vibronic bands and a value for the two-photon absorption cross section (Tam and Patel, 1979).

3.5 MISCELLANEOUS DETECTION METHODS

In this section we shall present other methods that have been proved to be sensitive enough for the detection of a two-photon process.

One of these methods is thermal blooming, which is also based on energy loss mechanisms in the excited states (Long et al., 1976). The transfer of energy from the excited states into heat causes a change in the refractive index of the material under investigation. This is then detected by a weak focusing or defocusing of the exciting laser beam due to the changed optical behavior of the material. The thermal blooming technique has been applied to the test molecule benzene. Two-photon spectra of liquid benzene were measured point by point in the range between 360 and 530 nm (Twarowski and Kligler, 1977; Vaida et al., 1978). A spectrum established by this technique yielded additional information about the position of the $^1E_{1g}$ state of benzene (Vaida et al., 1978).

Another detection method for two-photon absorption is based on the change in the susceptibility of the material under investigation in the presence of the light field. In a strong light field there are higher order contributions $\chi^{(n)}$ to the susceptibility (Bloembergen, 1965; see also Section 2.4). It has been known for a long time (Maker and Terhune, 1965) that the real part of the third-order susceptibility Re $\chi^{(3)}$ produces an intensity-dependent absorption coefficient that increases if there is a resonance at the two-photon energy. On the other hand, the imaginary part of the nonlinear susceptibility Im $\chi^{(3)}$ produces an intensity-dependent index of refraction that may turn the polarization vector of the incoming light wave. These two-photon resonances in the third-order susceptibility $\chi^{(3)}$ may be detected in several ways.

One possibility was demonstrated by Hochstrasser et al. (1978). They detected two-photon resonances of sulfur dioxide and nitric oxide gases in a three-wave mixing experiment. Since this method creates no real population of the resonant state, the detection of the resonances does not directly depend on the dynamic pathway followed by the excited state. The dynamics of the resonant state enters only through a damping parameter, thereby limiting the magnitude of the resonance term. It has been shown that this method is suitable for obtaining absolute two-photon cross sections by comparison of two-photon resonances with coherent anti-Stokes Raman resonances of $\chi^{(3)}$ (Kramer and Bloembergen, 1976; Lynch and Lotem, 1977; Anderson et al., 1977). Two-photon cross sections are then given in terms of the accurately known Raman cross sections (Schrötter and Klöckner, 1979).

Anderson et al. (1978) demonstrated for sodium vapor that the change of polarization produced by the imaginary part of the third-order susceptibility $\chi^{(3)}$ can be detected for observation of the two-photon spectrum. An extension of this method to the case of molecules seems possible even though the sensitivity is not expected to be better than that of the other methods discussed above. The general virtue of two-photon absorption detection via $\chi^{(3)}$ is the calibration of two-photon cross sections on the basis of Raman cross sections.

REFERENCES

Anderson, R. J. M., Holtom, G. R., and McClain, W. M. (1977). *J. Chem. Phys.* **66**, 3332.
Anderson, R. J. M., Stachelek, T. M., and McClain, W. M. (1978). *Chem. Phys. Lett.* **59**, 100.
Bergman, A., and Jortner, J. (1972). *Chem. Phys. Lett.* **15**, 309.
Bergman, A., and Jortner, J. (1974). *Chem. Phys. Lett.* **26**, 323.
Bloembergen, N. (1965). "Nonlinear Optics." Benjamin, New York.
Boesl, U., Neusser, H. J., and Schlag, E. W. (1976). *Chem. Phys.* **15**, 167.
Boesl, U., Neusser, H. J., and Schlag, E. W. (1978). *Z. Naturforsch., Teil A:* **33**, 1546.
Boesl, U., Neusser, H. J., and Schlag, E. W. (1980). *J. Chem. Phys.* **72**, 4327.
Boesl, U., Neusser, H. J., and Weinkauf, R., and Schlag, E. W. (1982). *J. Phys. Chem.* **86**, 4857.
Bray, R. G., and Hochstrasser, R. M. (1976). *Mol. Phys.* **31**, 412.
Bray, R. G., and Hochstrasser, R. M., and Wessel, J. E. (1974). *Chem. Phys. Lett.* **27**, 167.
Bray, R. G., and Hochstrasser, R. M., and Sung, H. N. (1975). *Chem. Phys. Lett.* **33**, 1.
Bredikhin, V. I., and Galanin, M. D. (1973). *Sov. Phys—Usp.* (*Engl. Transl.*) **16**, 299.
Danyluk, M. D., and King, G. W. (1976). *Chem. Phys. Lett.* **44**, 440.
Drucker, R. P., and McClain, W. M. (1974a). *J. Chem. Phys.* **61**, 2609.
Drucker, R. P., and McClain, W. M. (1974b). *Chem. Phys. Lett.* **28**, 255.
Eisenthal, K. B., Dowley, M. W., and Peticolas, W. L. (1968). *Phys. Rev. Lett.* **20**, 93.
Esherick, P., Zinsli, P., and El-Sayed, M. A. (1975). *Chem. Phys.* **10**, 415.
Friedrich, D. M., and McClain, W. M. (1975). *Chem. Phys. Lett.* **32**, 541.
Fröhlich, D., and Mahr, H. (1966). *Phys. Rev. Lett.* **16**, 895.
Genkin, V. N., and Kitai, M. S. (1975). *Opt. Spectrosc.* (*Engl. Transl.*) **39**, 184.
M. Göppert-Mayer (1929). *Naturwissenschaften* **17**, 932.
Hänsch, T. W. (1972) *Appl. Opt.* **11**, 898.
Hänsch, T. W., Schawlow, A. L., and Toschek, P. E. (1972). *J. Quant. Electron.* **QE-8**, 802.
Hochstrasser, R. M., Sung, H. N., and Wessel, J. E. (1973). *J. Chem. Phys.* **58**, 4694.
Hochstrasser, R. M., Sung, H. N., and Wessel, J. E. (1974a). *Chem. Phys. Lett.* **24**, 7.
Hochstrasser, R. M., Sung, H. N., and Wessel, J. E. (1974b). *Chem. Phys. Lett.* **24**, 168.
Hochstrasser, R. M., Wessel, J. E. and Sung, H. N. (1974c). *J. Chem. Phys.* **60**, 317.
Hochstrasser, R. M., Meredith, G. R., and Trommsdorff, H. P. (1978). *Chem. Phys. Lett.* **53**, 423.
Hopfield, J., Worlock, J., and Park, K. (1963). *Phys. Rev. Lett.* **11**, 414.
Johnson, P. M. (1975). *J. Chem. Phys.* **62**, 4562.
Johnson, P. M. (1976). *J. Chem. Phys.* **64**, 4638.
Kaiser, W., and Garrett, C. G. B. (1961). *Phys. Rev. Lett.* **7**, 229.
Kleinschmidt, J., Tottleben, W., and Rentsch, S. (1974). *Exp. Tech. Phys.* **22**, 191.
Knight, A. E. W., and Parmenter, C. S. (1976). *Chem. Phys. Lett.* **43**, 399.
Kramer, S. D., and Bloembergen, N. (1976). *Phys. Rev.* **114**, 4654.
Lineberger, W. C., and Patterson, T. A. (1972). *Chem. Phys. Lett.* **13**, 40.
Long, M. E., Swofford, R. L., and Albrecht, A. C. (1976). *Science* **191**, 183.

Lynch, R. T., Jr., and Lotem, M. (1977). *J. Chem. Phys.* **66**, 1905.

Maker, P. D., and Terhune, R. W. (1965). *Phys. Rev.* **137**, A801.

Mikami, N., and Ito, M. (1975). *Chem. Phys. Lett.* **31**, 472.

Monson, P. R., and McClain, W. M. (1970). *J. Chem. Phys.* **53**, 29.

Monson, P. R., and McClain, W. M. (1972). *J. Chem. Phys.* **56**, 4817.

Peterson, N. C., Kurylo, M. J., Braun, W., Bass, A. M., and Keller, R. A. (1971). *J. Opt. Soc. Am.* **61**, 746.

Peticolas, W. L., Goldsborough, I. P., and Rieckhoff, K. E. (1963). *Phys. Rev. Lett.* **10**, 43.

Rosencwaig, A. (1978). *Adv. Electron. Electron Phys.* **46**, 207.

Rousseau, D. L., and Williams, P. F. (1974). *Phys. Rev. Lett.* **33**, 1368.

Schröder, H., Neusser, H. J., and Schlag, E. W. (1975). *Opt. Commun.* **74**, 395.

Schrötter, H. W., and Klöckner, H. W. (1979). *Top. Curr. Phys.* **11** 123.

Staginnus, B., Fröhlich, D., and Caps, T. (1968). *Rev. Sci. Instrum.* **39**, 1129.

Swofford, R. L., and McClain, W. M. (1975). *Rev. Sci. Instrum.* **46**, 246.

Tam, A. C., and Patel, C. K. N. (1979). *Nature (London)* **280**, 304.

Twarowski, A. J., and Kligler, D. S. (1977). *Chem. Phys. Lett.* **50**, 36.

Vaida, V., Robin, M. B., and Kuebler, N. A. (1978). *Chem. Phys. Lett.* **58**, 557.

Webman, J., and Jortner, J. (1969). *J. Chem. Phys.* **50**, 2706.

Wieland, K., Tellinghuisen, J. B., and Nobs, A. (1972). *J. Mol. Spectrosc.* **41**, 69.

Worlock, J. M. (1972). In "Laser Handbook" (E. O. Schultz-Dubois and F. T. Arecchi, eds.), p. 1323. North-Holland Publ., Amsterdam.

Wunsch, L., Neusser, H. J., and Schlag, E. W. (1975). *Chem. Phys. Lett.* **31**, 433.

Wunsch, L., Neusser, H. J., and Schlag, E. W. (1981). *Z. Naturforsch.* **36**, 1340.

Wunsch, L., Metz, F., Neusser, H. J., and Schlag, E. W. (1977). *J. Chem. Phys.* **66**, 386.

Zandee, L., and Bernstein, R. B. (1979). *J. Chem. Phys.* **71**, 1359.

CHAPTER
4
Characteristics of Multiphoton Spectroscopy

In this chapter some characteristics of molecular multiphoton spectroscopy will be discussed. The dependence of the process on laser intensity is described first. Measurement of the laser-intensity dependence is very important to understanding the mechanism and order of the multiphoton process. We shall discuss saturation phenomena that appear in resonant multiphoton processes as well and then review polarization behavior of multiphoton spectroscopy, concentrating on the two-photon absorption process.

4.1 INTENSITY DEPENDENCE

The laser intensity dependence observed in multiphoton transitions can be classified into two kinds. One type originates from the intrinsic laser intensity dependence, which is involved in the expression of the transition probability, and the other type from geometrical effects of the focused laser beam in the region of molecule–photon interaction. In some cases the measured intensity dependence is governed by both effects.

As stated in Chapter 2, the transition rate constant for the n-photon process is proportional to the nth order of laser intensity: $k^{(n)} = (\sigma^{(n)} I^n)/(\hbar \omega_r)^n$, where $\sigma^{(n)}$, I, and ω_r are the nth order transition cross section (strength) in units of $cm^{2n} s^{n-1}$, the laser intensity in units of photons $cm^{-2} s^{-1}$, and the laser frequency, respectively. This I^n-dependence of the transition probability for the n-photon process is called the formal intensity law.

In Section 4.1.1 the formal intensity law will be discussed. This law is found to hold well for nonresonant multiphoton processes and to hold sometimes for resonant multiphoton processes. Deviations from the I^n-dependence have been observed for many multiphoton processes via resonant states. In Section 4.1.2, by using the rate equation approach, it will be shown that the deviation is interpreted by means of saturation between the initial and resonant states.

The intensity dependence originating from geometrical effects in focused laser experiments is sometimes called the $\frac{3}{2}$ power law. The origin of the $\frac{3}{2}$ power law will be discussed in Section 4.1.4.

4.1.1 The Formal Intensity Law

The formal intensity law, I^n-intensity dependence of the observed quantities, has been utilized to determine the orders of multiphoton processes such as excitation, ionization, and/or dissociation of molecules. Fig. 4-1 shows log–log plots of the ion yield versus laser intensity for two-photon ionization of aniline observed by Brophy and Rettner (1979). The laser pulse with a 293.9 nm wavelength and a pulse duration $t_p \sim 1$ μs excites the first excited singlet state $^1B_2$ of aniline. In Fig. 4.1b the I^2-intensity dependence of the ion yield can be seen at the lowest intensities corresponding to the unfocused Nd–YAG pumped laser. The estimated values of the cross sections are $\sigma_1 = (1.0 \pm 0.2) \times 10^{-17}$ cm^2 and $\sigma_2 = (3.5 \pm 0.8) \times 10^{-17}$ cm^2 for the absorption from the $^1A_1$ ground to the $^1B_2$ resonant state and that from the resonant to the ionized state, respectively. Assuming a 1-kW laser pulse, the absorption rate constants $k_{B_1 A_1}^{(1)}$ and $k_{f B_1}^{(1)}$ are found to be 10^5 s^{-1}, and the condition $k_{B_1 A_1}^{(1)} t_p$ and $k_{f B_1}^{(1)} t_p \simeq 10^{-1} < 1$ is satisfied. Under this condition, the ion yield $R_f(t_p)$ can be safely expressed as

$$R_f(t_p) = \sigma_{f B_1}^{(1)} \sigma_{B_1 A_1}^{(1)} t_p^2 I^2 / 2(\hbar \omega_r)^2,$$

which represents the formal intensity law for $n = 2$ and has been derived by using a simple kinetic equation (see Eq. (4.1.14)).

The linear plot of the ion yield with a slope of 1.5 (Fig. 4.1) has been measured by using a high-power, strongly focused laser beam. This is a typical example of the $\frac{3}{2}$ power law originating from the geometric effects of the focused laser in the interaction region.

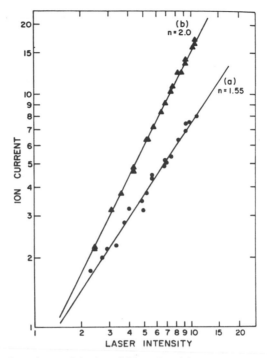

Fig. 4.1 Intensity dependence of the ion yield produced by resonant two-photon ionization via the $^1B_2$ state of aniline: (a) with a focused high-power laser, (b) with an unfocused laser. [From Brophy and Rettner (1979).]

4.1.2 Saturation Phenomena

Note that I^n-dependence generally holds for cases of low-intensity laser experiments, long-lived intermediate states for short pulse times, and before the steady-state condition is satisfied for resonant multiphoton processes. The use of high-intensity lasers may result in saturation of the population between the resonant and ground states and make it easy to reach the steady-state condition.

In Fig. 4.2 is shown the intensity dependence of the ion number for two-photon ionization via the resonant 6^1 vibronic state of S_1 in benzene (Boesl *et al.*, 1981). The measured intensity dependences are presented in Fig. 4.2 for bulk benzene gas at a pressure of 35 mTorr with an unfocused parallel light beam and focused laser light, respectively. A pure quadratic intensity dependence that obeys the formal intensity law is observed for laser intensities below 10^7 W cm^{-2}. Above this threshold value the ion number changes from a quadratic to a roughly linear intensity dependence.

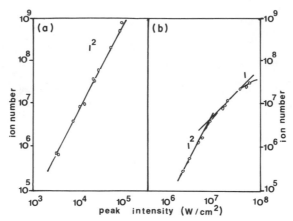

Fig. 4.2 Intensity dependence of the ion number produced by two-photon ionization via the resonant 6^1 state in benzene: (a) with an unfocused parallel light beam in a bulk gas experiment, (b) with focused light in a molecular beam experiment. [From Boesl *et al.* (1981).]

The geometric effect of the focused laser beam in this experiment has been successfully eliminated by setting up the laser in the crossed molecular beam (Fig. 4.2b). The observed linear intensity dependence thus represents the intrinsic intensity dependence and takes saturation into account.

Theoretical considerations for the deviations of transition probability, ion current, and yield from I^n-dependence have been reported by several authors (Lambropoulos, 1974; Kafri and Kimel, 1974; Parker *et al.*, 1978; Zakheim and Johnson, 1980.) Lambropoulos (1974) has taken into account the intensity dependence for the near-resonant multiphoton ionization transition rate of atoms by using the Green's function method: He has evaluated the linewidth and the shift that originates from the intensity-dependent self-energy part associated with induced emission and ionization.

4.1.3   The Rate Equation Approach

One of the methods used to study the deviation from I^n-dependence is the rate equation approach (Parker *et al.*, 1978; Zakheim and Johnson, 1980). For simplicity, let us consider the resonant two-photon ionization shown in Fig. 4.3 by using the rate equation approach and see the conditions under which the deviation takes place.

From the master equation, Eq. (2.3.87), the rate equations associated with two-photon ionization for states a to n through a resonant state m can be

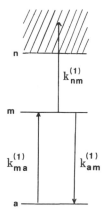

Fig. 4.3   A simple model for resonant two-photon ionization.

expressed as

$$dp_a(t)/dt = -k^{(1)}_{aa}\rho_a(t) + k^{(1)}_{am}\rho_m(t), \tag{4.1.1}$$

$$dp_m(t)/dt = -k^{(1)}_{mm}\rho_m(t) + k^{(1)}_{ma}\rho_a(t), \tag{4.1.2}$$

$$dp_n(t)/dt = k^{(1)}_{nm}\rho_m(t), \tag{4.1.3}$$

where $\rho_a(t) \equiv \rho_{aa}(t)$ is the density matrix element for the initial state, and $k^{(1)}_{am} = k^{(1)}_{aa:mm}$ the radiative rate constant associated with the transition from states m to a. The rate constants satisfy $k^{(1)}_{mm} = k^{(1)}_{am} + k^{(1)}_{nm}$ and $k^{(1)}_{aa} = k^{(1)}_{ma}$. In this treatment effects of the simultaneous two-photon process, specified by $k^{(2)}_{am}$ and $k^{(2)}_{aa}$, and relaxation have been omitted. Note that the radiative rate constants $k^{(n)}_{am}$, etc., have been derived in Chapter 2. The Laplace transformation of Eqs. (4.1.1) and (4.1.2) yields

$$[p + k^{(1)}_{aa}]\rho_a(p) = N_0 + k^{(1)}_{am}\rho_m(p), \tag{4.1.4}$$

$$[p + k^{(1)}_{mm}]\rho_m(p) = k^{(1)}_{ma}\rho_a(p), \tag{4.1.5}$$

where

$$\rho(p) = \int_0^\infty dt\, \rho(t)e^{-pt}. \tag{4.1.6}$$

Here the initial conditions $\rho_a(t = 0) = N_0$ and $\rho_m(t = 0) = \rho_n(t = 0) = 0$ have been assumed. The solution is written as

$$\rho_m(p) = \frac{k^{(1)}_{ma}N_0}{(p - \alpha_1)(p - \alpha_2)} = \frac{k^{(1)}_{ma}N_0}{\alpha_1 - \alpha_2}\left(\frac{1}{p - \alpha_1} - \frac{1}{p - \alpha_2}\right), \tag{4.1.7}$$

$$\rho_a(p) = \frac{N_0}{\alpha_1 - \alpha_2}\left[\frac{(k^{(1)}_{mm} + \alpha_1)}{p - \alpha_1} - \frac{(k^{(1)}_{mm} + \alpha_2)}{p - \alpha_2}\right], \tag{4.1.8}$$

where α_1 and α_2 are the solution for the equation

$$[p + k_{aa}^{(1)}][p + k_{mm}^{(1)}] - k_{am}^{(1)}k_{ma}^{(1)} = 0 \qquad (4.1.9)$$

and are given by

$$\alpha_{1,2} = \tfrac{1}{2}\{-(k_{ma}^{(1)} + k_{am}^{(1)} + k_{nm}^{(1)}) \pm [(k_{ma}^{(1)} + k_{am}^{(1)} + k_{nm}^{(1)})^2 - 4k_{ma}^{(1)}k_{nm}^{(1)}]^{\frac{1}{2}}\}. \qquad (4.1.10)$$

Applying the inverse Laplace transformation to Eqs. (4.1.7) and (4.1.8) yields

$$\rho_m(t) = \frac{k_{ma}^{(1)}N_0}{\alpha_1 - \alpha_2}(e^{\alpha_1 t} - e^{\alpha_2 t}) \qquad (4.1.11)$$

and

$$\rho_a(t) = \frac{N_0}{\alpha_1 - \alpha_2}[(\alpha_1 + k_{am}^{(1)} + k_{nm}^{(1)})e^{\alpha_1 t} - (\alpha_2 + k_{am}^{(1)} + k_{nm}^{(1)})e^{\alpha_2 t}]. \qquad (4.1.12)$$

The ionization rate is proportional to $d\rho_n(t)/dt$, which is given by

$$\frac{d\rho_n(t)}{dt} = \frac{k_{nm}^{(1)}k_{ma}^{(1)}N_0}{\alpha_1 - \alpha_2}(e^{\alpha_1 t} - e^{\alpha_2 t}). \qquad (4.1.13)$$

The ion yield in the case of a square laser pulse can be obtained by integrating Eq. (4.1.13) over t and dividing the resulting expression by N_0 as

$$R_n(t_p) = \frac{\rho_n(t_p)}{N_0} = \frac{k_{nm}^{(1)}k_{ma}^{(1)}}{\alpha_1 - \alpha_2}\left(\frac{e^{\alpha_1 t_p} - 1}{\alpha_1} - \frac{e^{\alpha_2 t_p} - 1}{\alpha_2}\right), \qquad (4.1.14)$$

where t_p is the pulse duration. The emission rate $f_m(t)$ from the resonant state m is proportional to $d\rho_m(t)/dt$:

$$f_m(t) = d\rho_m(t)/dt \qquad (4.1.15)$$

Let us consider a case of weak laser intensity, in which the effect of stimulated emission is negligible and the contribution of $k_{am}^{(1)}$ to the intensity dependence may be neglected [see Eq. (2.1.90)]. In this case the solution of Eq. (4.1.10) can be approximated by $\alpha_1 = -k_{nm}^{(1)}$ and $\alpha_2 = -k_{ma}^{(1)}$. The ion current is proportional to

$$\frac{d\rho_n(t)}{dt} = \frac{k_{nm}^{(1)}k_{ma}^{(1)}N_0}{k_{nm}^{(1)} - k_{ma}^{(1)}}(e^{-k_{ma}^{(1)}t} - e^{-k_{nm}^{(1)}t}) \qquad (4.1.16)$$

for $k_{nm}^{(1)} > k_{ma}^{(1)}$. For the case in which $k_{nm}^{(1)} < k_{ma}^{(1)}$, $k_{nm}^{(1)}$ and $k_{ma}^{(1)}$ should be transposed in Eq. (4.1.16). In the time regions of $k_{ma}^{(1)}t \leq 1$, Eq. (4.1.16) can be expressed approximately as

$$\frac{d\rho_n(t)}{dt} \simeq k_{nm}^{(1)}k_{ma}^{(1)}N_0 t e^{-k_{ma}^{(1)}t}$$

$$\simeq k_{nm}^{(1)}k_{ma}^{(1)}N_0 t$$

$$= \frac{\sigma_{nm}^{(1)}\sigma_{ma}^{(1)}}{(\hbar\omega_r)^2}N_0 I^2 t. \tag{4.1.17}$$

The ion yield $R_n(t_p)$ is expressed as

$$R_n(t_p) = \frac{k_{nm}^{(1)}k_{ma}^{(1)}t_p^2}{2} = \frac{\sigma_{nm}^{(1)}\sigma_{ma}^{(1)}}{2(\hbar\omega_r)^2} I^2 t_p^2. \tag{4.1.18}$$

In the case of $k_{ma}^{(1)} \simeq k_{nm}^{(1)}$, the quadratic intensity dependence can be also derived as

$$d\rho_n(t)/dt \simeq \tfrac{1}{2}(k_{nm}^{(1)}k_{ma}^{(1)})N_0 t. \tag{4.1.19}$$

In other words, in the case of the weak laser field, in which the laser pulse duration satisfies $k_{ma}^{(1)}t_p < 1$, the formal intensity law holds for the ion current and yield.

Next, we consider the case in which a strong laser intensity is applied and the stimulated emission process cannot be neglected. In this case, the spontaneous emission can safely be omitted $k_{am}^{(1)} = k_{ma}^{(1)}$ and

$$\alpha_1 = \tfrac{1}{2}[-(2k_{ma}^{(1)} + k_{nm}^{(1)}) + \sqrt{4(k_{ma}^{(1)})^2 + (k_{nm}^{(1)})^2}], \tag{4.1.20}$$

and

$$\alpha_2 = \tfrac{1}{2}[-(2k_{ma}^{(1)} + k_{nm}^{(1)}) - \sqrt{4(k_{ma}^{(1)})^2 + (k_{nm}^{(1)})^2}]. \tag{4.1.21}$$

Equation (4.1.13) is rewritten as

$$\frac{d\rho_n(t)}{dt} = \frac{k_{nm}^{(1)}k_{ma}^{(1)}N_0}{\alpha_1 - \alpha_2} e^{\alpha_1 t}[1 - e^{-(\alpha_1 - \alpha_2)t}]. \tag{4.1.22}$$

Here α_1 and α_2 are linear functions of the laser intensity. We notice that the coefficient of the intensity of α_1 is very small compared to the difference between α_1 and α_2. For the time scale $t < (\alpha_1 - \alpha_2)^{-1} = [4(k_{ma}^{(1)})^2 + (k_{nm}^{(1)})^2]^{-1/2}$,

$$d\rho_n(t)/dt \simeq k_{nm}^{(1)}k_{ma}^{(1)}N_0 t, \tag{4.1.23}$$

and for the time scale $t > (\alpha_1 - \alpha_2)^{-1}$,

$$d\rho_n(t)/dt \simeq k_{nm}^{(1)}k_{ma}^{(1)}N_0/[4(k_{ma}^{(1)})^2 + (k_{nm}^{(1)})^2]^{1/2}, \tag{4.1.24}$$

Equations (4.1.23) and (4.1.24) indicate quadratic and linear intensity dependences, respectively. The deviation of the intensity from I^2 dependence takes place at $t > (\alpha_1 - \alpha_2)^{-1}$, and

$$\frac{d\rho_n(t)}{dt} \simeq \frac{k_{nm}^{(1)} N_0}{2} \qquad \text{for} \quad k_{ma}^{(1)} \doteq k_{am}^{(1)} > k_{nm}^{(1)}, \tag{4.1.25}$$

$$\frac{d\rho_n(t)}{dt} \simeq k_{ma}^{(1)} N_0 \qquad \text{for} \quad k_{am}^{(1)} < k_{nm}^{(1)}. \tag{4.1.26}$$

For the former case, an equilibrium between the initial and resonant states has been achieved (saturation), and we can observe the process from the resonant to the final state as the apparent transition.

Note that absorption rate constants are proportional to the cross section as well as to the laser intensity. Therefore, an appreciable difference between $\sigma_{ma}^{(1)}$ and $\sigma_{nm}^{(1)}$ may change the I^n intensity dependence of multiphoton processes, even in the presence of a weak laser field. As an example, let us consider a molecular system in which the transition from initial to resonant states is forbidden and the next absorption step is allowed: $\sigma_{nm}^{(1)} \gg \sigma_{ma}^{(1)}$. In this case, Eq. (4.1.16) can be reduced to

$$d\rho_n(t)/dt \simeq k_{ma}^{(1)} N_0\, e^{-k_{ma}^{(1)} t} \tag{4.1.27}$$

under the condition $(k_{nm} - k_{ma})t > k_{ma}t$.

Equation (4.1.27) can be derived simply by invoking the steady-state condition for the resonant state $d\rho_m/dt = 0$. Under the steady-state condition, the ion current can be expressed as

$$\frac{d\rho_n(t)}{dt} = \frac{k_{nm}^{(1)} k_{ma}^{(1)} N_0}{k_{am}^{(1)} + k_{nm}^{(1)}} \exp\left[-\frac{k_{nm}^{(1)} k_{ma}^{(1)} t}{k_{am}^{(1)} + k_{nm}^{(1)}} \right], \tag{4.1.28}$$

which yields Eq. (4.1.27) for $k_{nm}^{(1)} > k_{ma}^{(1)}$.

In deriving Eq. (4.1.27) by using Eq. (4.1.16), a restrictive condition, as compared with $k_{nm}^{(1)} > k_{ma}^{(1)}$, is required.

Equation (4.1.27) indicates that the excitation spectrum of the ion current reflects the vibronic structure related to the initial and resonant states.

4.1.4 The $\frac{3}{2}$ Power Law

Focused laser beam techniques have frequently been utilized to obtain high power intensity for many multiphoton processes, for multiphoton dissociation by using IR lasers and multiphoton ionization, and/or dissociation by using visible or UV lasers. The resultant product yields are found to be proportional to $I^{3/2}$ or in some cases to noninteger powers of the laser intensity that are not relevant to the intrinsic intensity dependence. The $I^{3/2}$ dependence has been theoretically investigated by Arutyunyan et al. (1970), Speiser and Kimel (1970), and Speiser and Jortner (1976). The appearance of

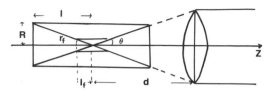

Fig. 4.4 Dog bone conical geometry of laser light.

$I^{3/2}$ dependence has been explained by the laser intensity change due to conical focusing in the interaction region where the intensity exceeds a critical value for saturation effects.

Let us consider the geometrical effect of the laser beam on the ion yield for multiphoton ionization by making use of a "dog bone" conically focused laser beam that takes into account diffraction effects. The dog bone geometry is shown in Fig. 4.4, in which d is the focal length and 2θ is the aperture angle of the lens.

The intrinsic ionization yield R_n is a function of the intensity I, which depends on position z in the interaction region. Let us express R_n as

$$R_n = \begin{cases} C_{n0}I^n(z) & \text{for} \quad E < E^*, & (4.1.29) \\ C_{nk}I^{n-k}(z) & \text{for} \quad E \geq E^*, & (4.1.30) \end{cases}$$

where $0 \leq k \leq n$ and E^* is the critical laser field amplitude for the onset of saturation effects. Let us consider an optical sample that is thin enough for us to neglect decay of the laser intensity, and assume uniform distribution of the sample in the interaction region. Under these conditions the intensity (field amplitude) satisfies $I(z)r^2(z) = \text{const}$ $[E(z)r(z) = \text{const}]$, where $r(z)$ is the radius of the cross section of the laser beam at z. It is convenient to divide the interaction region in terms of a characteristic length for the saturation effects as

$$l^* = E_f l_f / E^* = E_0 d / E^*, \qquad (4.1.31)$$

where E_f is the field amplitude within the focal region, which is characterized by the radius r_f and the length $2l_f$, and E_0 is the incident field amplitude at $z = d$. The observed ion yield should be averaged over the interaction volume as

$$\langle R_n \rangle = \int R_n \, dv, \qquad (4.1.32)$$

where the volume element dv is given by $dv = \pi r^2 \, dz = \pi \tan^2 \theta z^2 \, dz$. Three cases can be considered for the evaluation of Eq. (4.1.32).

Case 1. The saturation condition is not satisfied over all of the interaction region.

Case 2. The saturation condition is satisfied in the interaction volume; that is, the characteristic length for the saturation effect is $l_f < l^* < l$.

Case 3. The saturation condition is satisfied over all of the interaction region.

For Case 1, Eq. (4.1.32) is given by

$$\langle R_n \rangle = C_{n0} I_0^n \left\{ 1 + \frac{1 - (l_f/l)^{2n-3}}{2n-3} \right\} \left(\frac{d}{l_f} \right)^{2n} V_f, \tag{4.1.33}$$

where I_0 is the incident intensity at $z = d$ and V_f is the focal volume, which is given by $V_f = 2\pi r_f^2 l_f$. For Cases 2 and 3 the ion yields are given by

$$\langle R_n \rangle = \left[C_n k I_0^{n-k} \left(1 - \frac{1}{3 - 2n + 2k} \right) \left(\frac{d}{l_f} \right)^{2n-2k} \right.$$

$$+ I_0^{3/2} \left(\frac{C_{nk} E^{*-2k}}{3 - 2n + 2k} - \frac{C_{n0}}{3 - 2n} \right) \left(\frac{d}{l_f} \right)^3 E^{*2n-3}$$

$$\left. + \frac{C_{n0} I_0^n}{3 - 2n} \left(\frac{d}{l} \right)^{2n} \left(\frac{l}{l_f} \right)^3 \right] V_f \tag{4.1.34}$$

and

$$\langle R_n \rangle = C_{nk} I_0^{n-k} \left(1 - \frac{1 - (l_f/l)^{2n-2k-3}}{3 - 2n + 2k} \right) \left(\frac{d}{l_f} \right)^{2n-2k} V_f, \tag{4.1.35}$$

respectively. Here one should note that the term proportional to I_0^{n-k} in Eq. (4.1.34) vanishes for multiphoton processes in the case of $k = n - 1$. From the above expressions, the conclusions that can be drawn for the geometrical effects of the focused laser beam in a thin sample cell are that

(1) at low or high laser intensities the intrinsic intensity dependence holds, and, on the other hand,

(2) $\frac{3}{2}$ power dependence can be expected for cases in which $E_0 < E^*$, and onset of saturation is located at a cross section in the interaction volume.

The intensity dependence that has been discussed is based on the simplified model of dog bone geometry, assuming uniform radial distribution of the laser beam. A theoretical treatment that takes into account a smooth radial distribution in terms of a Gaussian distribution $E = E_0(z) \exp[-r^2/2a(r)]$ has been given by Arutyunyan *et al.* (1970). The asymptotic behavior was found to be the same as that mentioned above.

Noninteger power dependence for acetaldehyde ions used in multiphoton ionization/dissociation processes has been reported by Fisanick *et al.* (1980). They have claimed that ion-to-ion photofragmentation processes, in addition to photoionization/dissociation during a single pulse, can result in non-integer power dependence.

So far, the intensity dependence of multiphoton processes has been discussed by using simple rate equations that neglect coherent multiphoton effects, i.e., simultaneous terms. For resonant multiphoton processes, coherent multiphoton effects are usually neglected because of their small contribution to the total transition rate.

In our treatment the rate equations describing the dynamics of resonant multiphoton excitation are derived from the master equation in Section 2.3. Another method, called the adiabatic approximation (Sargent et al., 1974), is to set the time derivatives of off-diagonal density matrices to zero. Note that there have been many papers on the coherent effects of multiphoton transitions (Ackerhalt and Shore, 1977; de Meijere and Eberly, 1978; Milonni and Eberly, 1978; Shore and Johnson, 1978; Eberly and O'Neil, 1979).

4.2 POLARIZATION BEHAVIOR OF TWO-PHOTON PROCESSES

The cross section of the multiphoton transition of molecules in liquids and gases has polarization dependence; for example, the cross section depends on whether linearly or circularly polarized laser light is applied. Measurement of the polarization dependence has been used to assign the excited state symmetry. This polarization dependence is one of the important characteristics of multiphoton spectroscopy. In one-photon spectroscopy, on the other hand, the cross section is independent of polarization because the polarization vectors λ and κ satisfy $\lambda \cdot \kappa^* = 1$. In this section we restrict ourselves to the polarization behavior of two-photon absorption processes of molecules. The polarization dependence of the two-photon absorption cross section of randomly oriented, nonrotating molecules has been investigated in detail by Monson and McClain (1970), McClain (1971), and Nascimento (1983).

In Section 4.2.1, following mainly the work of Monson and McClain (1970), we show how polarization dependence can be seen for two-photon absorption of randomly oriented nonrotating molecules. Polarization dependence of rotating molecules, which appears in the spectra of gases via moderately well-resolved two-photon spectroscopy, is described in Section 4.2.2.

4.2.1 Polarization Behavior of Nonrotating Molecules

We shall first consider the polarization behavior of two-photon absorption for randomly oriented nonrotating molecules. Here two photons with frequencies ω_λ and ω_κ and polarizations λ and κ are assumed to excite the molecules. In this case, from Eq. (2.1.105) the transition probability from states i to f is given by

$$W^{(2)}_{i \to f} = \frac{2}{\hbar^2} \left(\frac{2\pi e^2 n_\lambda \omega_\lambda}{L^3} \right) \left(\frac{2\pi e^2 n_\kappa \omega_\kappa}{L^3} \right) |S^{\kappa\lambda}_{fi}|^2 \delta(\omega_{fi} - \omega_\lambda - \omega_\kappa), \quad (4.2.1)$$

where

$$S^{\kappa\lambda}_{fi} = \sum_m \left[\frac{(\kappa \cdot \mathbf{R}_{fm})(\lambda \cdot \mathbf{R}_{mi})}{\omega_{mi} - \omega_\kappa} + \frac{(\lambda \cdot \mathbf{R}_{fm})(\kappa \cdot \mathbf{R}_{mi})}{\omega_{mi} - \omega_\lambda} \right], \quad (4.2.2)$$

in which m denotes the intermediate states. The polarization vectors κ and λ are expressed in a laboratory coordinate system (X, Y, Z) and satisfy $\kappa \cdot \kappa^* = \lambda \cdot \lambda^* = 1$. These vectors take real form for linearly polarized laser light and complex form for circularly or elliptically polarized light; in the right-handed coordinate system, the vectors of right- and left-circularly polarized light propagating along the Z axis are represented by

$$\lambda^R = (\tfrac{1}{2})^{1/2}(1, -i, 0) \quad \text{and} \quad \lambda^L = (\tfrac{1}{2})^{1/2}(1, i, 0), \quad (4.2.3)$$

respectively. For example, this means that λ^R is viewed as moving in a clockwise manner by an observer at the detector.

The projection of the matrix element of the electronic dipole moment operator $\mathbf{R}$, which is expressed in the molecular coordinate system along the laboratory fixed direction of the polarization of light, is given by

$$\lambda \cdot \mathbf{R} = \sum_A \lambda_A R^A, \quad (4.2.4)$$

where $A = (X, Y, Z)$. The column matrix elements R^A are connected with the transition dipole moment matrix elements R^a, expressed in terms of molecular fixed coordinates, through the transformation

$$R^A = \sum_a \xi_{Aa}(\alpha, \beta, \gamma) R^a, \quad (4.2.5)$$

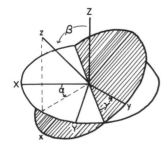

Fig. 4.5 Rotation of the molecular coordinate system (x, y, z) with respect to the laboratory coordinate system (X, Y, Z) specified by the Euler angles α, β, and γ.

where $a = (x, y, z)$ represents molecular fixed coordinates, and α, β, and γ denote the Euler angles shown in Fig. 4.5.

In terms of the Euler angles, the transformation matrix is expressed as [for example, see Rose (1957)]

$$
\xi = \begin{bmatrix}
\cos\alpha\cos\beta\cos\gamma & \sin\alpha\cos\beta\cos\gamma & -\sin\beta\cos\gamma \\
-\sin\alpha\sin\gamma & +\cos\alpha\sin\gamma & \\
-\cos\alpha\cos\beta\sin\gamma & -\sin\alpha\cos\beta\sin\gamma & \sin\beta\sin\gamma \\
-\sin\alpha\cos\gamma & +\cos\alpha\cos\gamma & \\
\cos\alpha\sin\beta & \sin\alpha\sin\beta & \cos\beta
\end{bmatrix}. \quad (4.2.6)
$$

Substituting Eq. (4.2.4) with Eq. (4.2.5) into Eq. (4.2.2) yields

$$
S_{fi}^{\kappa\lambda} = \sum_A \sum_B \sum_a \sum_b \kappa_B \lambda_A \xi_{Aa} \xi_{Bb} \sum_m \left[\frac{R_{fm}^b R_{mi}^a}{\omega_{mi} - \omega_\kappa} + \frac{R_{fm}^a R_{mi}^b}{\omega_{mi} - \omega_\lambda} \right]. \quad (4.2.7)
$$

The two-photon absorption transition probability is given by

$$
\begin{aligned}
W_{i\to f}^{(2)} &= \frac{2}{\hbar^2} \left(\frac{2\pi e^2 n_\lambda \omega_\lambda}{L^3} \right) \left(\frac{2\pi e^2 n_\kappa \omega_\kappa}{L^3} \right) |S_{fi}^{\kappa\lambda}|^2 \delta(\omega_{fi} - \omega_\lambda - \omega_\kappa) \\
&= \frac{2}{\hbar^2} \left(\frac{2\pi e^2 n_\lambda \omega_\lambda}{L^3} \right) \left(\frac{2\pi e^2 n_\kappa \omega_\kappa}{L^3} \right) \\
&\quad \times \sum_A \sum_B \sum_C \sum_D \sum_a \sum_b \sum_c \sum_d (\lambda_A \kappa_B \lambda_C^* \kappa_D^*) \\
&\quad \times \xi_{Aa} \xi_{Bb} \xi_{Cc} \xi_{Dd} |S_{fi}^{ba}|^2 \delta(\omega_{fi} - \omega_\lambda - \omega_\kappa), \quad (4.2.8)
\end{aligned}
$$

where S_{fi}^{ba}, the component of the two-photon transition tensor, is given by

$$
S_{fi}^{ba} = \sum_m \left[\frac{R_{fm}^b R_{mi}^a}{\omega_{mi} - \omega_\kappa} + \frac{R_{fm}^a R_{mi}^b}{\omega_{mi} - \omega_\lambda} \right]. \quad (4.2.9)
$$

In Eq. (4.2.8) all polarization information is included in the first factor, all orientation information is in the middle factor with ξ, and all molecular information is in the last factor $|S_{fi}^{ba}|^2$.

To obtain an expression for the transition probability of randomly oriented nonrotating molecules, we have to average $|S_{fi}^{\kappa\lambda}|^2$ over all the molecular orientations,

$$
\langle W_{i\to f}^{(2)} \rangle = \frac{2}{\hbar^2} \left(\frac{2\pi e^2 n_\lambda \omega_\lambda}{L^3} \right) \left(\frac{2\pi e^2 n_\kappa \omega_\kappa}{L^3} \right) \langle |S_{fi}^{\kappa\lambda}|^2 \rangle \delta(\omega_{fi} - \omega_\lambda - \omega_\kappa). \quad (4.2.10)
$$

This averaging can be accomplished by integrating over the Euler angles α, β, and γ and by dividing the resulting expression by the total volume $8\pi^2$;

that is,

$$
\begin{aligned}
\overline{|S_{fi}^{\kappa\lambda}|^2} &= \frac{1}{8\pi^2} \int_0^{2\pi} d\alpha \int_0^{\pi} d\beta \int_0^{2\pi} d\gamma |S_{fi}^{\kappa\lambda}|^2 \sin\beta \\
&= \sum_A \sum_B \sum_C \sum_D \sum_a \sum_b \sum_c \sum_d (\lambda_A \kappa_B \lambda_C^* \kappa_D^*) \\
&\quad \times S_{fi}^{dc*} S_{fi}^{ba} \frac{1}{8\pi^2} \int_0^{2\pi} d\alpha \int_0^{\pi} d\beta \int_0^{2\pi} d\gamma \, \xi_{Aa}\xi_{Bb}\xi_{Cc}\xi_{Dd} \sin\beta.
\end{aligned} \quad (4.2.11)
$$

Angular integrations of this type have already been evaluated and tabulated (Cyvin *et al.*, 1965; Power and Thirunamachandran, 1974). Some of them are given by

$$
I_{AB,ab}^{(2)} = \frac{1}{8\pi^2} \int \xi_{Aa}\xi_{Bb} \sin\beta \, d\alpha \, d\beta \, d\gamma = \frac{1}{3} \delta_{AB}\delta_{ab} \quad (4.2.12)
$$

for one-photon processes and

$$
\begin{aligned}
I_{ABCD,abcd}^{(4)} &= \frac{1}{8\pi^2} \int \xi_{Aa}\xi_{Bb}\xi_{Cc}\xi_{Dd} \sin\beta \, d\alpha \, d\beta \, d\gamma \\
&= \frac{1}{30} \big[\delta_{AC}\delta_{BD}(4\delta_{ac}\delta_{bd} - \delta_{ad}\delta_{bc} - \delta_{ab}\delta_{cd}) \\
&\quad + \delta_{AB}\delta_{CD}(-\delta_{ac}\delta_{bd} + 4\delta_{ab}\delta_{cd} - \delta_{ad}\delta_{bc}) \\
&\quad + \delta_{AD}\delta_{CB}(-\delta_{ab}\delta_{cd} - \delta_{ac}\delta_{bd} + 4\delta_{ad}\delta_{bc}) \big]
\end{aligned} \quad (4.2.13)
$$

for two-photon processes. Substituting Eq. (4.2.13) into Eq. (4.2.11) yields

$$
\begin{aligned}
\langle |S_{fi}^{\kappa\lambda}|^2 \rangle &= \frac{1}{30} \sum_A \sum_B \Big[\lambda_A \lambda_A^* \kappa_B \kappa_B^* \sum_a \sum_b (4 S_{fi}^{ba} S_{fi}^{ba*} \\
&\quad - S_{fi}^{aa} S_{fi}^{bb*} - S_{fi}^{ba} S_{fi}^{ab*}) + \lambda_A \kappa_A \lambda_B^* \kappa_B^* \sum_a \sum_b (-S_{fi}^{ba} S_{fi}^{ba*} \\
&\quad + 4 S_{fi}^{aa} S_{fi}^{bb*} - S_{fi}^{ba} S_{fi}^{ab*}) + \lambda_A \kappa_A^* \lambda_B^* \kappa_B \\
&\quad \times \sum_a \sum_b (-S_{fi}^{ba} S_{fi}^{ba*} - S_{fi}^{aa} S_{fi}^{bb*} + 4 S_{fi}^{ba} S_{fi}^{ab*}) \Big]
\end{aligned} \quad (4.2.14)
$$

$$
\begin{aligned}
&= \frac{1}{30} \sum_a \sum_b \big[4 S_{fi}^{ba} S_{fi}^{ba*} - S_{fi}^{aa} S_{fi}^{bb*} - S_{fi}^{ba} S_{fi}^{ab*} \\
&\quad + |\boldsymbol{\lambda}\cdot\boldsymbol{\kappa}|^2 (-S_{fi}^{ba} S_{fi}^{ba*} + 4 S_{fi}^{aa} S_{fi}^{bb*} - S_{fi}^{ba} S_{fi}^{ab*}) \\
&\quad + |\boldsymbol{\lambda}\cdot\boldsymbol{\kappa}^*|^2 (-S_{fi}^{ba} S_{fi}^{ba*} - S_{fi}^{aa} S_{fi}^{bb*} + 4 S_{fi}^{ba} S_{fi}^{ab*}) \big]
\end{aligned} \quad (4.2.15)
$$

By defining

$$F = 4|\lambda \cdot \kappa|^2 - 1 - |\lambda \cdot \kappa^*|^2, \qquad (4.2.16)$$

$$G = -|\lambda \cdot \kappa|^2 + 4 - |\lambda \cdot \kappa^*|^2, \qquad (4.2.17)$$

and

$$H = -|\lambda \cdot \kappa|^2 - 1 + 4|\lambda \cdot \kappa^*|^2, \qquad (4.2.18)$$

which are experimentally controllable polarization variables, Eq. (4.2.15) can be rewritten as

$$|S_{fi}^{\kappa\lambda}|^2 = \frac{1}{30} \sum_a \sum_b [S_{fi}^{aa}S_{fi}^{bb*}F + S_{fi}^{ba}S_{fi}^{ba*}G + S_{fi}^{ba}S_{fi}^{ab*}H]. \qquad (4.2.19)$$

Finally, we obtain the transition probability averaged over all the orientations as

$$\langle W_{i \to f}^{(2)} \rangle = \delta_F F + \delta_G G + \delta_H H. \qquad (4.2.20)$$

This expression indicates that by making three measurements of the transition probability with linearly independent values of the experimentally controllable polarization variables F, G, and H, one may determine each of the components δ_F, δ_G, and δ_H, which are characterized by the molecular quantity and laser intensity. These components are given by

$$\delta_F = \frac{1}{15\hbar^2} \left(\frac{2\pi e^2 n_\lambda \omega_\lambda}{L^3} \right) \left(\frac{2\pi e^2 n_\lambda \omega_\kappa}{L^3} \right) \sum_a \sum_b S_{fi}^{aa}S_{fi}^{bb*}\delta(\omega_{fi} - \omega_\lambda - \omega_\kappa), \quad (4.2.21)$$

$$\delta_G = \frac{1}{15\hbar^2} \left(\frac{2\pi e^2 n_\lambda \omega_\lambda}{L^3} \right) \left(\frac{2\pi e^2 n_\kappa \omega_\kappa}{L^3} \right) \sum_a \sum_b S_{fi}^{ba}S_{fi}^{ba*}\delta(\omega_{fi} - \omega_\lambda - \omega_\kappa), \quad (4.2.22)$$

and

$$\delta_H = \frac{1}{15\hbar^2} \left(\frac{2\pi e^2 n_\lambda \omega_\lambda}{L^3} \right) \left(\frac{2\pi e^2 n_\kappa \omega_\kappa}{L^3} \right) \sum_a \sum_b S_{fi}^{ba}S_{fi}^{ab*}\delta(\omega_{fi} - \omega_\lambda - \omega_\kappa). \quad (4.2.23)$$

Since Eq. (4.2.21) is expressed in terms of the absolute square of the trace of the two-photon transition tensor, $\delta_F \neq 0$ is satisfied only for the case of a transition to a totally symmetric state from a totally symmetric ground state.

Depending on the combination of polarizations that are used, some easily obtainable experimental cases can be considered:

(1)   two linearly polarized photons with parallel polarization;
(2)   two linearly polarized photons with perpendicular polarization;

TABLE 4.1

Values of the Polarization Variables F, G, and H, for Eight Two-Photon Transitions[a]

Polarization variable	Case[b]							
	1	2	3	4	5	6	7	8
F	2	-1	-1	$-\dfrac{1}{4}$	$\dfrac{1}{2}$	$\dfrac{1}{2}$	-2	3
G	2	4	4	$\dfrac{7}{2}$	3	3	3	3
H	2	-1	-1	$-\dfrac{1}{4}$	$\dfrac{1}{2}$	$\dfrac{1}{2}$	3	-2

[a] From Monson and McClain (1970).
[b] Cases 1–8 described in text.

(3) one linear and one circular with linear polarization perpendicular to the plane of the circular polarization;

(4) both circular, in either the same or opposite sense, with perpendicular propagation;

(5) both linear, with $\theta = 45°$ between the two polarization vectors;

(6) one linear and one circular, with linear polarization in the plane of the circular polarization;

(7) both circular in the same sense, with parallel propagation; and

(8) both circular in the opposite sense, with parallel propagation.

In Table 4.1, the values of F, G, and H that correspond to these cases are presented (Monson and McClain, 1970).

As a simple example of polarization dependence, consider two cases of a two-photon transition from the totally symmetric ground state to a non-totally symmetric state: in one case the transition is excited by using two linearly polarized photons, i.e., case (1), and in the other case the transition is excited by using two circularly polarized photons with parallel propagation, i.e., case (7). From Table 4.2, one can see that the ratio of the latter transition probability to the former, $\langle W^{(2)}_{i \to f} \rangle^{\circlearrowright\circlearrowright} / \langle W^{(2)}_{i \to f} \rangle^{\uparrow\uparrow}$, becomes

$$\langle W^{(2)}_{i \to f} \rangle^{\circlearrowright\circlearrowright} / \langle W^{(2)}_{i \to f} \rangle^{\uparrow\uparrow} = 3/2. \tag{4.2.24}$$

It is important to know the tensor patterns when we assign the two-photon transition of molecules. The tensor patterns depend only on the symmetry of the molecular states relevant to the transition. In Table 4.2, the tensor patterns that have been constructed by McClain (1971) and McClain and Harris (1977) are tabulated. It has been assumed that the initial state is a totally symmetric state A. The tabulated quantity is

$$S^{ba}_{fi} = \langle f^J |{}^A \hat{S}^{ba} + {}^B \hat{S}^{ba} + {}^C \hat{S}^{ba} + \cdots |i^A \rangle$$

where A, B, $\ldots$, J represent the name of the symmetry species (A, B, E, T, etc.).

4.2.2 Polarization Behavior of Rotating Molecules

In Section 4.2.1 we restricted ourselves to the polarization behavior of randomly oriented nonrotating molecules. Development of a narrow band tunable dye laser system has made it possible to observe well-resolved rotational spectra of molecules (Wunsch *et al.*, 1976; Lombardi *et al.*, 1976; Hampf *et al.*, 1977; Bray *et al.*, 1975). They have reported the effects of two-photon excitation on rotational contours, i.e., the different rotational branches of the same band differ in their polarization behavior.

In this section polarization behavior of rotating molecules is treated with concentration on two-photon absorption processes. The theoretical study of

TABLE 4.2

Cartesian Tensor Patterns for Two-Photon Processes[a]

1. Groups C_1 and C_i

$$A = \begin{bmatrix} s_1 & s_2 & s_3 \\ s_4 & s_5 & s_6 \\ s_7 & s_8 & s_9 \end{bmatrix}$$

2. Groups C_2, C_3, and C_{2h}

$$A = \begin{bmatrix} s_1 & s_4 & 0 \\ s_5 & s_2 & 0 \\ 0 & 0 & s_3 \end{bmatrix}, \qquad B = \begin{bmatrix} 0 & 0 & s_6 \\ 0 & 0 & s_7 \\ s_8 & s_9 & 0 \end{bmatrix}$$

3. Groups C_{2v}, D_2, and D_{2h}

$$A_1 = A = \begin{bmatrix} s_1 & 0 & 0 \\ 0 & s_2 & 0 \\ 0 & 0 & s_3 \end{bmatrix}, \qquad A_2 = B_1 = \begin{bmatrix} 0 & s_4 & 0 \\ s_5 & 0 & 0 \\ 0 & 0 & 0 \end{bmatrix}$$

$$B_1(C_{2v}) = B_2 = \begin{bmatrix} 0 & 0 & s_6 \\ 0 & 0 & 0 \\ s_7 & 0 & 0 \end{bmatrix}, \qquad B_2(C_{2v}) = B_3 = \begin{bmatrix} 0 & 0 & 0 \\ 0 & 0 & s_8 \\ 0 & s_9 & 0 \end{bmatrix}$$

4. Groups C_4, C_{4h}, and S_4

$$A = \begin{bmatrix} s_1 & s_3 & 0 \\ -s_3 & s_1 & 0 \\ 0 & 0 & s_2 \end{bmatrix}, \qquad B = \begin{bmatrix} s_4 & s_3 & 0 \\ s_5 & -s_4 & 0 \\ 0 & 0 & 0 \end{bmatrix}$$

$$E = \begin{bmatrix} 0 & 0 & s_6 \\ 0 & 0 & -is_6 \\ s_7 & -is_7 & 0 \end{bmatrix} \quad \text{and} \quad \begin{bmatrix} 0 & 0 & s_6^* \\ 0 & 0 & is_6^* \\ s_7^* & is_7^* & 0 \end{bmatrix}$$

5. Groups C_{4v}, D_4, D_{2d}, and D_{4h}

$$A_1 = \begin{bmatrix} s_1 & 0 & 0 \\ 0 & s_1 & 0 \\ 0 & 0 & s_2 \end{bmatrix}, \qquad A_2 = \begin{bmatrix} 0 & s_3 & 0 \\ -s_3 & 0 & 0 \\ 0 & 0 & 0 \end{bmatrix}$$

$$B_1 = \begin{bmatrix} s_4 & 0 & 0 \\ 0 & -s_4 & 0 \\ 0 & 0 & 0 \end{bmatrix}, \qquad B_2 = \begin{bmatrix} 0 & s_5 & 0 \\ s_5 & 0 & 0 \\ 0 & 0 & 0 \end{bmatrix}$$

$$E = \begin{bmatrix} 0 & 0 & s_6 \\ 0 & 0 & -is_6 \\ s_7 & -is_7 & 0 \end{bmatrix} \quad \text{and} \quad \begin{bmatrix} 0 & 0 & s_6^* \\ 0 & 0 & is_6^* \\ s_7^* & is_7^* & 0 \end{bmatrix}$$

6. Groups C_3 and $S_6 = C_{3h}$

$$A = \begin{bmatrix} s_1 & s_3 & 0 \\ -s_3 & s_1 & 0 \\ 0 & 0 & s_2 \end{bmatrix}$$

$$E = \begin{bmatrix} s_4 & is_4 & s_5 \\ is_4 & -s_4 & -is_5 \\ s_6 & -is_6 & 0 \end{bmatrix} \quad \text{and} \quad \begin{bmatrix} s_4^* & -is_4^* & s_5^* \\ -is_4^* & -s_4^* & is_5^* \\ s_6^* & is_6^* & 0 \end{bmatrix}$$

TABLE 4.2   *(cont.)*

7. Groups C_{3v}, D_3, and D_{3d}

$$A_1 = \begin{bmatrix} s_1 & 0 & 0 \\ 0 & s_1 & 0 \\ 0 & 0 & s_2 \end{bmatrix}, \qquad A_2 = \begin{bmatrix} 0 & s_3 & 0 \\ -s_3 & 0 & 0 \\ 0 & 0 & 0 \end{bmatrix}$$

$$E = \begin{bmatrix} s_4 & is_4 & s_5 \\ is_4 & -s_4 & -is_3 \\ s_6 & -is_6 & 0 \end{bmatrix} \quad \text{and} \quad \begin{bmatrix} s_4^* & -is_4^* & s_5^* \\ -is_4^* & -s_4^* & is_5^* \\ s_6^* & is_6^* & 0 \end{bmatrix}$$

8. Groups C_{3h}, C_6, and C_{6h}

$$A = \begin{bmatrix} s_1 & s_3 & 0 \\ -s_3 & s_1 & 0 \\ 0 & 0 & s_2 \end{bmatrix}$$

$$E_1 = \begin{bmatrix} 0 & 0 & s_4 \\ 0 & 0 & -is_4 \\ s_5 & -is_5 & 0 \end{bmatrix} \quad \text{and} \quad \begin{bmatrix} 0 & 0 & s_4^* \\ 0 & 0 & is_4^* \\ s_5^* & is_5^* & 0 \end{bmatrix}$$

$$E_2 = \begin{bmatrix} s_6 & -is_6 & 0 \\ -is_6 & -s_6 & 0 \\ 0 & 0 & 0 \end{bmatrix} \quad \text{and} \quad \begin{bmatrix} s_6^* & is_6^* & 0 \\ is_6^* & -s_6^* & 0 \\ 0 & 0 & 0 \end{bmatrix}$$

9. Groups C_{6v}, D_{3h}, D_6, and D_{6h}; Groups $C_{\infty v}$ and $D_{\infty h}$

$$A_1 = \Sigma^+ = \begin{bmatrix} s_1 & 0 & 0 \\ 0 & s_1 & 0 \\ 0 & 0 & s_2 \end{bmatrix}, \qquad A_2 = \Sigma^- = \begin{bmatrix} 0 & s_3 & 0 \\ -s_3 & 0 & 0 \\ 0 & 0 & 0 \end{bmatrix}$$

$$E_1 = \Pi = \begin{bmatrix} 0 & 0 & s_4 \\ 0 & 0 & -is_4 \\ s_5 & -is_5 & 0 \end{bmatrix} \quad \text{and} \quad \begin{bmatrix} 0 & 0 & s_4^* \\ 0 & 0 & is_4^* \\ s_5^* & is_5^* & 0 \end{bmatrix}$$

$$E_2 = \Delta = \begin{bmatrix} s_6 & -is_6 & 0 \\ -is_6 & -s_6 & 0 \\ 0 & 0 & 0 \end{bmatrix} \quad \text{and} \quad \begin{bmatrix} s_6^* & is_6^* & 0 \\ is_6^* & -s_6^* & 0 \\ 0 & 0 & 0 \end{bmatrix}$$

10. Groups T and T_h $\omega = \exp(2\pi i/3)$

$$A = \begin{bmatrix} s_1 & 0 & 0 \\ 0 & s_1 & 0 \\ 0 & 0 & s_1 \end{bmatrix}$$

$$E = \begin{bmatrix} s_2 & 0 & 0 \\ 0 & \omega s_2 & 0 \\ 0 & 0 & \omega^* s_2 \end{bmatrix} \quad \text{and} \quad \begin{bmatrix} s_2^* & 0 & 0 \\ 0 & \omega^* s_2^* & 0 \\ 0 & 0 & \omega s_2^* \end{bmatrix}$$

$$T = \begin{bmatrix} 0 & 0 & 0 \\ 0 & 0 & s_3 \\ 0 & s_4 & 0 \end{bmatrix} \quad \text{and} \quad \begin{bmatrix} 0 & 0 & s_4 \\ 0 & 0 & 0 \\ s_3 & 0 & 0 \end{bmatrix} \quad \text{and} \quad \begin{bmatrix} 0 & s_3 & 0 \\ s_4 & 0 & 0 \\ 0 & 0 & 0 \end{bmatrix}$$

(continued)

TABLE 4.2 *(cont.)*

11. Groups O, O_h, and T_d $\omega = \exp(2\pi i/3)$

$$A = \begin{bmatrix} s_1 & 0 & 0 \\ 0 & s_1 & 0 \\ 0 & 0 & s_1 \end{bmatrix}$$

$$E = \begin{bmatrix} s_2 & 0 & 0 \\ 0 & \omega s_2 & 0 \\ 0 & 0 & \omega^* s_2 \end{bmatrix} \quad \text{and} \quad \begin{bmatrix} s_2^* & 0 & 0 \\ 0 & \omega^* s_2^* & 0 \\ 0 & 0 & \omega s_2^* \end{bmatrix}$$

$$T_1 = \begin{bmatrix} 0 & s_2 & 0 \\ -s_2 & 0 & 0 \\ 0 & 0 & 0 \end{bmatrix} \quad \text{and} \quad \begin{bmatrix} 0 & 0 & -s_3 \\ 0 & 0 & 0 \\ s_3 & 0 & 0 \end{bmatrix} \quad \text{and} \quad \begin{bmatrix} 0 & 0 & 0 \\ 0 & 0 & s_3 \\ 0 & -s_3 & 0 \end{bmatrix}$$

$$T_2 = \begin{bmatrix} 0 & s_4 & 0 \\ s_4 & 0 & 0 \\ 0 & 0 & 0 \end{bmatrix} \quad \text{and} \quad \begin{bmatrix} 0 & 0 & s_4 \\ 0 & 0 & 0 \\ s_4 & 0 & 0 \end{bmatrix} \quad \text{and} \quad \begin{bmatrix} 0 & 0 & 0 \\ 0 & 0 & s_4 \\ 0 & s_4 & 0 \end{bmatrix}$$

[a] From McClain and Harris (1977). The tabulated quantity is $S^{ba}(A \rightarrow J)$. The table is divided into 11 sets of groups. Within each set the groups are either isomorphic or differ from each other only by the inclusion of the center of inversion element. The tensors are labeled by the symbol J only, and that symbol is often simplified by dropping primes and subscripts, due to the variability of nomenclature among different groups within the same isomorphic set. We write "and" between tensors that belong to the different parts of a degenerate transition; such pairs must always be used together, for example, for a twofold degeneracy, according to $W^{(2)} = |\mathbf{e}_1 \cdot S^I \cdot \mathbf{e}_2|^2 + |\mathbf{e}_1 \cdot S^{II} \cdot \mathbf{e}_2|^2$. The basis sets for symmetry species A and B are always unambiguous. When the group has one E species, the basis set is $[x + iy, x - iy]$, except in the tetrahedral and octahedral groups (sets 10 and 11) where the basis is $[u + iv, u - iv]$, with $u = 2z^2 - x^2 - y^2$ and $v = 3^{1/2}(x^2 - y^2)$. When the group has two E species, the basis of E_1 is $[x + iy, x - iy]$ and the basis of E_2 is $[(x + iy)^2, (x - iy)^2]$. In groups T and T_h (set 10) the basis of species T is (x, y, z). In groups O, O_h, and T_d the basis of T_1 is (x, y, z) and the basis of T_2 is (yz, zx, xy). In sets 10 and 11 note that $1 + \omega + \omega^* = 0$.

polarization behavior of rotating molecules has been done by Bray and Hochstrasser (1976), McClain and Harris (1977), Metz *et al.* (1978), and Chen and Yeung (1978). The basic idea of polarization dependence can be seen in the theory of Raman scattering developed by Placzek and Teller (1933) and Altmann and Strey (1972).

In order to take into account the effects of molecular rotation, it is convenient to use spherical coordinates rather than Cartesian coordinates, because the rotational wave function of a symmetric top molecule can be well described in terms of spherical coordinates and because the matrix elements of the dipole transition operator are easily evaluated with the aid of the operation algebra of rotation matrices expressed in spherical coordinates.

Let us first derive an expression for the two-photon absorption probability of a rotating molecule, primarily following the treatment given by Metz *et al.* (1978). Next, it will be shown how the polarization dependence that appeared in highly resolved rotational spectra of two-photon excited vibronic bands of some molecules was analyzed by using the expression for the derived transition probability.

Before we begin our detailed derivation of the transition probability for a rotating molecule, it is instructive to briefly summarize the fundamentals of molecular rotation. The rotational Hamiltonian $\hat{H}_{R}$ is expressed as

$$\hat{H}_{R} = \hat{L}_x^2/2I_x + \hat{L}_y^2/2I_y + \hat{L}_z^2/2I_z, \tag{4.2.25}$$

where I_i is the moment of inertia, and $\hat{L}^2 = \hat{L}_x^2 + \hat{L}_y^2 + \hat{L}_z^2$ is the operator of the total angular momentum whose eigenvalue is $\hbar^2 J(J + 1)$. Here J is the quantum number of the total angular momentum, and $\hat{L}_i$ is its projection on the principal molecular axis i. In this treatment, the principal molecular axes are identified with the molecular axes. Molecules can be classified in terms of the moment of inertia I_i as

(1) linear top molecules for $I_x = 0$, $I_y = I_z$,
(2) prolate symmetric top molecules for $I_x < I_y = I_z$,
(3) oblate symmetric top molecules for $I_x = I_y < I_z$,
(4) spherical symmetric top molecules for $I_x = I_y = I_z$,
(5) asymmetric top molecules for $I_x < I_y < I_z$,

where we have chosen $I_x \leq I_y \leq I_z$. For simplicity, we restrict ourselves to a symmetric molecule, specifically, an oblate symmetric one. The normalized rotational wave function $\eta_{JKM}(\Omega)$ can be expressed in terms of a rotation matrix $D_{MK}^{(J)}(\Omega)$ called the Wigner function, generalized spherical function, or D-function [see, for example, Carrington *et al.* (1970)],

$$\eta_{JKM}(\Omega) = \left(\frac{2J + 1}{8\pi^2}\right)^{1/2} D_{MK}^{(J)*}(\Omega)$$

$$= (-1)^{M-K}\left(\frac{2J + 1}{8\pi^2}\right)^{1/2} D_{-M\,-K}^{(J)}(\Omega), \tag{4.2.26}$$

where Ω denotes Euler angles α, β, and γ with respect to the laboratory coordinates. The variables J, K, and M represent the total angular momentum mentioned above, the component of J along the molecular-fixed z-axis, and that of J along Z-axis of the laboratory coordinates.

The corresponding rotational energy $E_{J,K}$ is given by

$$E_{J,K} = \frac{\hbar^2 J(J + 1)}{2I_x} + \frac{\hbar^2}{2}\left(\frac{1}{I_z} - \frac{1}{I_x}\right)K^2. \tag{4.2.27}$$

Some useful symmetric properties of the rotation matrix are summarized in Appendix II.A.

To derive an expression for the two-photon absorption probability, the wave functions of a symmetric top molecule are assumed to be expressed as the product of the electronic, vibrational, and rotational wave functions,

$$\Psi_i(r, q, \Omega) = \Phi_i(r, q)\chi_i(q)\eta_i(\Omega) \qquad (4.2.28)$$

for the i state, where $\Phi_i(r, q)$, $\chi_i(q)$, and $\eta_i(\Omega)$ are the wave functions of the electronic, vibrational, and rotational motions, respectively.

We consider a nonresonant two-photon transition from rovibronic state $|i\rangle = |\bar{i}, J_i, K_i, M_i\rangle$ to state $|f\rangle = |\bar{f}, J_f, K_f, M_f\rangle$, where $\bar{i}$ and $\bar{f}$ specify the quantum numbers of the vibronic states in the initial and final states, respectively. The probability of the transition $|i\rangle \rightarrow |f\rangle$ is given by

$$W_{i\rightarrow f}^{(2)} = \frac{2}{\hbar^2}\left(\frac{2\pi e^2 n\omega}{L^3}\right)^2 |\langle J_f K_f M_f|S_{\bar{f}\bar{i}}^{\kappa\lambda}|J_i K_i M_i\rangle|^2 \delta(\omega_{fi} - 2\omega), \quad (4.2.29)$$

where κ and λ are the photon polarization vectors, $\langle \cdots \rangle$ denotes integration over the Euler angles $d\Omega$, and $S_{\bar{f}\bar{i}}^{\kappa\lambda}$ is the two-photon transition matrix element, which is assumed to be expressed as

$$S_{\bar{f}\bar{i}}^{\kappa\lambda} = \sum_m \left[\frac{(\kappa \cdot \mathbf{R}_{\bar{f}\bar{m}})(\lambda \cdot \mathbf{R}_{\bar{m}\bar{i}})}{\omega_{\bar{m}\bar{i}} - \omega} + \frac{(\lambda \cdot \mathbf{R}_{\bar{f}\bar{m}})(\kappa \cdot \mathbf{R}_{\bar{m}\bar{i}})}{\omega_{\bar{m}\bar{i}} - \omega}\right]. \qquad (4.2.30)$$

In other words, rotational quantum number dependences of the matrix element have been neglected. This assumption is valid for nonresonant transitions.

In the absence of a magnetic field, each JK level is $(2J + 1)$-fold degenerate. We may then take summations over M_i and M_f to yield

$$\begin{aligned}
W_{\bar{i}J_iK_i \rightarrow \bar{f}J_fK_f}^{(2)} &= \frac{2}{\hbar^2}\left(\frac{2\pi e^2 n\omega}{L^3}\right)^2 \frac{1}{2J_i + 1} \\
&\quad \times \sum_{m_i}\sum_{m_f} |\langle J_f K_f M_f|S_{\bar{f}\bar{i}}^{\kappa\lambda}|J_i K_i M_i\rangle|^2 \delta(\omega_{fi} - 2\omega) \\
&\equiv W_{J_iK_i \rightarrow J_fK_f}^{(2)}.
\end{aligned} \qquad (4.2.31)$$

In order to study polarization behavior of rotating molecules it is convenient to use the spherical coordinate basis set for vectors $\mathbf{R}^A$, $A = (1, 0, -1)$, and $\mathbf{R}^a$, $a = (1, 0, -1)$ where $\mathbf{R}^A$ and $\mathbf{R}^a$ are expressed in terms of the laboratory and molecular fixed coordinates, respectively. These are connected through the $D^{(1)}$ function as

$$\mathbf{R}^a = \sum_A D^{(1)}_{Aa}(\Omega)\mathbf{R}^A \tag{4.2.32}$$

or

$$\mathbf{R}^A = \sum_a D^{(1)*}_{Aa}(\Omega)\mathbf{R}^a = \sum_a (-1)^{A-a}D^{(1)}_{-A-a}(\Omega)\mathbf{R}^a. \tag{4.2.33}$$

Noting that the scalar product of λ and $\mathbf{R}$ is then given by

$$\lambda \cdot \mathbf{R} = \sum_A \sum_a \lambda_A(-1)^{A-a}D^{(1)}_{-A-a}(\Omega)\mathbf{R}^a, \tag{4.2.34}$$

Eq. (4.2.30) can be expressed as

$$S^{\kappa\lambda}_{fi} = \sum_A \sum_B \sum_a \sum_b \lambda_A \kappa_B(-1)^{A+B-a-b}D^{(1)}_{-A-a}(\Omega)D^{(1)}_{-B-b}(\Omega)S^{ba}_{fi}, \tag{4.2.35}$$

where S^{ba}_{fi}, expressed in terms of the molecular fixed coordinates, is given by

$$S^{ba}_{fi} = \sum_{\bar{m}} \left[\frac{R^b_{f\bar{m}}R^a_{\bar{m}i}}{\omega_{\bar{m}i} - \omega} + \frac{R^a_{f\bar{m}}R^b_{\bar{m}i}}{\omega_{\bar{m}i} - \omega} \right]. \tag{4.2.36}$$

Applying the combination theorem for the rotation matrices (see Appendix III.A) to Eq. (4.2.35) yields

$$S^{\kappa\lambda}_{fi} = \sum_A \sum_B \sum_a \sum_b \sum_J \sum_M \sum_k \lambda_A \kappa_B(-1)^{A+B-a-b}(2J+1)$$

$$\times \begin{bmatrix} 1 & 1 & J \\ -A & -B & M \end{bmatrix}\begin{bmatrix} 1 & 1 & J \\ -a & -b & k \end{bmatrix} D^{(J)*}_{Mk}(\Omega)S^{ba}_{fi}, \tag{4.2.37}$$

Here the

$$\begin{bmatrix} J_1 & J_2 & J_3 \\ M_1 & M_2 & M_3 \end{bmatrix},$$

called the Wigner $3j$ symbols, have the properties

$$M_1 + M_2 + M_3 = 0 \quad \text{and} \quad |J_1 - J_2| \le J_3 \le |J_1 + J_2|. \tag{4.2.38}$$

From the preceding restriction, we find that $J = 0, 1, 2$. The symmetric properties and theorems of the Wigner $3j$ symbols relevant to our derivation of the transition probability are given in Appendix II.B.

Equation (4.2.37) can be rewritten as

$$S^{\kappa\lambda}_{fi} = \sum_J \sum_M \sum_R L^{(J)}_M M^{(J)}_k D^{(J)*}_{Mk}(\Omega). \tag{4.2.39}$$

Here

$$L_M^{(J)} = (2J + 1)^{1/2} \sum_A \sum_B (-1)^{A+B} \lambda_A \kappa_B \begin{bmatrix} 1 & 1 & J \\ -A & -B & M \end{bmatrix}, \quad (4.2.40)$$

which is purely geometric and depends only on the direction of the polarization vectors λ and κ, and

$$M_k^{(J)} = (2J + 1)^{1/2} \sum_a \sum_b \begin{bmatrix} 1 & 1 & J \\ -a & -b & k \end{bmatrix} S_{fi}^{ba}. \quad (4.2.41)$$

This term is free from the experimental conditions and is mainly determined by the symmetry of the vibronic wave functions. Using Eqs. (4.2.39) and (4.2.28), the rotational matrix element in Eq. (4.2.29) can be expressed as

$$\langle J_f K_f M_f | S_{fi}^{\kappa\lambda} | J_i K_i M_i \rangle$$
$$= \sum_J \sum_M \sum_k (2J_f + 1)^{1/2} (2J_i + 1)^{1/2} L_M^{(J)} M_k^{(J)}$$
$$\times \int d\Omega \, D_{M_f K_f}^{(1)}(\Omega) D_{Mk}^{(J)*}(\Omega) D_{M_i K_i}^{(1)*}(\Omega). \quad (4.2.42)$$

Integration over the Euler angles can easily be performed (see Appendix II.A) to give

$$\langle J_f K_f M_f | S_{fi}^{\kappa\lambda} | J_i K_i M_i \rangle$$
$$= \sum_J \sum_M \sum_k L_M^{(J)} M_k^{(J)} (2J_f + 1)^{1/2} (2J_i + 1)^{1/2}$$
$$\times (-1)^{M_i + M - K_i - k} \begin{bmatrix} J_i & J_f & J \\ -M_i & M_f & -M \end{bmatrix} \begin{bmatrix} J_i & J_f & J \\ -K_i & K_f & -k \end{bmatrix}. \quad (4.2.43)$$

From the symmetry property of the Wigner 3j symbols,

$$|J_i - J_f| \le J \le |J_i + J_f|, \quad \text{with} \quad J = 0, 1, 2,$$

we find that the selection rule for the two-photon absorption is $\Delta J = J_f - J_i = 0, \pm 1, \pm 2$, that is, transitions for two-photon absorptions are classified into five branches as

transition	$\Delta J = J_f - J_i$
O branch,	$\Delta J = -2,$
P branch,	$\Delta J = -1,$
Q branch,	$\Delta J = 0,$
R branch,	$\Delta J = 1,$
S branch,	$\Delta J = 2.$

Taking the absolute square of Eq. (4.2.43) and summing over M_i and M_f, we obtain the transition probability as

$$
W^{(2)}_{J_iK_i \to J_fK_f} = \frac{2}{\hbar^2} \left(\frac{2\pi e^2 n\omega}{L^3}\right)^2
$$

$$
\times \sum_{M_i}\sum_{M_f}\sum_{J}\sum_{J'}\sum_{M}\sum_{M'}\sum_{k}\sum_{k'}(-1)^{M+M'-k-k'}
$$

$$
\times (2J_f+1)L^{(J)*}_M L^{(J)}_{M'} M^{(J)*}_k M^{(J)}_{k'}
$$

$$
\times \begin{bmatrix} J_i & J_f & J \\ -M_i & M_f & -M \end{bmatrix}\begin{bmatrix} J_i & J_f & J \\ -K_i & K_f & -k \end{bmatrix}
$$

$$
\times \begin{bmatrix} J_i & J_f & J' \\ -M_i & M_f & -M' \end{bmatrix}\begin{bmatrix} J_i & J_f & J' \\ -k_i & k_f & -k' \end{bmatrix}\delta(\omega_{fi}-2\omega)
$$

$$\tag{4.2.44}$$

Applying the theorem for summation of the Wigner $3j$ symbols [Eq. (II.B.6) in Appendix II.B] to Eq. (4.2.44) yields

$$
W^{(2)}_{J_iK_i \to J_fK_f} = \frac{2}{\hbar^2}\left(\frac{2\pi e^2 n\omega}{L^3}\right)^2
$$

$$
\times \sum_{J}\sum_{J'}\sum_{M}\sum_{M'}\sum_{k}\sum_{k'}(-1)^{M+M'-k-k'}
$$

$$
\times (2J_f+1)(2J+1)^{-1}L^{(J)*}_M L^{(J')}_{M'} M^{(J)*}_k M^{(J')}_{k'}
$$

$$
\times \begin{bmatrix} J_i & J_f & J \\ -K_i & K_f & -k \end{bmatrix}\begin{bmatrix} J_i & J_f & J' \\ -K_i & K_f & -k' \end{bmatrix}\delta_{MM'}\delta_{JJ'}\delta(\omega_{fi}-2\omega)
$$

$$
= \frac{2}{\hbar^2}\left(\frac{2\pi e^2 n\omega}{L^3}\right)^2 \sum_{J}\sum_{M}\sum_{k}(2J_f+1)(2J+1)^{-1}
$$

$$
\times |L^{(J)}_M|^2 |M^{(J)}_k|^2 \begin{bmatrix} J_i & J_f & J \\ -K_i & K_f & -k \end{bmatrix}^2.
$$

$$\tag{4.2.45}$$

By defining

$$
C_J = (2J+1)^{-1}\sum_{M}|L^{(J)}_M|^2, \tag{4.2.46}
$$

Eq. (4.2.45) can be rewritten as

$$
W^{(2)}_{J_iK_i \to J_fK_f} = \frac{2}{\hbar^2}\left(\frac{2\pi e^2 n\omega}{L^3}\right)^2 (2J_f+1)
$$

$$
\times \sum_{J=0}^{2} C_J|M^{(J)}_{\Delta K}|^2 \begin{bmatrix} J_i & J_f & J \\ -K_i & K_f & -\Delta K \end{bmatrix}^2 \delta(\omega_{fi}-2\omega), \quad (4.2.47)
$$

with $|\Delta K| \le J$. This is the desired expression for the two-photon transition, taking into account rotational effects. If we sum over J_f and K_f in Eq.

TABLE 4.3

The Geometrical Factors C_0, C_1, and C_2 of the General Expression in Eq. (4.2.50) for Two-Photon Absorption (TPA) in Rotating Molecules for Different Polarizations of the Two Exciting Photons

Polarization of incident photons $(e_r, e_s)^a$	Geometrical factor		
	C_0	C_1	C_2
↕ $(x,x)(y,y)(0,0)(v,v)$	$\frac{1}{3}$	0	$\frac{2}{15}$
↔ $(x,y)(v,h)(0,\pm)$	0	$\frac{1}{6}$	$\frac{1}{10}$
↕ $(x,\pm)(y,\pm)(v,\pm)$	$\frac{1}{6}$	$\frac{1}{12}$	$\frac{7}{60}$
$(\pm,\pm)$	0	0	$\frac{1}{5}$
$(\pm,\mp)$	$\frac{1}{3}$	$\frac{1}{6}$	$\frac{1}{30}$
$(h,h)_\phi$	$\frac{1}{3}\cos^2\phi$	$\frac{1}{6}\sin^2\phi$	$\frac{1}{10}(1+\frac{1}{3}\cos^2\phi)$
$(h,\pm)_\phi$	$\frac{1}{6}\cos^2\phi$	$\frac{1}{6}(1-\frac{1}{2}\cos^2\phi)$	$\frac{1}{10}(1+\frac{1}{6}\cos^2\phi)$
$(\pm,\pm)_\phi^b$	$\frac{1}{12}(1-\cos\phi)^2$	$\frac{1}{24}(4-(1+\cos\phi)^2)$	$\frac{1}{120}(13+10\cos\phi+\cos^2\phi)$

^{a} The polarization is expressed in Cartesian and spherical coordinates, where r_0 corresponds to z and $r_\pm$ to $2^{-1/2}(x \pm iy)$. The k-vector of the light is usually assumed to be in the z or 0 direction. For photons coming from different directions at an angle ϕ we also distinguish between the polarization vertical (v) and horizontal (h) to the scattering plane. [From Metz et al., (1978).]

^{b} The other possible case $(\pm, \mp)$ can be obtained by replacing ϕ by $(\phi + \pi/2)$.

(4.2.47), with $J_i = K_i = 0$, we obtain

$$\sum_{J_f} \sum_{K_f} W^{(2)}_{0_i 0_i \to J_f K_f} = \frac{2}{\hbar^2} \left(\frac{2\pi e^2 n\omega}{L^3}\right)^2 \sum_{J=0}^{2} C_J |M^{(J)}_{\Delta K}|^2 \delta(\omega_{fi} - 2\omega). \quad (4.2.48)$$

This is equivalent to Eq. (4.2.20), which was derived by neglecting rotational effects. Equation (4.2.47) can be expressed in the form of the products of the geometrical factor C_J, the molecular factor M_J, and the rotational factors R_J, as

$$W^{(2)}_{J_i K_i \to J_f K_f} = C_0 M_0 R_0 + C_1 M_1 R_1 + C_2 M_2 R_2, \quad (4.2.49)$$

where the geometrical factors are a function of the polarization vectors κ and λ only and are given by

$$C_0 = \tfrac{1}{3}|\lambda \cdot \kappa|^2,$$
$$C_1 = \tfrac{1}{6}|\lambda \times \kappa|^2 = \tfrac{1}{6}(1 - |(\lambda \cdot \kappa^*)|^2), \quad (4.2.50)$$
$$C_2 = \tfrac{1}{5}(1 - C_0 - 3C_1),$$

which are derived by applying the symmetric properties of the Wigner $3j$ symbols to Eq. (4.2.46).

Table 4.3 contains values of the geometrical factors for linearly and/or circularly polarized light (Metz *et al.*, 1978). The molecular factor M_J, which is defined in spherical coordinates as

$$M_J = |M^{(J)}_{\Delta K}|^2 = (2J + 1) \left| \sum_a \sum_b \begin{bmatrix} 1 & 1 & J \\ -a & -b & \Delta K \end{bmatrix} S^{ba}_{\bar{f}\bar{i}} \right|^2, \quad (4.2.51)$$

with $|\Delta K| \leq J$, can be evaluated with the aid of the symmetry property of the Wigner $3j$ symbols. All possible values of the molecular factors are tabulated in Table 4.4. The transformation of M_J in spherical coordinates into that in Cartesian coordinates can be accomplished by noting that $R^{+1} = -(x + iy)/\sqrt{2}$, $R^0 = z$, and $R^{-1} = (x - iy)/\sqrt{2}$. The rotational factor R_J is defined as

$$R_J = (2J_f + 1)(2J_i + 1) \begin{bmatrix} J_i & J_f & J \\ -K_i & K_f & \Delta K \end{bmatrix}^2. \quad (4.2.52)$$

Each factor has the form

$$R_0 = (2J_f + 1)\delta_{J_i, J_f}\delta_{K_i, K_f},$$
$$R_1 = (2J_f + 1)(2J_i + 1) \begin{bmatrix} J_i & J_f & 1 \\ -K_i & K_f & \Delta K \end{bmatrix}^2, \quad (4.2.53)$$
$$R_2 = (2J_f + 1)(2J_i + 1) \begin{bmatrix} J_i & J_f & 2 \\ -K_i & K_f & \Delta K \end{bmatrix}^2.$$

TABLE 4.4

The Molecular Factors M_J of the General Expression in Eq. (4.2.51) for TPA
in Rotating Molecules in Spherical and Cartesian Coordinates[a]

		Molecular factors $M_J = |M_k^{(J)}|^2$ with $|\Delta K| \leq J$	
J	ΔK	In spherical coordinates	In Cartesian coordinates
2	± 2	$\|M_{\pm\pm}\|^2$	$(M_{xx} - M_{yy})^2/4 + (M_{xy} + M_{yx})^2/4$
	± 1	$\|M_{\pm 0} + M_{0\pm}\|^2/2$	$(M_{xz} + M_{zx})^2/4 + (M_{yz} + M_{zy})^2/4$
	0	$\|M_{+-} + M_{-+} + 2M_{00}\|^2/6$	$(2M_{zz} - M_{xx} - M_{yy})^2/6$
1	± 1	$\|M_{\pm 0} - M_{0\pm}\|^2/2$	$(M_{xz} - M_{zx})^2/4 + (M_{yz} - M_{zy})^2/4$
	0	$\|M_{+-} - M_{-+}\|^2/2$	$(M_{xy} - M_{yx})^2/2$
0	0	$\|M_{+-} + M_{-+} - M_{00}\|^2/3$	$(M_{xx} + M_{yy} + M_{zz})^2/3$

[a] The molecular factors are listed for different ΔK and given in the molecular
frame system. The spherical coordinates r_+, r_0, and r_- correspond to $-2^{-1/2}(x + iy)$,
z, and $2^{-1/2}(x - iy)$, respectively. The matrix elements M_{ba} correspond to S_{fi}^{ba} in
Eq. (4.2.51). [From Metz et al. (1978).]

From Eq. (4.2.53), we obtain the rotational selection rules.

$R_0 \neq 0$ for $\Delta J = 0$ and $\Delta K = 0$;   Q branch only;

$R_1 \neq 0$ for $\Delta J = 0, \pm 1$,   and $\Delta K = 0, \pm 1$,

P, Q, R branches;   (4.2.54)

$R_2 \neq 0$ for $\Delta J = 0, \pm 1, \pm 2$,   and $\Delta K = 0, \pm 1, \pm 2$;

O, P, Q, R, S branches.

Now we consider the polarization effects of two-photon absorption for a
rotating molecule. Let us calculate the ratio of the linearly to the circularly
polarized two-photon transition probabilities. From Eq. (4.2.49) and Table
4.3 we find that the ratio has the form

$$\frac{W^{(2)}_{J_iK_i \to J_f^n K_f}}{W^{(2)}_{J_iK_i \to J_f^n K_f}} = \frac{2}{3} + \frac{5}{3}\frac{M_0 R_0}{M_2 R_2}. \tag{4.2.55}$$

This expression shows that for nontotally symmetric transitions characterized by $M_0 = 0$, the ratio is independent of the rotational quantum numbers J, K and the value is $\frac{2}{3}$, which has already been noted in Section 4.2.1
and Eq. (4.2.24). The same behavior of the ratio is also predicted for the
rotational lines of the O, P, R, and S branches of a totally symmetric transition, because $M_0 \neq 0$ and $R_0 = 0$. On the other hand, the Q branch of the
totally symmetric transition, in which $M_0 \neq 0$ and $R_0 \neq 0$, shows a different
behavior in that its ratio deviates from $\frac{2}{3}$; the magnitude of the deviation
depends strongly on $M_0 R_0/M_2 R_2$, in which usually $M_0 > M_2$ and $R_0 > R_2$.

So far we have restricted ourselves to the two-photon transition theory of a symmetric top molecule. The extension to asymmetric top molecules has been done by Metz *et al.* (1978). Let us briefly mention the effects of rotational asymmetry, following their treatment. For the asymmetric top molecule, the projection of the total angular momentum on the molecular axis is no longer conserved. There are no general analytical expressions for the rotational states and their energies. The rotational wave function in this case is usually expressed by expansion of the wave functions of the symmetric top molecule as

$$\eta_{J\tau M}(\Omega) = \sum_K A_{\tau K}^J \eta_{JKM}(\Omega), \tag{4.2.56}$$

where τ specifies the eigenstates of the asymmetric top molecule.

If we neglect the detailed derivation of the transition probability, the resulting expression can be written as

$$W_{J_i\tau_i \to J_f\tau_f}^{(2)} = \sum_{J=0}^{2} C_J |M_{\tau_i\tau_f}^{J_i J_f J}|^2, \tag{4.2.57}$$

where C_J is the same geometrical factor as that given by Eq. (4.2.46), and the factor M takes the form

$$M_{\tau_i\tau_f}^{J_i J_f J} = (2J_f + 1)^{1/2}(2J_i + 1)^{1/2}$$

$$\times \sum_K \sum_{K'} \sum_m A_{\tau_f K'}^{J_f *} A_{\tau_i K}^{J_i} \begin{bmatrix} J_i & J_f & J \\ -K_i & K_f & -m \end{bmatrix} M_m^{(J)}, \tag{4.2.58}$$

which involves both molecular and rotational factors, and which cannot be separated from each other as they can in the case of the symmetric top molecule. This is the essential difference from the symmetric molecule case. Equation (4.2.57) indicates that the polarization effect is not fundamentally affected by the introduction of rotational asymmetry.

We just note the measurement of the polarization behavior of rotating molecules. Rovibronic-resolved two-photon excitation spectra have been observed by Lombardi *et al.* (1976) and Wunsch *et al.* (1977), who have measured the two-photon excitation spectra of C_6H_6 and C_6D_6 for linear and circular polarizations of the exciting laser radiation in the region of the $S_1 \to S_0$ and have analyzed the rotational structures. Comparison of the band profiles predicted from theory with the experimental shapes yields good agreement for band shape and polarization behavior.

To demonstrate polarization behavior we present experimental results for the totally symmetric two-photon transition 14_0^1 of C_6H_6 for different photon polarizations. Of importance here is the strong polarization behavior of the Q branch, according to Eq. (4.2.55), whereas the intensity ratio $W^{(2)\uparrow\uparrow}/W^{(2)\curvearrowright\curvearrowright}$ of the R and S branches is found to be $\frac{2}{3}$, in agreement with

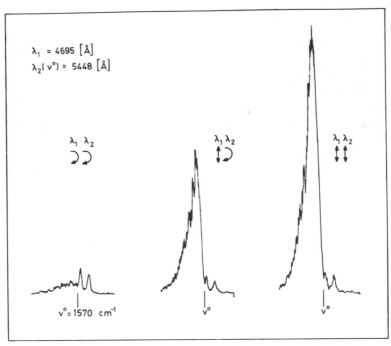

Fig. 4.6. The strongest absorption band of benzene h_6, which corresponds to a totally symmetric two-photon transition induced by the $v_{14}(b_{2u})$ vibration. The variation of the rotational structure of this band is shown for three different photon polarizations obtained by simultaneous absorption of two unlike photons from two collinear laser beams. [From Hampf *et al.* (1977).]

Eq. (4.2.55), where $R_0 = 0$. The spectral shape of the R and S branches is *not* changed with polarization, since polarization behavior is independent of J, K, and therefore the shape is the same for all rotational lines within the R and S branches.

However, for the Q branch ($R_0 \neq 0$), according to Eq. (4.2.55), the polarization behavior is governed by the Placzek–Teller factor R_2, which depends strongly on J, K. This produces not only a large change in the shape of the total vibronic band (see Figs. 4.6 and 4.7) but also a variation in the rotational structure of the Q branch when the polarization is changed. In Fig. 4.8 the very blue edge of the Q branch is shown under Doppler-free resolution for two different photon polarizations (Riedle *et al.*, 1982). Therefore, single rotational lines can be observed. Comparison of parts (a) and (b) of Fig. 4.8 shows that the typical spectral pattern changed for the two different photon polarizations. The reason for this is the polarization behavior, which in the isotropic Q branch is different for different lines. For countercircularly

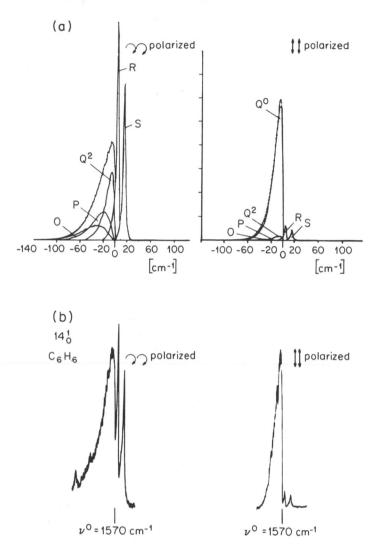

Fig. 4.7.   Comparison of the computer plot (a) with the experimental spectra (b) of the strongest totally symmetric two-photon excitation band b_{2u} 14_0^1 in benzene h_6. In the calculated profiles the rotational branches are separately plotted; Q^0 and Q^2 correspond to the isotropic and anisotropic contributions to the transition. The wave numbers in (a) and (b) are drawn on the same scale. The band in the circularly polarized light is enlarged by a factor of 10 in (a) and of $\sim$9 in (b). [From Wunsch *et al.* (1977).]

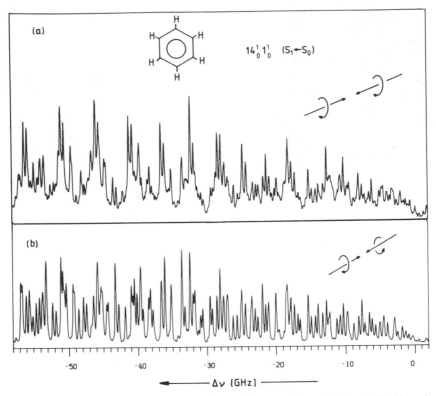

Fig. 4.8. Parts of the two-photon electronic $^1B_{2u} \leftarrow {}^1A_{1g}$ transition of C_6H_6. (a) The blue edge of the Q branch ($\Delta J = 0$) of the totally symmetric $14^1_0 1^1_0$ transition, which is excited with two laser beams polarized in the same direction. (b) The same frequency range of the two-photon spectrum as in (a); countercircularly polarized photons are used. [From Riedle *et al.* (1982).]

polarized ($\circlearrowleft\circlearrowright$) photons, R_0 is independent of J, K, whereas for two photons circularly polarized in the same sense R_2 varies with J, K and is largest for $J = K$ lines [for rotational factors see Broderson (1979) or Papousek (1982)].

In Section 4.2, theoretical aspects of polarization effects on two-photon transitions have been described. Polarization dependence of more than three-photon processes in molecules was omitted. For readers who are interested in polarization behavior we give the following references on the polarization of three- and four-photon processes. McClain (1972, 1973) has formulated the polarization dependence of three-photon transitions in randomly oriented nonrotating molecules. An expression for the intensity of two-photon excited fluorescence from a symmetric top molecule has been derived by using the density matrix formalism by Chen and Yeung (1979). A polarization study on three- and four-photon absorption of nonrotating molecules has been done by Andrews and Ghoul (1981). Selection rules for multiphoton transitions of rotating molecules have been discussed by Bialkowski and Guillory (1981).

REFERENCES

Ackerhalt, J. R., and Shore, B. W. (1977). *Phys. Rev. A* **16**, 277.
Altman, K., and Strey, G. (1972). *J. Mol. Spectrosc.* **44**, 571.
Andrews, D. L., and Ghoul, W. A. (1981). *J. Chem. Phys.* **75**, 530.
Arutyunyan, N., Askar'yan, G. A., and Pogosyan, V. A. (1970). *Solv. Phys.–JETP (Engl. Transl.)* **3**, 548.
Bialkowski, S. E., and Guillory, W. A. (1981). *Chem. Phys.* **55**, 229.
Boesl, U., Neusser, H. J., and Schlag, E. W. (1981). *Chem. Phys.* **55**, 193.
Bray, R. G., and Hochstrasser, R. M. (1976). *Mol. Phys.* **31**, 1199.
Bray, R. G., Hochstrasser, R. M., and Sung, H. N. (1975). *Chem. Phys. Lett.* **33**, 1.
Broderson, S. (1979). *In* "Raman Spectroscopy of Gases and Liquids." (A. Weber, ed.), p. 11. Springer-Verlag, Berlin and New York.
Brophy, J. H., and Rettner, C. T. (1979). *Chem. Phys. Lett.* **67**, 351.
Carrington, A., Levy, D. H., and Miller, T. A. (1970), *Adv. Chem. Phys.* **18**, 149.
Chen, K. M., and Yeung, E. S. (1978). *J. Chem. Phys.* **69**, 43.
Chen, K., and Yeung, E. S. (1979). *J. Chem. Phys.* **70**, 1312.
Cyvin, S. J., Rauch, J. E., and Decius, J. C. (1965). *J. Chem. Phys.* **43**, 4083.
de Meijere, J. L. F., and Eberly, J. H. (1978). *Phys. Rev. A* **17**, 1416.
Eberly, J. H., and O'Neil, S. V. (1979). *Phys. Rev. A* **19**, 1161.
Fisanick, G. J., Eichelberger, T. S., IV, Heath, B. A., and Robin, M. B. (1980). *J. Chem. Phys.* **72**, 5571.
Hampf, W., Neusser, H. J., and Schlag, E. W. (1977). *Chem. Phys. Lett.* **46**, 406.
Kafri, O., and Kimel, S. (1974). *Chem. Phys.* **5**, 448.
Lambropoulos, P. (1974). *Phys. Rev. A* **9**, 1992.
Lombardi, J. R., Wallenstein, R., T. W., Hänsch, T. W., and Friedrich, D. M. (1976). *J. Chem. Phys.* **65**, 2357.
McClain, W. M. (1971). *J. Chem. Phys.* **55**, 2789.
McClain, W. M. (1972). *J. Chem. Phys.* **57**, 2264.
McClain, W. M. (1973). *J. Chem. Phys.* **58**, 324.
McClain, W. M., and Harris, R. A. (1977). *Excited States* **3**, 2.
Metz, F., Howard, W. E., Wunsch, L., Neusser, H. J., and Schlag, E. W. (1978). *Proc. R. Soc. London, Ser. A*, **363**, 381.
Milonni, P. W., and Eberly, J. H. (1978). *J. Chem. Phys.* **68**, 1602.
Monson, P. R., and McClain, W. M. (1970). *J. Chem. Phys.* **53**, 29.
Nascimento, M. A. C. (1983). *Chem. Phys.* **74**, 51.
Papousek, D., and Aliev, M. R. (1982). "Molecular Vibrational-Rotational Spectra," p. 130. Elsevier, Amsterdam.
Parker, D. H., Berg, J. O., and El-Sayed, M. A. (1978) *Adv. Laser Chem.* **3**, 320.
Placzeck, G., and Teller, E. (1933). *Z. Phys.* **81**, 209.
Power, E. A., and Thirunamachandran, T. (1974). *J. Chem. Phys.* **60**, 3695.
Riedle, E., Moder, R., Neusser, H. J. (1982). *Opt. Commun.* **43**, 388.
Rose, M. E. (1957). "Elementary Theory of Angular Momentum." Wiley, New York.

Sargent, M., III, Scully, M. O., and Lamb, W. E., Jr. (1974). "Laser Physics." Addison-Wesley Reading, Massachusetts.
Shore, B. W., and Johnson, M. A. (1978). *J. Chem. Phys.* **68**, 5631.
Speiser, S., and Jortner, J. (1976). *Chem. Phys. Lett.* **44**, 399.
Speiser, S., and Kimel, S. (1970) *Chem. Phys. Lett* **7**, 19.
Wunsch, L., Neusser, H. J., and Schlag, E. W. (1976). *Chem. Phys. Lett.* **38**, 216.
Wunsch, L., Metz, F., Neusser, H. J., and Schlag, E. W. (1977). *J. Chem. Phys.* **66**, 386.
Zakheim, D. S., and Johnson, P. M. (1980) *Chem. Phys.* **46**, 263.

CHAPTER
5

Spectral
Properties
of Multiphoton
Transitions

New spectroscopic techniques, such as two-photon excitation and the multiphoton ionization methods discussed in Chapter 3, and polarization measurements have made it possible to assign vibrational modes in the electronically excited states of molecules. In this chapter, we are concerned with the spectroscopic properties of multiphoton transitions. Effects of the vibrations on multiphoton transitions in the nonresonant and resonant cases will be treated in Sections 5.1 and 5.2, respectively. In Section 5.1 the mechanisms of vibronic coupling in the forbidden two-photon transitions are presented. In Section 5.2, we investigate how the mechanism of the resonant multiphoton process affects the vibronic structure that appears in the spectra. Analytical expressions for the probability of simultaneous and sequential two-photon absorption are presented in the displaced harmonic oscillator model. In Section 5.3, we shall describe basic principles of Doppler-free multiphoton spectroscopy, which is a powerful method for observation of rotational lines. Next, we shall present a light source and the experimental setup for Doppler-free two-photon spectroscopy of molecules. Rovibronic bands appearing in the two-photon excitation spectrum of benzene are analyzed, and the origin of a new nonradiative channel, called the "third channel," is discussed in terms of an intramolecular vibrational relaxation.

5.1   THE NONRESONANT INTERMEDIATE STATE

It can easily be understood that two-photon absorption (TPA) takes place between the g and g or u and u electronic states of molecules with inversion symmetry (McClain, 1974). However, transitions between the g and u electronic states are observed as well, though their intensity is very weak, for example, the $^1B_{2u} \leftarrow {}^1A_g$ transition in benzene. In this case, the u $\leftrightarrow$ g transition, called the forbidden TPA, can be allowed by the coupling of the nuclear vibration of u symmetry with electrons (vibronic coupling). The mechanisms of the forbidden TPA in aromatic hydrocarbons have been studied by Honig *et al.* (1967), Hochstrasser and Wessel (1974), Metz (1975), Friedrich and McClain (1975), Wunsch *et al.* (1975), and Rava *et al.* (1981).

We shall consider the mechanisms of vibronic coupling in the nonresonant TPA case. Here the expression for the transition probability can be approximated by the simultaneous term in Eq. (2.1.105) or (2.3.145): The cross section $\sigma_{na}^{(2)}$ of the transition from the a to the n states is then expressed as

$$\sigma_{na}^{(2)} \propto \sum_a \sum_n \frac{\rho_a \Gamma_{na}}{(E_n - E_a - 2\hbar\omega_l)^2 + \Gamma_{na}^2} |S_{na}^{\kappa\lambda}|^2, \tag{5.1.1}$$

where two photons with identical frequency have been assumed, Γ and E are the damping constants associated with the a and n states and the vibronic energy, respectively, and S_{na} is the TPA matrix element expressed in terms of the laboratory fixed coordinate system, which can be transformed into that expressed in terms of the molecular fixed coordinates (x, y, z), as has already been seen in Chapter 4; that is, it takes the form

$$S_{na}^{\beta\alpha} = \sum_m \frac{R_{nm}^{\beta} R_{ma}^{\alpha} + R_{nm}^{\alpha} R_{ma}^{\beta}}{E_m - E_a - \hbar\omega_l}. \tag{5.1.2}$$

Here $R_{nm}^{\beta} = \langle n|r_\beta|m \rangle$ are the transition moments, the subscripts specify both electronic and vibrational quantum states, and $\beta = x, y, z$. For the nonresonant case, $E_m - E_a > \hbar\omega_l$, the vibrational quantum number dependence of the energy denominator can safely be neglected. For the resonant case, on the other hand, the nuclear coordinate-dependent terms of the energy denominator make an essential contribution to the cross section.

We adopt the Born–Oppenheimer approximation for the molecular eigenstates:

$$|a\rangle = |\Phi_a\rangle|\theta_a\rangle, \tag{5.1.3}$$

where $\Phi_a(r, Q)$ and $\theta_a(Q)$ are the electronic and vibrational wave functions, respectively, and r and Q denote the coordinates of the electrons and the nuclei, respectively. In this approximation the molecular Hamiltonian $\hat{H}$ satisfies

$$\hat{H}|a\rangle = E_a|a\rangle \tag{5.1.4}$$

and

$$\langle \Phi_a | \hat{H} | \Phi_a \rangle = (\hat{T}_N + \varepsilon_a) | \theta_a \rangle = E_a | \theta_a \rangle, \tag{5.1.5}$$

where $\hat{T}_N$ is the nuclear kinetic energy operator and $\varepsilon_a \equiv \varepsilon_a(Q)$ represents the adiabatic potential energy of electronic state a. By applying the completeness of the vibrational wave functions θ_m, that is, $\sum_m |\theta_m\rangle\langle\theta_m| = 1$, to Eq. (5.1.2), we obtain

$$S_{na}^{\beta\alpha} = \left\langle \theta_n \left| \sum_m^{\text{elec}} \frac{R_{nm}^\beta(Q)R_{ma}^\alpha(Q) + R_{nm}^\alpha(Q)R_{ma}^\beta(Q)}{\varepsilon_m(Q_0) - \varepsilon_a(Q_0) - \hbar\omega_l} \right| \theta_a \right\rangle, \tag{5.1.6}$$

where summation over m means the summation over the electronic states, and $R_{nm}(Q) = \langle \Phi_n(r, Q) | r | \Phi_m(r, Q) \rangle$.

The two-photon transition moments of Eq. (5.1.6) can be expanded around the equilibrium nuclear configuration of the ground state as

$$R_{nm}^\beta(Q)R_{ma}^\alpha(Q) = R_{nm}^\beta(0)R_{ma}^\alpha(0) + \sum_j [R_{nm}^\beta(0)R_{ma}^\alpha(j) + R_{nm}^\beta(j)R_{ma}^\alpha(0)]Q_j + \cdots, \tag{5.1.7}$$

where

$$R_{nm}^\beta(0) = R_{nm}^\beta(Q)|_{Q=Q_0} \quad \text{and} \quad R_{ma}^\alpha(j) = \frac{\partial}{\partial Q_j} R_{ma}^\alpha(Q)|_{Q=Q_0}.$$

Just as for the case of the one-photon transition, the two-photon transition can be classified as the dipole-allowed or the vibronically induced transition, depending on whether the transition moments are independent of the nuclear coordinates. Substituting Eq. (5.1.7) into Eq. (5.1.6) yields

$$S_{na}^{\beta\alpha} = S_{na}^{\beta\alpha}(\text{C-TPA}) + S_{na}^{\beta\alpha}(\text{VI-TPA}), \tag{5.1.8}$$

where the matrix elements denoted by C-TPA and VI-TPA are those of the TPA in the Condon approximation and the vibronically-induced TPA, respectively, and they are given by

$$S_{na}^{\beta\alpha}(\text{C-TPA}) = \left\langle \theta_n \left| \frac{R_{nm}^\beta(0)R_{ma}^\alpha(0) + R_{nm}^\alpha(0)R_{ma}^\beta(0)}{\varepsilon_{ma}^0 - \hbar\omega_l} \right| \theta_a \right\rangle \tag{5.1.9}$$

and

$$S_{na}^{\beta\alpha}(\text{VI-TPA}) = \sum_j \left\langle \theta_n \left| \sum_m \left(\frac{\begin{array}{c} R_{nm}^\beta(0)R_{ma}^\alpha(j) + R_{nm}^\beta(j)R_{ma}^\alpha(0) \\ + R_{nm}^\alpha(0)R_{ma}^\beta(j) + R_{nm}^\alpha(j)R_{ma}^\beta(0) \end{array}}{\varepsilon_{ma}^0 - \hbar\omega_l} \right) Q_j \right| \theta_a \right\rangle, \tag{5.1.10}$$

with $\varepsilon_{ma}^0 = \varepsilon_m(Q_0) - \varepsilon_a(Q_0)$.

Equation (5.1.9), in which the operator is independent of the nuclear co-ordinates (Condon approximation), expresses the allowed TPA and indicates that the geometrical change between the initial and final electronic states is reflected in the vibronic structure (progressions) appearing in the TPA spectrum. Equation (5.1.10) expresses the vibronically induced TPA. In deriving Eq. (5.1.10), terms involving second and higher orders of Q_j have been omitted.

To qualitatively discuss the mechanisms of the vibronically induced TPA, we introduce the conventional Herzberg–Teller approximation, which was originally used by Albrecht (1960) to explain the forbidden one-photon transitions. In the approximation the electronic wave function Φ_m is expanded to the first order of nuclear displacement as

$$\Phi_m(r, Q) \simeq \Phi_m(r, Q_0) + \sum_j \sum_k V_{mk}(j)\Phi_k(r, Q_0)Q_j, \qquad (5.1.11)$$

where $V_{mk}(j)$ is the vibronic coupling matrix element including the energy denominator:

$$V_{mk}(j) = \sum_j \frac{\langle \Phi_m(r, Q_0)|(\partial \hat{H}_{e-n}/\partial Q_j)_{Q_j = Q_{j0}}|\Phi_k(r, Q_0)\rangle}{\varepsilon_k^0 - \varepsilon_m^0} Q_j. \qquad (5.1.12)$$

Here $\hat{H}_{e-n}$ represents the interaction between the electron and the nucleus. Substituting Eq. (5.1.11) into the transition moment $R_{ma}^\alpha (Q)$ and differentiating the resulting equation over Q_j, we find that

$$R_{ma}^\alpha(j) = \sum_k [R_{mk}^\alpha(0)V_{ka}(j) - V_{mk}(j)R_{ka}^\alpha(0)]. \qquad (5.1.13)$$

The matrix element of the vibronically induced two-photon transition, Eq. (5.1.10), is then written as

$$\begin{aligned}
S_{na}^{\beta\alpha}(\text{VI} - \text{TPA}) = \sum_j \sum_k \sum_m \Bigg(& \frac{[R_{nm}^\beta(0)R_{mk}^\alpha(0) + R_{nm}^\alpha(0)R_{mk}^\beta(0)]V_{ka}(j)}{\varepsilon_{ma}^0 - \hbar\omega_l} \\
& + \frac{[R_{nm}^\beta(0)R_{ka}^\alpha(0) + R_{nk}^\beta(0)R_{ma}^\alpha(0)]V_{km}(j)}{\varepsilon_{ma}^0 - \hbar\omega_l} \\
& + \frac{[R_{nm}^\alpha(0)R_{ka}^\beta(0) + R_{nk}^\alpha(0)R_{ma}^\beta(0)]V_{km}(j)}{\varepsilon_{ma}^0 - \hbar\omega_l} \\
& - \frac{V_{nk}(j)[R_{km}^\beta(0)R_{ma}^\alpha(0) + R_{km}^\alpha(0)R_{ma}^\beta(0)]}{\varepsilon_{ma}^0 - \hbar\omega_l} \Bigg), \qquad (5.1.14)
\end{aligned}$$

where the first two terms involving $V_{ka}(j)$ express the vibronically induced TPA originating from the vibronic couplings in the electronic ground state,

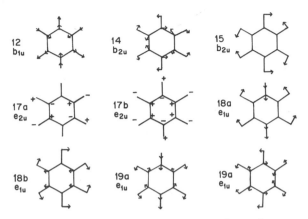

Fig. 5.1 The active ground-state modes in the two-photon $^1B_{2u} \leftarrow \, ^1A_{1g}$ spectra of benzene. [From Rava *et al.* (1981).]

the last two terms involving $V_{nk}(j)$ correspond to that in the final electronic state, and the other terms correspond to that in the intermediate states.

As an example of the vibronically induced TPA, let us consider the TPA from the ground state $^1A_{1g}$ to the first excited singlet state $^1B_{2u}$ in benzene. For the TPA of D_{6h} benzene, using two photons with identical frequency, the tensor pattern belongs to the A_{1g}, E_{1g} and E_{2g} irreducible representations of the D_{6h} point group (see Table 4.2.2). The species of vibrations that induce forbidden TPA ($^1B_{2u} \rightarrow \, ^1A_{1g}$) can be specified from symmetry consideration, $\Gamma_j \otimes B_{2u} \otimes A_{1g} = \{A_{1g}, E_{1g}, E_{2g}\}$; that is, the inducing modes belong to the b_{2u}, e_{2u}, and e_{1u} species. The normal modes, ν_{14} and ν_{15} of the b_{2u} species, ν_{17a} and ν_{17b} of the e_{2u}, species, and ν_{18a}, ν_{18b}, ν_{19a}, and ν_{19b} of the e_{1u} species, are shown in Fig. 5.1. Normal modes of b_{1u} symmetry can induce TPA in the case in which a different polarization is used. There have been several discussions about which modes are dominant for inducing TPA $^1B_{2u} \leftarrow \, ^1A_{1g}$. Honig *et al.* (1967), and Hochstrasser and Wessel (1974) predicted that the e_{1u} mode is the inducing mode and that the main vibronic coupling is in the intermediate states. Experiments and theoretical calculations, however, have confirmed that the dominant inducing mode is the bond-alternating vibration ν_{14} (Wunsch *et al.*, 1975a, b; Friedrich and McClain, 1975; Metz, 1975; Bray *et al.*, 1975; Rava *et al.*, 1981).

Friedrich and McClain (1975) have reported the polarization dependence on the two-photon fluorescence excitation spectrum of benzene in the vapor phase at 60 Torr for the energy region from 38,086 to 42,441 cm^{-1} by using both linearly and circularly polarized light; the two-photon excitation spectra that were observed are shown in Fig. 5.2. The upper and lower traces of

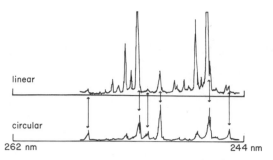

Fig. 5.2 Effect of linear and circular polarization on the TPA spectrum of benzene vapor. The same sensitivity was used for both spectra; 262 nm marks the $^1B_{2u}$ electronic origin and 244 nm marks the short-wavelength limit of the dye solution. Corresponding features are marked with arrows. The two tallest peaks in the upper trace are 14_1^0 and $14_0^1 1_0^1$. [From Friedrich and McClain, (1975.)]

Fig. 5.2 represent the spectra obtained using linearly and circularly polarized light, respectively. The prominent feature in the spectra is that the two tallest peaks, 14_0^1 and $14_0^1 1_0^1$ ($v_1 = 923$ cm^{-1} in the excited state), in the upper trace are missing in the lower trace. This is characteristic of the Q branch of a totally symmetric transition ($A_{1g} \leftarrow A_{1g}$), and the $^1B_{2u}$ electronic excited state is coupled with the b_{2u} vibration [See Section 4.2, and Friedrich and McClain (1980)]. As discussed in Section 4.2.2, anisotropic R and S branches of the 14_0^1 transition can be seen in Fig. 5.2.

To determine the inducing modes for the TPA of benzene in gases, Wunsch *et al.* (1975a, b) measured TPA from the hot bands of the ground state to the vibrationless level of the excited state $^1B_{2u}$. Because the frequencies of the normal modes are well known in the ground state, the technique of Wunsch *et al.* can be expected to uniquely identify the inducing modes. The results of their hot-band analysis are summarized in Table 5.1. Their conclusion is that the vibrations v_{14}, v_{15}, and $v_8(e_{1u})$ induce TPA, and the strongest line is brought about by the v_{14} vibration.

Rava *et al.* (1981) have made the intensity measurement of TPA ($^1B_{2u} \leftarrow$ $^1A_{1g}$) spectra of benzene using the multiphoton ionization (MPI) technique. Their measurement supports the v_{14} vibration as the most effective inducing mode. The important role of the v_{14} vibration the MPI spectra of the $^1B_{2u} \leftarrow {}^1A_{1g}$ transition has been recognized by Murakami *et al.* (1980) as well.

The normal modes of the b_{2u} symmetry can induce TPA through the vibronic coupling in the final $^1B_{2u}$ and ground $^1A_{1g}$ states and not through the coupling in the intermediate states; that is, the first and last two terms in Eq. (5.1.14) contribute to the TPA of benzene.

It seems that the failure of the earlier vibronic theory of TPA to predict the correct inducing modes originates from using approximate electronic

TABLE 5.1

The Hot-Band Frequencies below the $^1B_{2u}$ Origin in Gas-Phase Benzene $(h_6)^a$

v (cm^{-1})	Intensity	Transition	Excess energy (cm^{-1})	IR and Raman data in the ground state[b]	Boltzmann factor
36,447	vw	$14_1^0 16_2^0$	2,125	—	4×10^{-5}
36,599	m	$14_1^0 16_1^1$	1,720	—	3×10^{-4}
36,681	w	$14_1^0 6_1^1$	1,921	—	1×10^{-4}
36,771	s	14_1^0	1,315	1,309	2×10^{-3}
36,848	w	$15_1^0 6_1^1$	1,764	—	2×10^{-4}
36,928	m	15_1^0	1,158	1,146	4×10^{-3}
37,047	m	18_1^0	1,039	1,037	7×10^{-3}
37,530	w	$14_1^0 16_1^1 1_0^1$	1,720	—	3×10^{-4}

^{a} The intensity of the lines is described by vw (very weak), w (weak), m (medium), and s (strong). The 0–0 difference of the ground and upper states $v_0 = 38,089$ cm^{-1} [from Wunsch et al. (1975a,b)].
^{b} From Paramenter (1972).

wave functions that were far from the adiabatic wave functions. It is necessary to properly evaluate the transition matrix elements by using the adiabatic wave functions. There have been two attempts to explain the inducing modes on the basis of semiempirical calculations for all-valence electron molecular orbitals (MO) (Metz, 1975; Rava et al., 1981). Metz (1975) has applied the modified Herzberg–Teller approximation proposed by Roche and Jaffe (1974) to the adiabatic calculation of the transition moments. The zeroth-order wave functions are obtained from a combined neglect of differential overlap in spectroscopy (CNDO/S) calculation. This calculation is of the type of a self-consistent (SCF) LCAO MO configuration interaction (CI), with atomic orbitals that follow the nuclei. In this approximation, the transition moment $R_{ma}^\alpha(j)$ is expressed as

$$R_{ma}^\alpha(j) = \frac{\partial}{\partial Q_j} \langle \Phi_m | r_\alpha | \Phi_a \rangle + \sum_k [R_{mk}^\alpha(0)\tilde{V}_{ka}(j) - \tilde{V}_{mk}(j)R_{ka}^\alpha(0)]. \quad (5.1.16)$$

Equation (5.1.16) involves terms similar to those in Eq. (5.1.13), but the coupling matrix element $\tilde{V}_{mk}(j)$ takes the form

$$\tilde{V}_{mk}(j) = \frac{\partial \langle \Phi_m | \hat{H}_{e-e} | \Phi_k \rangle}{\partial Q_j (\varepsilon_k^0 - \varepsilon_m^0)}. \quad (5.1.17)$$

Here $\hat{H}_{e-e}$ is the electron–electron interaction Hamiltonian, and the wave functions Φ_m are constructed from $\Phi_m(r, Q_0)$ by keeping MO and CI coefficients constant and changing the atomic orbitals to those following the nuclei. The matrix elements $\partial/\partial Q_j \langle \Phi_m | \hat{H}_{e-e} | \Phi_k \rangle$, restricting the configurations to singly excited ones, have the form

$$\partial/\partial Q_j \langle \Phi_m(i \to \alpha) | \hat{H}_{e-e} | \Phi_k(l \to \beta) \rangle$$
$$= \partial/\partial Q_j [\delta_{il} F_{\alpha\beta} - \delta_{\alpha\beta} F_{il} + 2(i\alpha | l\beta) - (il | \alpha\beta)], \qquad (5.1.18)$$

where $F_{\alpha\beta}$ are the F-matrix elements and $(i\alpha | l\beta)$ are the two-electron repulsion integrals. The result of this vibronic calculation is consistent with the experimental findings of Wunsch *et al.* (1975a, b).

Rava *et al.* (1981) have calculated the values of the transition tensors for benzene two-photon vibronic transitions by using the all-valence INDO/S MO method, taking into account the floating orbital basis performed by Ziegler and Albrecht (1974). Their calculation shows that the most important mechanism for TPA is the vibronic coupling in the final state $^1B_{2u}$, which corresponds to the last two terms in Eq. (5.1.14): Its intermediate state is $^1E_{1u}$, the lending state being the A_{1g} ground state. It is certain that the discovery of the vibronic coupling of the $^1B_{2u}$ electronic state with the ground state through the b_{2u} mode explains the large frequency change of the v_{14} vibration (1566 and 1309 cm^{-1} in the $^1B_{2u}$ and ground states, respectively).

So far we have restricted ourselves to vibronically induced TPA of benzene. The importance of vibronic coupling in the excited state with the ground state has been observed for two-photon excitation spectra of the lowest $^1B_{2u}$ state of crystalline napthalene by Mikami and Ito (1977). These authors have reported that the 1535-cm^{-1} vibration of the b_{2u} symmetry is the most prominent inducing vibration in the excited state, which corresponds to the vibration with 1361 cm^{-1} in the ground state.

Work on p-difluorobenzene with inducing modes has been published by Robey and Schlag (1978). Allowed and forbidden (vibronic coupling) characters in the two-photon spectra of various substituted benzenes have been investigated by Rava and Goodman (1982) and Rava *et al.* (1982). They have shown that the two-photon spectral intensities can be understood in terms of the tendency for an electron of the substituent group to enter into conjugation with the benzene electrons. Interesting review articles on the two-photon spectra of perturbed benzenes have been published by Goodman and Rava (1982, 1983). Two-photon rotational contours of the $14(^1B_{2u}{}^1A_{1g})$ band in benzene and its derivatives have been reported by Thakur and Goodman (1983). Interesting two-photon work on pyrazine crystal has been published by Esherick *et al.* (1975), and on toluene vapor by Vasudev and Brand (1979). Vibronically induced two-photon transitions have been also found in pyrazine vapor and in pyrimidine by Knoth *et al.* (1979, 1982).

Quantum-mechanical calculations of nonresonant TPA probability within the semiempirical molecular orbital method have been reported by several authors. Evleth and Peticolas (1964) have calculated the TPA cross sections for pyrene and 3,4-benzpyrene using the SCF-CI MO method within the π-electron approximation. Calculations of nonresonant TPA probabilities within a semiempirical all-valence electrons MO framework combined with the CNDO/S method have been carried out by Hohlneicher and Dick (1979) and Marchese *et al.* (1980).

5.2   THE RESONANT INTERMEDIATE STATE

As is well known, multiphoton processes are said to be in resonance if the energy of one or several quanta of laser photons is close to the transition energy from the initial to an intermediate excited state or to that between the intermediate states. When a resonant condition is satisfied, not only is the multiphoton absorption greatly enhanced, but also the vibronic structure, which cannot be explained by the simple Frank–Condon overlap of the vibrational wave functions of the initial and final states, appears in the multiphoton spectrum. Let us consider multiphoton ionization of a molecular system. In the near-resonant or resonant case, there may exist two types of ionizing channels for multiphoton ionization: One is direct multiphoton ionization from the vibrational level in the ground electronic state, and the other is a stepwise ionization via the resonant state. The vibrational structure in the multiphoton spectrum is related to the difference among the potential energy surfaces involved in the multiphoton transition, and the potential energy differences between the initial and resonant states and/or between the resonance and final states, as well as those between the initial and final electronic states, may be reflected in the spectra.

To see how the mechanism of multiphoton ionization relates to the ionization spectrum, we consider a simple resonant two-photon ionization (Fig. 5.3). Let us derive an expression for the ion yield by using the kinetic

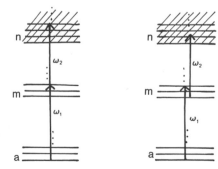

Fig. 5.3 Model for (left) direct and (right) stepwise two-photon ionization. In the case in which effects of the pure dephasing are negligibly small, the direct process has the same meaning as the simultaneous one discussed in Chapter 2. Energy conservation during the transition is satisfied for the simultaneous process. For the stepwise process, on the other hand, there is no need to satisfy energy conservation.

equation approach introduced in Chapter 4. Here we focus our attention on both the direct ionization, which was neglected in Chapter 4, and the stepwise ionization at the same time. The kinetic equations are expressed as

$$\frac{d\rho_a(t)}{dt} = -k_{aa}\rho_a(t), \tag{5.2.1}$$

$$\frac{d\rho_m(t)}{dt} = \sum_a k_{ma}^{(1)}\rho_a(t) - k_{mm}\rho_m(t), \tag{5.2.2}$$

$$\frac{d\rho_n(t)}{dt} = \sum_a k_{na}^{(2)}\rho_a(t) + \sum_m k_{nm}^{(1)}\rho_m(t), \tag{5.2.3}$$

where the superscripts (1) and (2) of k represent the one- and two-photon processes and $k_{aa} = k_{aa}^{(1)} + k_{aa}^{(2)}$ and $k_{mm} = \sum_n k_{nm}^{(1)}$. The structure of $k_{na}^{(2)}$ has been given in Section 2.3.3.

Along with the derivation given in Chapter 4, the ion yield $R_n(t_p)$ for an ideal experiment using a square laser pulse of duration t_p is expressed as

$$R_n(t_p) = \sum_a \sum_n \frac{k_{na}^{(2)}}{k_{aa}} (1 - e^{-k_{aa}t_p})\rho_a(0)$$

$$+ \sum_a \sum_n \sum_m \frac{k_{nm}^{(1)}k_{ma}^{(1)}}{k_{mm} - k_{aa}} \left(\frac{1 - e^{-k_{aa}t_p}}{k_{aa}} - \frac{1 - e^{-k_{mm}t_p}}{k_{mm}} \right) \rho_a(0), \tag{5.2.4}$$

where the averaging over the initial distribution of molecules and the summation over the final vibronic states have been carried out, and the first and the second terms represent the ion yields of the direct and stepwise processes, respectively. The ion yield depends on the laser pulse duration t_p and the radiative rate constants. For the time scale with $k_{aa}t_p \ll 1$ and $k_{mm}t_p \ll 1$, ion yields up to the order of t_p^2 are expressed as

$$R_n(t_p) \simeq \sum_a \sum_n \left(k_{na}^{(2)}t_p + \frac{1}{2}\sum_m k_{nm}^{(1)}k_{ma}^{(1)}t_p^2 \right) \rho_a(0). \tag{5.2.5}$$

That is, for a short time pulse the direct two-photon process that appears in the first term of Eq. (5.2.5) dominates the multiphoton transition. When the pulse duration is increased, on the other hand, the stepwise process will make a dominant contribution to the transition.

An interesting situation can be seen in the case in which the steady-state condition in the resonant state is satisfied. In this case, $d\rho_m(t)/dt = 0$, the rate of ionization is given by

$$\frac{d\rho_n(t)}{dt} = \sum_a \left(\sum_n k_{na}^{(2)} + \sum_m k_{ma}^{(1)} \right) \rho_a(t). \tag{5.2.6}$$

For a pulse of $k_{aa}t_p < 1$, the ion yield takes the form

$$R_n(t_p) \simeq \sum_a \left(\sum_n k_{na}^{(2)} + \sum_m k_{ma}^{(1)} \right) \rho_a(0)t_p. \qquad (5.2.7)$$

Equations (5.2.6) and (5.2.7) indicate that when the first-order laser power dependence dominates, the vibrational structure associated with the transition from the initial to the resonant state appears in the spectrum.

We restricted ourselves to resonant two-photon ionization, but it is straightforward to extend the preceding discussion to $i + j$ ionization. The ion yield corresponding to Eq. (5.2.7) can be expressed as

$$R_n \simeq \sum_a \left(\sum_n k_{na}^{(i+j)} + \sum_m k_{ma}^{(i)} \right) \rho_a(0)t_p. \qquad (5.2.8)$$

Resonant multiphoton ionization spectra of molecules have been reported by several authors (Johnson, 1980, and references therein). Murakami *et al.* (1980) have observed resonant four-photon ionization spectra of benzene and halogen-substituted benzenes. Figure 5.4 shows the ionization spectrum of benzene. The vibronic structure is very similar to the two-photon fluorescence excitation spectrum of the vibronically induced $^1B_{2u} \leftarrow {}^1A_{1g}$ transition observed by Wunsch *et al.* (1977): Its vibrational structure consists of the ν_1 progression that starts from the vibronically induced 14_0^1 band. Comparing the intensities observed by Murakami *et al.* with those observed by Wunsch *et al.*, there exists a sudden decrease of the two-photon fluorescence excitation spectrum when the excitation to $14_0^1 1_0^2$, which is located 3415 cm^{-1} above 0–0, is performed. The origin of the observed discrepancy between two-photon excitation and four-photon ionization spectra has been explained by the existence of the third channel, which is found 3000 cm^{-1} above 0–0 and which is directly related to the two-photon excitation spectra. Murakami *et al.* have concluded that the resonant four-photon ionization

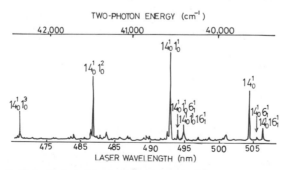

Fig. 5.4. Normalized multiphoton ionization spectrum of benzene resonant with S_1. [From Murakami *et al.* (1980).]

spectrum reflects the two-photon process of the $S_1 \leftarrow S_0$ transition. This suggests that the rate process $k_{S_1 S_0}^{(1)}$ from the ground to the resonant state makes a dominant contribution to the resonant ionization, and, on the other hand, the direct four-photon processes associated with the coherent interaction are unimportant.

Use of a highly powerful laser will cause the direct process to dominate the transition, because of its short time pulse and $I^{(i+j)}$ versus $I^{(i)}$ intensity dependence. This phenomenon is widely recognized in the field of multiphoton ionization decomposition reactions (Bernstein, 1982, and references therein).

From the spectroscopic viewpoint, a significant contribution of the direct process has been thought to complicate the analysis of the spectrum. In fact, the transition probability of the direct process consists of the simultaneous, sequential, and mixed terms, as shown in Chapter 2 [see, for example, Eq. (2.3.144) for two-photon processes]. Furthermore, for the resonant case the same approximation as that used in the nonresonant case, regardless of the nuclear coordinate dependence on the energy denominator of the transition operator, cannot be valid. Fujimura and Lin (1981) have derived an analytical expression for the simultaneous two-photon transition in the harmonic oscillator approximation. The resulting expression may be used to analyze the spectra originating from the resonant, simultaneous transitions.

5.2.1 An Analytical Expression for the Simultaneous Two-Photon Transition

Let us now derive an analytical expression for the simultaneous two-photon transition. The simultaneous term of the resonant two-photon excitation process, as we have already noticed, is similar to that of resonant Raman scattering (RRS). The difference between the two terms is that the resonant two-photon excitation process involves two potential displacements and distortions associated with the ground and resonant states and with the resonant and final states; on the other hand, RRS involves one potential displacement and distortion between the ground and resonant states. In this treatment, we adopt the displaced harmonic oscillator model as shown in Fig. 5.5, in which the electronic energy gaps are assumed to be $\varepsilon_m^0 - \varepsilon_a^0 > \varepsilon_n^0 - \varepsilon_m^0$.

In order to derive the expression for the transition probability of the multilevel molecular system, we shall start with the expression

$$W_{\text{sim}}^{(2)}(\omega_1, \omega_2) = \frac{2}{\hbar 4} \text{Re} \sum_a \sum_n \frac{\rho_a}{i\omega_{na} + \Gamma_{na}} \left| \sum_m \frac{(\hat{H}_R')_{nm}(\hat{H}_R')_{ma}}{i\omega_{ma} + \Gamma_{ma}} \right|^2, \quad (5.2.9)$$

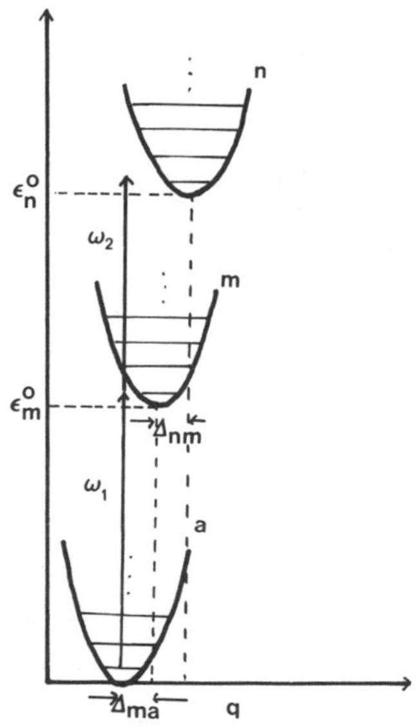

Fig. 5.5 Model for a two-photon transition. Here a, m, and n specify the electronic ground, resonant, and final states, respectively. ε_n^0 and ε_m^0 represent the energy gap between the final and ground states and that between the resonant and ground states, respectively. The terms Δ_{ma} and Δ_{nm} denote the dimensionless displacements between the resonant and ground states and the final and resonant states, respectively. In the one-dimensional displaced harmonic oscillator model $\Delta_{na} = \Delta_{nm} + \Delta_{ma}$.

which is equivalent to the first or last term of Eq. (2.3.149). Here $\rho_a = \rho_a(0)$, $\omega_{na} = (\varepsilon_n - \varepsilon_a)/\hbar - \omega_1 - \omega_2$, and $\omega_{ma} = (\varepsilon_m - \varepsilon_a)/\hbar - \omega_1$. The laser frequencies ω_1 and ω_2 are assumed to be $\omega_1 > \omega_2$, and ω_{ma} satisfies the near-resonance or resonance condition. A contribution of the nonresonant term has been omitted. It is convenient to use the following transformations:

$$\frac{\gamma}{x^2 + \gamma^2} = \frac{1}{2} \int_{-\infty}^{\infty} dt \, \exp(-ixt - \gamma|t|) \qquad (5.2.10)$$

and

$$\frac{1}{ix + \gamma} = \int_0^{\infty} dt \, \exp[-\tau(ix + \gamma)]. \qquad (5.2.11)$$

In the Born–Oppenheimer approximation [see Eq. (5.1.3)], Eq. (5.2.9) is then expressed in integral form as

$$W_{\text{sim}}^{(2)}(\omega_1, \omega_2) = \frac{|M_{nm}M_{ma}|^2}{\hbar^4} \int_{-\infty}^{\infty} dt \int_0^{\infty} d\tau \int_0^{\infty} d\tau' \sum_{an} \exp(i\omega_{na}t - \Gamma_{na}|t|)$$

$$\times \sum_{mm'} \langle \theta_a | \rho_a \exp[-\tau'(i\omega_{ma} + \Gamma_{ma})] | \theta_m \rangle \langle \theta_m | \theta_n \rangle \langle \theta_n |$$

$$\times \exp[-\tau(i\omega_{ma} + \Gamma_{ma})] | \theta_{m'} \rangle \langle \theta_{m'} | \theta_a \rangle, \qquad (5.2.12)$$

where the Condon approximation has been used to evaluate the transition moment $M_{nm} = \langle \Phi_n | er | \Phi_m \rangle$; that is, effects of vibronic coupling have been omitted. Equation (5.2.12) can be rewritten as

$$W^{(2)}_{sim}(\omega_1, \omega_2) = \frac{|M_{nm} M_{ma}|^2}{\hbar^4} \int_{-\infty}^{\infty} dt \int_0^{\infty} d\tau \int_0^{\infty} d\tau'$$

$$\times \sum_{an} \sum_{mm'} \exp[-\Gamma_{na}|t| - it(\omega_1 + \omega_2) - i\omega_1 \tau' + i\omega_1 \tau]$$

$$\times \langle \theta_a | \rho_a \exp[-i\varepsilon_a(t + \tau' - \tau)/\hbar]$$

$$\times \exp(i\varepsilon_m \tau'/\hbar - \Gamma_{ma}\tau') |\theta_m\rangle \langle \theta_m| \exp(i\varepsilon_n t/\hbar) |\theta_n\rangle \langle \theta_n|$$

$$\times \exp(-i\varepsilon_{m'} \tau/\hbar - \Gamma_{m'a}\tau) |\theta_{m'}\rangle \langle \theta_{m'} | \theta_a \rangle. \tag{5.2.13}$$

By using the closure relation of the vibronic states and carrying out the summations in Eq. (5.2.13), we obtain

$$W^{(2)}_{sim}(\omega_1, \omega_2) = \frac{|M_{nm} M_{ma}|^2}{\hbar^4} \int_{-\infty}^{\infty} dt \int_0^{\infty} d\tau \int_0^{\infty} d\tau'$$

$$\times \exp[-\Gamma_{na}|t| - \Gamma_{ma}(\tau + \tau') - it(\omega_1 + \omega_2)$$

$$- i\omega_1(\tau' - \tau)]G(\tau, \tau', t), \tag{5.2.14}$$

where

$$G(\tau, \tau', t) = \sum_a \left\langle \theta_a \left| \rho_a \exp\left[-\frac{i\hat{H}_a(t + \tau' - \tau)}{\hbar} \right] \right. \right.$$

$$\times \exp\left(\frac{i\hat{H}_m \tau'}{\hbar} \right) \exp\left(\frac{i\hat{H}_n t}{\hbar} \right) \exp\left(\frac{-i\hat{H}_m \tau}{\hbar} \right) \left| \theta_a \right\rangle, \tag{5.2.15}$$

and $\hat{H}_m$ represents the vibronic Hamiltonian of the resonant electronic state. In deriving Eq. (5.2.14), the dependence of Γ on the vibrational quantum number was neglected. Equation (5.2.14) expresses the transition probability of simultaneous TPA in the Born–Oppenheimer basis set.

In the displaced harmonic oscillator model, the vibronic Hamiltonians of the initial, resonant, and final electronic states are written as

$$\hat{H}_a = \sum_{i=1}^{N} \hat{H}_{ai}, \qquad \hat{H}_{ai} = \frac{\hbar\omega_i}{2}(\hat{p}_i^2 + \hat{q}_i^2), \tag{5.2.16}$$

$$\hat{H}_m = \sum_{i=1}^{N} \hat{H}_{mi} + \varepsilon_m^0, \qquad \hat{H}_{mi} = \frac{\hbar\omega_i}{2}[\hat{p}_i^2 + (\hat{q}_i - \Delta_{mai})^2], \tag{5.2.17}$$

and

$$\hat{H}_n = \sum_{i=1}^{N} \hat{H}_{ni} + \varepsilon_n^0, \qquad \hat{H}_{ni} = \frac{\hbar\omega_i}{2}[\hat{p}_1^2 + (\hat{q}_i - \Delta_{nai})^2], \qquad (5.2.18)$$

respectively, where N represents the number of the vibrational modes, ω_i is the frequency, and $\hat{q}_i$ and $\hat{p}_i$ are the dimensionless nuclear coordinate and conjugate momentum of the ith vibrational mode, respectively; Δ_{mai} represents the dimensionless displacement between the equilibrium points of the ith vibrational mode in the resonant and ground electronic states, and ε_m^0 the electronic energy gap between the bottoms of the two electronic states (Fig. 5.5). Assuming that the density matrix of the initial molecular state ρ_a is given by the Boltzmann distribution

$$\rho_a = \prod_i \rho_{ai} = \prod_i \left[1 - \exp\left(-\frac{\hbar\omega_i}{kT} \right) \right] \exp\left(-\frac{a_i\hbar\omega_i}{kT} \right), \qquad (5.2.19)$$

Eq. (5.2.14) can be expressed in the analytical form as (Appendix III.A)

$$W_{\text{sim}}^{(2)}(\omega_1, \omega_2)$$

$$= \frac{2|M_{nm}M_{ma}|^2}{\hbar^4} \exp\left[-\sum_{i=1}^{N} (2n_i + 1)\left(\frac{\Delta_{mai}^2}{2} + \frac{\Delta_{nmi}^2}{2} \right) \right]$$

$$\times \sum_{k_1=0}^{\infty} \sum_{k_2=0}^{\infty} \cdots \sum_{k_N=0}^{\infty} \sum_{l_1=0}^{\infty} \sum_{l_2=0}^{\infty} \cdots \sum_{l_N=0}^{\infty} \left[\prod_{i=1}^{N} \frac{(n_i + 1)^{k_1} n_i^{-l_i}}{k_i! \, l_i!} \right]$$

$$\times \frac{\Gamma_{na}}{[\omega_1 + \omega_2 - \varepsilon_n^0/\hbar - \sum_{i=1}^{N} (k_i - l_i)\omega_l]^2 + \Gamma_{na}^2}$$

$$\times \left| \sum_{p_1=0}^{k_1} \sum_{p_2=0}^{k_2} \cdots \sum_{p_N=0}^{k_N} \sum_{q_1=0}^{l_1} \sum_{q_2=0}^{l_2} \cdots \sum_{q_N=0}^{l_N} \sum_{r_1=0}^{\infty} \sum_{r_2=0}^{\infty} \cdots \sum_{r_N=0}^{\infty} \sum_{s_1=0}^{\infty} \sum_{s_2=0}^{\infty} \right.$$

$$\cdots \sum_{s_N=0}^{\infty} \left\{ \prod_{i=1}^{N} \binom{k_i}{p_i}\binom{l_i}{q_i} \left[-(n_i + 1)\frac{\Delta_{nmi}\Delta_{mai}}{2} \right]^{r_i} \left(-\bar{n}_i \frac{\Delta_{nmi}\Delta_{mai}}{2} \right)^{s_i} \right.$$

$$\times \left. \left(\frac{\Delta_{nmi}}{\sqrt{2}} \right)^{k_i - p_i + l_i - q_i} \frac{(\Delta_{mai}/\sqrt{2})^{p_i + q_i}}{(r_i! \, s_i!)} \right\}$$

$$\left. \times \frac{1}{i[\varepsilon_m^0/\hbar - \omega_1 + \sum_{i=1}^{N} (p_i - q_i + r_i - s_i)\omega_i] + \Gamma_{ma}} \right|^2, \qquad (5.2.20)$$

where Δ_{nmi} represents the displacement between the equilibrium points of the ith vibrational mode in the final and resonant states, $\binom{k}{p} = k!/\{p!(k-p)!\}$, and $\bar{n}_i$ is the occupation number of the ith vibrational mode given by

$$\bar{n}_i = [\exp(\hbar\omega_i/kT) - 1]^{-1}. \qquad (5.2.21)$$

The details of the derivation of Eq. (5.2.20), using the boson operator technique, are given in Appendix III.A. For a simple case in which the molecular system is characterized by a single mode in the low temperature limit, $\bar{n}_i = 0$, Eq. (5.2.20) reduces to

$$W^{(2)}_{sim}(\omega_1, \omega_2) = \frac{2|M_{nm}M_{ma}|^2}{\hbar^4} \exp\left(-\frac{\Delta^2_{ma}}{2} - \frac{\Delta^2_{nm}}{2}\right)$$

$$\times \sum_{l=0}^{\infty} \frac{\Gamma_{na}}{l!\{[\varepsilon^0_n/\hbar] + l\omega - \omega_1 - \omega_2]^2 + \Gamma^2_{na}\}}$$

$$\times \left| \sum_{j=0}^{l} \binom{l}{j} \left(\frac{\Delta_{nm}}{\sqrt{2}}\right)^j \left(\frac{\Delta_{ma}}{\sqrt{2}}\right)^{l-j} \right.$$

$$\times \left. \sum_{k=0}^{\infty} \frac{(-1)^k(\Delta_{ma}\Delta_{nm}/2)^k}{k!\{i[(\varepsilon^0_m/\hbar) - \omega_1 - j\omega + l\omega + k\omega] + \Gamma_{ma}\}} \right|^2. \qquad (5.2.22)$$

The derivation of the analytical expression for sequential TPA, $W^{(2)}_{seq}$, on the other hand, is simple because the transition probability can be expressed approximately as the product of the line shape function associated with the transition from the initial to the resonant states, $I_{a\to m}(\omega_1)$, and that from the resonant to the final states, $I_{m\to n}(\omega_2)$ [see Eq. (2.3.148)]. That is, in the Born–Oppenheimer approximation, the transition probability is given by

$$W^{(2)}_{seq}(\omega_1, \omega_2) = \frac{2|M_{nm}M_{ma}|^2(2\Gamma_{ma} - \Gamma_{mm})}{\hbar^4\Gamma_{mm}\Gamma_{ma}} \sum_m I_{a\to m}(\omega_1)I_{m\to n}(\omega_2), \qquad (5.2.23)$$

where the line shape functions take the form

$$I_{a\to m}(\omega_1) = \sum_a \frac{\rho_a|\langle\theta_a|\theta_m\rangle|^2\Gamma_{ma}}{\omega^2_{ma} + \Gamma^2_{ma}} \qquad (5.2.24)$$

and

$$I_{m\to n}(\omega_2) = \sum_n \frac{|\langle\theta_m|\theta_n\rangle|^2\Gamma_{mn}}{\omega^2_{mn} + \Gamma^2_{mn}}, \qquad (5.2.25)$$

respectively. The analytical form of Eq. (5.2.24) in the harmonic approximation has been derived to study the vibronic structure in one-photon processes (Lin, 1980a). In the displaced harmonic oscillator model for a single mode, an analytical expression for $W^{(2)}_{seq}$ in the low temperature limit can be obtained as

$$
W_{\text{seq}}^{(2)}(\omega_1, \omega_2) = \frac{2|M_{nm}M_{ma}|^2}{\hbar^4} \frac{(2\Gamma_{ma} - \Gamma_{mm})\Gamma_{ma}}{\Gamma_{mm}}
$$

$$
\times \sum_{k=0}^{\infty} \frac{\exp(-\Delta_{ma}^2/2)(\Delta^2{}_{ma}/2)^k}{k![\varepsilon_m^0/\hbar + k\omega - \omega_1)^2 + \Gamma_{ma}^2]}
$$

$$
\times \sum_{j=0}^{k} \frac{\exp(-\Delta_{nm}^2/2)k!(\Delta_{ma}^2/2)^j}{(k-j)!(j!)^2} \sum_{p=0}^{2j} \binom{2j}{p}(-1)^p
$$

$$
\times \sum_{l=0}^{\infty} \frac{(\Delta_{nm}^2/2)^l}{l!\{[\varepsilon_n^0 - \varepsilon_m^0)/\hbar - \omega_2 + j\omega - p\omega + l\omega]^2 + \Gamma_{mn}^2\}}.
$$

$$(5.2.26)$$

The analytical expressions derived here may serve as useful tools for analyzing the vibrational structures, for studying temperature effects, and for determining the geometries of the resonant and final states. A model calculation of the vibrational structure in TPA spectra has been performed to illustrate the difference of the Franck–Condon effect (i.e., the effect of displacements Δ_{ma} and Δ_{nm}) between the simultaneous and sequential processes for a model system in which both the simultaneous and sequential TPA processes $W_{\text{sim}}^{(2)}$ and $W_{\text{seq}}^{(2)}$ may make a significant contribution (Fujimura and Lin, 1981). Finally, the theoretical treatment of the molecular multiphoton transition probability presented here is still in an early stage of development, because a simple displaced harmonic oscillator model in the Born–Oppenheimer approximation has been adopted. Effects of frequency change, anharmonicity, breakdown of the Born–Oppenheimer approximation, and so on, will have to be elucidated in the near future.

5.3 DOPPLER-FREE MULTIPHOTON SPECTROSCOPY

5.3.1 Basic Principles

One of the important advantages of two-photon spectroscopy is the possible elimination of Doppler broadening, which makes a resolution higher than 10^6 impossible under ambient gas conditions. Even though there are a number of Doppler-free methods known, Doppler-free two-photon spectroscopy presents some basic features that are not present in other methods. The classical method of obtaining Doppler-free spectra is to use a collimated molecular beam. However, this technique suffers from small particle densities. Moreover, most molecular transitions lie in the UV spectra where monochromatic laser light sources are hardly obtainable. Both disadvantages are also responsible for the limited application of saturation spectroscopy in molecular spectroscopy, which is known to provide Doppler-free molecular spectra (Letokhov and Chebotayev, 1977). The important merits of Doppler-free two-photon spectroscopy are that narrow-band visible light can be used for observation of UV transitions, which is readily obtainable from commercial cw dye lasers, and that molecules independent of their velocity components within the interaction volume contribute to the Doppler-free signal. This is particularly different from saturation spectroscopy, in which only molecules with a zero velocity component in the propagation direction of the light beam contribute to the Doppler-free signal. One has to bear in mind, however, that the spectral information obtained from both methods is different and complementary for symmetric molecules due to the different selection rules.

5.3.1.1 Doppler Broadening

Even at low pressures the observed linewidth of molecular transitions in a gas is not always given by the lifetime of the states, that is the observed linewidth is not due to homogeneous broadening. In particular, for visible and UV transitions under high resolution, in most cases inhomogeneous Doppler broadening is observed, since the Doppler shift is proportional to the transition frequency.

For an ambient molecular gas there is an isotropic velocity distribution that produces different shifts for molecules with different velocity components in the direction of light propagation. The average of these shifts results in a Doppler broadening of the transition. For a Maxwell–Boltzmann velocity distribution we obtain a Gaussian line profile whose full width at half-maximum (FWHM) is given by (Letokhov and Chebotayev, 1977)

$$\Delta v_D = (2v_0/C)[2\ln(2KT/m)]^{1/2} = 7.163 \times 10^{-7}\sqrt{T/M}\, v_0. \quad (5.3.1)$$

Here v_0 is the transition frequency, T is the temperature in degrees K, and m is the mass in kilograms, and M is the molecular weight of the molecules in atomic mass units. Typically, for a polyatomic molecule such as benzene (C_6H_6) with $M = 78$, we obtain $\Delta v_D = 1.67$ GHz for $v_0 = 40,000$ cm^{-1} and room temperature. Therefore, the highest resolution possible in ambient benzene vapor at room temperature is $v_0/\Delta v_0 = 7.2 \times 10^5$ for the $S_1 \leftarrow S_0$ transition.

If molecular transitions are so close in frequency that their spectral line shapes overlap due to Doppler broadening, they cannot be resolved in conventional spectroscopy and Doppler-free methods are necessary. This situation is mainly given for rovibronic transitions in polyatomic molecules even

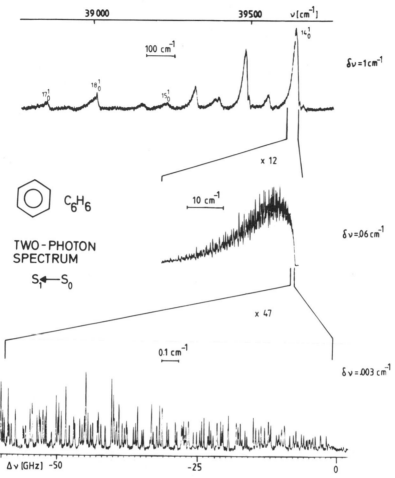

Fig. 5.6 Part of the two-photon spectrum of C_6H_6 as measured with different spectral resolution. The middle trace represents the highest resolution possible in Doppler-limited spectroscopy. Only in the Doppler-free spectrum (lower part) are single rotational lines resolved.

though their Doppler width is smaller than that of light molecules. In large molecules the average spacing of rovibronic transitions is several times smaller than the Doppler width. As a consequence, it is not possible to observe single Doppler-broadened lines, but the envelope of the lines produces a typical rotational contour of the vibronic band (see Fig. 5.6).

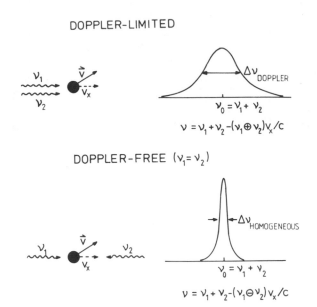

Fig. 5.7   Principle of Doppler-limited and Doppler-free two-photon absorption.

5.3.1.2   Elimination of Doppler Broadening in Two-Photon Spectroscopy

The principles of Doppler-limited and Doppler-free two-photon absorption are shown in Fig. 5.7. Here the interaction of a molecule with two monochromatic light beams having frequencies v_1 and v_2 is shown. In the upper part of Fig. 5.7 both light beams are parallel and propagate in the same direction. For each molecule with a velocity v and a component in the propagation direction of the light beams v_x, the transition frequency v_0 is shifted by $(v_1 + v_2)(v_x/c)$ since the Doppler shift of both light waves has the same sign. When an ensemble of molecules in thermal equilibrium is investigated, this yields a broadening of the transition line according to Eq. (5.3.1).

If, on the other hand, light beams with frequency v_1 and v_2 propagate in opposite directions and the molecule absorbs one photon from each beam, then the corresponding Doppler shifts have opposite signs and the residual Doppler shift of the transition frequency is $(v_1 - v_2)(v_x/c)$. The shifts cancel exactly to zero if the frequencies of both light beams are equal ($v_1 = v_2$). It is seen immediately from Fig. 5.7 that the Doppler shift is canceled independently of the direction and magnitude of the velocity v.[1]

---

[1] Actually, this is the case only for the linear Doppler effect. The much smaller quadratic Doppler effect, however, may be neglected for molecular spectroscopy.

This means that all molecules, independent of their velocity, may interact with light beams, and the signal strength is determined by the full molecular density within a given quantum state.

After elimination of the inhomogeneous Doppler broadening, in principle one is able to obtain information about the dynamic behavior of the excited quantum states. Since at low pressures and for final states with low excess energies the homogeneous width is usually much smaller than the Doppler width, single rovibronic lines that are normally hidden under their Doppler-broadened contour can be resolved.

For slightly different light frequencies a residual Doppler broadening occurs. This situation is well known from classical Raman experiments. Raman scattered light observed in the forward direction is nearly Doppler-free due to the decrease of the Doppler shift, as shown in Fig. 5.7. The residual Doppler width is given by the shift of the Raman scattered light.

5.3.1.3 History of Doppler-Free Two-Photon Spectroscopy

The possibility of eliminating the Doppler shift by two-photon absorption in a standing-wave light field was originally proposed by Vasilenko et al. (1970). Soon after the appearance of narrow-band tunable dye lasers, Doppler-free two-photon absorption was demonstrated independently by several groups for the case of sodium atoms (Biraben et al., 1974; Levenson and Bloembergen, 1974; Hänsch et al., 1974). In the meantime, Doppler-free two-photon measurements were performed for a number of atoms to study hyperfine splitting, collisional effects, and dynamic stark effects [for a review see Giacobino and Cagnac (1980)].

Doppler-free UV two-photon spectroscopy of molecules has long been limited to the small molecular systems Na_2 (Woerdman, 1976), CO and N_2 (Filseth et al., 1977), and NO (Timmermann and Wallenstein, 1981). In 1981, the first Doppler-free two-photon spectrum of a polyatomic molecule was published by Riedle et al. (1981). In this way rotational lines were resolved within the electronic spectrum of a polyatomic molecule yielding very precise structural information about the electronically excited state of this molecule (see Section 3.1.2).

5.3.2 Experimental Problems

5.3.2.1 The Light Source and Experimental Setup

In order to obtain high resolution in a two-photon experiment with counterpropagating light beams, the bandwidth of the exciting dye lasers has to be much smaller than the Doppler width of the molecular transition under investigation.

So far, the only laser system that provides continuously tunable light of a frequency width smaller than 1 GHz is a cw single-mode dye laser pumped by an Ar^+ or Kr^+ ion laser. The output power typically is several 100 mW for a single-mode ring laser. In the focus an intensity of about 10^4 W/cm^2 is obtained. This is smaller by about 5 orders of magnitude than the peak intensity of a pulsed nitrogen-pumped dye laser, which has proven to be a proper light source for two-photon measurements. However, it has been shown by Riedle et al. (1981) that this cw intensity is sufficient to observe Doppler-free two-photon signals not only in atoms but also in polyatomic molecules, even though the oscillator strength of the electronic transition is distributed over many thousands of rovibronic lines in the molecular spectrum.

For weakly fluorescing molecules, however, it is desirable to have an extremely narrow-band pulsed tunable laser source with a peak intensity higher by some orders of magnitude and thereby to increase the probability of the nonlinear two-photon process. Therefore, Riedle et al. (1982a) used a special amplification system for pulsed amplification of the narrow-band cw laser light and demonstrated that in this way highly resolved Doppler-free spectra can be obtained with higher sensitivity than with cw light.

The scheme of the experimental setup is shown in Fig. 5.8. The high resolution and the possibility for a precise linear scan are provided by a commercial cw ring laser system. The ring laser operates around 4900 Å with a coumarin 102 dye and is pumped by the violet lines of a Kr^+ ion laser. At this wavelength the output power is in excess of 100 mW. The wavelength of the cw light is continuously controlled by a wavemeter with an absolute wavelength accuracy of 0.03 cm^{-1} or 1 GHz.

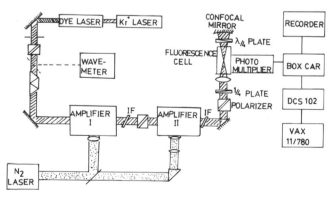

Fig. 5.8 Experimental setup for recording Doppler-free two-photon electronic spectra of molecules. The single-mode cw dye laser output is amplified in two stages and pumped by the pulsed light of the N$_2$ laser. The resulting light pulses are used for the two-photon excitation of the molecules. [From Riedle et al. (1982).]

The bandwidth of the ring laser is about 1 MHz, and therefore it would allow a resolution far below the Doppler width if it is used for two-photon experiments directly. For amplification a dye amplifier system is pumped by a nitrogen laser with a pump power of about 700 kW. The cw laser system is optically decoupled from the amplifier system by a direct vision prism. The general design of the amplifier system corresponds to that published by Salour (1977). Here, for molecular spectroscopy, narrow-band transmission interference filters (FWHM: 30 Å) are used for decoupling the two amplifier stages and suppressing stimulated emission from the amplifier stages. The transmission maximum can be easily tuned over a wavelength range of 300 Å by tilting the filter. For simplicity, only two amplifier stages (C 485 in dioxane) have been used, yielding up to 100-kW light pulses with a pulse length of 5 ns and a nearly Fourier transform limited bandwidth of 100 MHz.

In molecular spectroscopy, broad-band stimulated emission from the amplifiers would be a severe problem since the absorption spectrum of molecules is broad and two-photon absorption by the broad-band light would take place simultaneously and lead to a large background in the Doppler-free spectrum.

The total power of the broad stimulated emission is smaller than the power of the narrow-band amplified light by a factor of 20. This ratio is by far sufficient to guarantee that there be no background due to a two-photon absorption of the broad-band stimulated emission in the setup shown in Fig. 5.8.

The pulsed narrow-bandwidth laser light is focused into a fluorescence cell having a low stray light level and containing the molecular gas (C_6H_6) under a pressure of some Torr. The divergent light beam is then refocused by an adjustable confocal mirror so that a complete overlap of the oppositely propagating beams is obtained and a molecule placed within the common focus is able to absorb one photon each from the counterpropagating light beams. By inserting a Glan polarizer and a quarter-wave retardation plate the light beam is either linearly or circularly polarized. The direction of the circular polarization of the backward reflected light beam can be changed by inserting a second $\lambda/4$ retardation plate.

The UV fluorescence from the two-photon excited molecules is observed with a solar blind photomultiplier whose signal is fed into a boxcar integrator. The integrated signal is recorded on a strip chart recorder and at the same time transferred to a data processing system where it is digitized, stored on floppy discs, and then transferred to a larger computer. This procedure enables one to obtain the complete two-photon spectrum (made from several individual overlapping scans of the dye laser) with suitable scaling factors.

5.3.2.2   The Doppler-Broadened Background

A fundamental problem in Doppler-free two-photon spectroscopy is the Doppler-broadened background. A background appears if molecules absorb two photons propagating in the same direction, i.e., from one laser beam (Biraben *et al.*, 1974; Levenson and Bloembergen, 1974; Hänsch *et al.*, 1974). If the spacing of the molecular transitions is much larger than the Doppler width, as is the case for small molecules, then the Doppler-free transition is placed on a Gaussian background whose width is the Doppler width. For larger polyatomic molecules several transitions are located within the Doppler width and one observes the sum of all Doppler-broadened contributions. This is a severe problem since the intensity ratio of Doppler-free transition and Doppler-limited background a priori is not as good as for the case of only one line located within the Doppler width.

For two counterpropagating light beams having the same polarization, it is readily shown that the area of the Doppler-free line is twice the area of the corresponding Doppler-broadened contribution (Giacobino and Cagnac, 1980; Demtröder, 1981). Of course, the ratio of peak amplitudes is much better because of the drastically reduced width of the Doppler-free line. However, if 10 transitions of equal strength are located closely in a small frequency interval of 100 MHz, then at this point the amplitude of the Doppler-broadened (> 1 GHz) background is increased by a factor of 10 due to the superimposed 10 contributions to the background. Thus the amplitude ratio of Doppler-free and Doppler-broadened background has decreased by a factor of 10.

Therefore, in order to obtain a good signal-to-background ratio it is important to suppress the Doppler-broadened background for the dense spectra of polyatomic molecules, particularly when the resolution is not very high ($\gtrsim 50$ MHz).

5.3.2.2.1   Polarization Behavior and the Doppler-Broadened Background
The Doppler-broadened background can be suppressed if the absorption of two photons from one laser beam is less probable than the absorption of one photon from each laser beam. This can be achieved due to either different frequencies or different polarizations of the two oppositely propagating laser beams. If only one laser is used, as in Fig. 5.9, then only different photon polarizations are possible.

As an example, in Figs. 5.9 and 5.10 the high-resolution two-photon spectrum of C_6H_6 vapor is shown, as measured by using the pulsed light source shown in Fig. 5.8 under diverse polarization conditions. In both figures a small part of the $S_1 \leftarrow S_0$ spectrum is shown. This is the blue edge of the Q-branch ($\Delta J = 0$) of the totally symmetric $(A_{1g})14_0^1 1_0^1$ (Wunsch

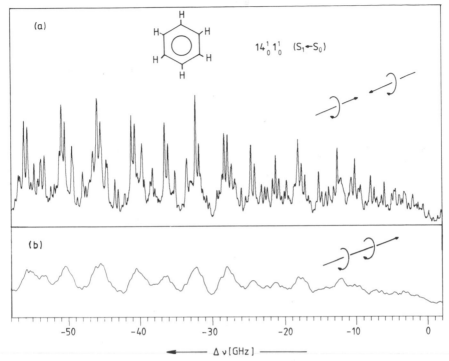

Fig. 5.9 Part of the two-photon electronic $S_1 \leftarrow S_0$ spectrum of benzene (C_6H_6). The blue edge of the Q-branch of the $14_0^1 1_0^1$ band ($A_{1g} \leftarrow A_{1g}$) is shown. If recorded (a) with one (circularly polarized) laser beam, a Doppler-broadened spectrum is seen, and (b) with two oppositely propagating beams, Doppler-free contributions are also seen. [From Riedle *et al.* (1982a).]

et al., 1977) transition induced by the vibration ν_{14} (b_{2u}) in the $^1B_{2u}$ electronic state (see also Section 4.2.2 and Fig. 4.7).

To record Fig. 5.9b only one (circularly polarized) laser beam was used. The two photons that were absorbed have the same direction of propagation and therefore a Doppler-broadened spectrum is seen (Doppler width = 1.7 GHz).

For Fig. 5.9a the light wave was reflected back into the fluorescence cell. Whereas the incoming wave has right-hand circular polarization, the reflected wave has left-hand circular polarization. However, in the molecular reference frame the two beams are polarized in the same sense (⟩⟩).

A comparison of Fig. 5.9a and b immediately reveals that the broad structure in Fig. 5.9a is a Doppler-broadened background due to the absorption of two photons from one laser beam. The sharp peaks on top of this broad background represent the Doppler-free spectrum of the molecule

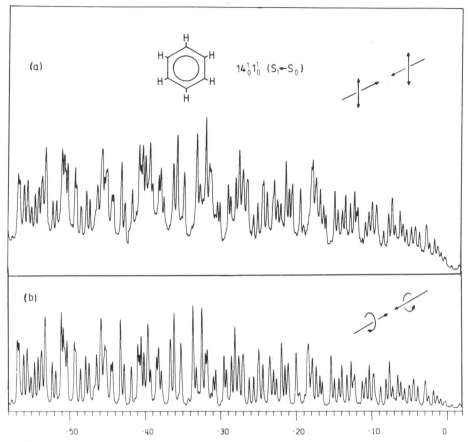

Fig. 5.10 Part of the Doppler-free two-photon spectrum of benzene (C_6H_6) ($A_{1g} \leftarrow A_{1g}$). Different photon polarizations of the two laser beams were used. (a) Linear polarization gives an isotropic spectrum with a strong Doppler-broadened background. (b) Use of countercircularly polarized photons eliminates this background. [From Riedle *et al.* (1982).]

obtained by the absorption of one photon from each of the two oppositely propagating light beams.

For this polarization the Doppler-broadened background has a strong intensity, and to obtain the correct relative heights of the Doppler-free peaks the background has to be subtracted from the measured spectrum.

A signal that is stronger by one order of magnitude is obtained when both photons are linearly polarized. The spectrum recorded in this way (↑↑) is shown in Fig. 5.10a, whereas in Fig. 5.10b both photons (in the molecular system) are circularly polarized but in opposite senses (↺ ↻). The reason for the stronger intensity is the appearance of the strong isotropic part of the spectrum, which has been explained in detail in Section 4.2.2. The spectra in Fig. 5.10a and b are identical, the only difference being an additional strong Doppler-broadened background in the case of the linear (↑↑) polarization.

Even though Figs. 5.9 and 5.10 show the same frequency range of the C_6H_6 spectrum, the measured Doppler-free and Doppler-broadened spectra are completely different under different polarization conditions (see Section 4.2.2. and Fig. 4.7). A Doppler-free spectrum that is not perturbed by the Doppler-broadened background is observed only for the case of counter-circularly polarized light (⟩ ⟨) shown in Fig. 5.10b.

For a brief explanation of these results, we consider the general expression for the two-photon absorption probability [Eq. (4.2.49)]. As in the Raman effect, there exist three tensor contributions to the two-photon transition probability: the isotropic, antisymmetric anisotropic and symmetric anisotropic contributions. The different contributions have been investigated theoretically and experimentally for liquids (Monson and McClain, 1970) and for gases (Metz et al., 1978; Hampf et al., 1977). For this discussion it is important that the isotropic contribution exist only for the Q-branch ($\Delta J = 0$) of totally symmetric two-photon transitions and be observed only for the photon polarizations ↑↑ and ⟩ ⟨ . On the other hand, symmetric anisotropic contributions exist for all allowed rotational branches $\Delta J = \pm 2, \pm 1, 0$ of all two-photon transition symmetries, whereas antisymmetric contributions do not exist for photons of equal frequency and have not been observed in benzene (Hampf et al., 1977). Symmetric anisotropic contributions do appear for all possible photon polarizations. From the results in Section 4.2.2 and Table 4.3 it is clear that the choice of photon polarizations allows for the selection of isotropic and anisotropic parts of the spectrum. The Doppler-broadened background can be eliminated if the contribution to the signal strength from the absorption of two photons from one laser beam is small. For s–s transitions in atoms this is easily obtained by using countercircularly polarized (⟩ ⟨) light (Biraben et al., 1974). The situation is more complicated in molecular spectroscopy, because the anisotropic component of the two-photon tensor leads to an absorption in the circularly polarized (⟩ ⟩) light of one laser beam. Fortunately, for C_6H_6 (Wunsch et al., 1977) and other aromatic molecules (Boesl et al., 1976; Robey and Schlag, 1978) this anisotropic component is smaller by nearly one order of magnitude than the corresponding isotropic component of the Q-branch. For the photon polarization in Fig. 5.10b the Doppler-broadened two-photon absorption from one laser beam can originate only from the anisotropic contribution and therefore is accordingly smaller than the isotropic contribution. The Doppler-broadened background is too small to be observed in Fig. 5.10b. In summary, the best experimental conditions for suppressing the Doppler-broadened background are found to be given for countercircularly polarized ($\circlearrowright$ $\circlearrowleft$) light in isotropic totally symmetric two-photon transitions (Q-branch of a $A_{1g} \leftarrow A_{1g}$ two-photon transition). On the other hand, for pure anisotropic two-photon absorption bands (O, P, R, S-branches of $A_{1g} \leftarrow A_{1g}$ transitions and all branches of $E_{1g} \leftarrow A_{1g}$ or $E_{2g} \leftarrow A_{1g}$ transitions in D_{6h}) the best ratio of Doppler-free signal to Doppler-broadened background is obtained for ($\circlearrowright$ $\circlearrowright$) polarization.

5.3.2.2.2   Two-Photon Absorption in Laser Beams of Different Wavelength.

Another possible way to eliminate the Doppler-broadened background is to allow the molecules to absorb only two photons with opposite propagation directions by choosing slightly different frequencies for the two light beams. Then, if the absorption of two equal photons from one laser beam is not possible, since the resonance condition for the two-photon energy is not fulfilled, no Doppler-broadened background will appear. But now, because of the slightly different Doppler shifts, a residual Doppler-broadening of the observed transition occurs. However, this is of importance only for ultrahigh resolution (< 10 MHz) experiments with cw lasers.

Doppler-free experiments with different photon energies were first demonstrated for the case of atoms (Liao and Bjorkholm, 1976). Here the reason for using two different wavelengths was to match the resonant intermediate states of the two-photon processs and to study the Stark effect and signal enhancement related to resonance. An application to molecular spectroscopy was demonstrated by Riedle et al. (1981) who used two cw lasers. One was a linear single-mode dye laser with an output power of about 10 mW and was tunable at about 4943 Å. This power level is too low for an efficient two-photon absorption in benzene. A cw Ar^+ ion laser of high power provided the second photon of fixed energy for the two-photon process. This Ar^+ ion laser was operated with a temperature-controlled internal etalon at the 5145 Å line in single mode with a linewidth of about 10 MHz. The output power was about 300 mW, which increases the two-photon absorption probability by a factor of 30 compared to a two-photon experiment with only the 10 mW single-mode dye laser.

The two counterpropagating laser beams were carefully aligned and brought to a common focus within a vacuum cell containing benzene vapor. In this experiment it is possible to eliminate the Doppler-broadened background if the resonance condition is not fulfilled for absorption of two photons from one laser beam. This is easily achieved for atoms because their spectra consist only of single sharp lines which lead in turn to sharp resonance conditions. For polyatomic molecules the spectrum is broad due to the many vibrational and rotational transitions and the resonance condition may be accidentally given for all photon combinations $v_1 + v_2$, $2v_1$, and $2v_2$ in a two-color experiment. In the experiment of Riedle et al. (1981) the wavelengths were selected for resonance at $v_1 + v_2$, i.e., the absorption of one photon from each laser beam gives a considerable contribution to the two photon signal.

Absorption of two photons from the tunable dye laser beam $(2v_1)$ generally is not critical because of the low power of 10 mW. A more severe problem is the absorption of two photons from the Ar^+ ion laser beam. When using the 5145 Å line, the energy of two photons is not sufficient for an electronic excitation starting from the zero ground-state level, and therefore two-photon absorption is in the very weak hot-band region of the two-photon spectrum (Wunsch et al., 1975). Hence there is only a constant 10% background in the observed spectrum due to this effect.

With this setup, the two two-photon bands 14_0^1 and $14_0^1 1_0^1$ located at 39,656 and 40,578 cm^{-1}, respectively, have been measured under high resolution, and single rotational transitions have been resolved (see Section 5.3.3.1.2). For a fixed Ar^+ ion laser frequency of 19,436 cm^{-1} the frequencies of the tunable dye laser have to be 20,220 and 21,142 cm^{-1}. The general expression for the residual Doppler width is

$$\Delta v_D = (v_1 - v_2)\bar{u}/c = 7.163 \times 10^{-7}\sqrt{T/M}(v_1 - v_2) \qquad (5.3.2)$$

where $\bar{u}$ is the mean velocity of the molecules, T the temperature in degrees Kelvin, and M the molecular weight in atomic mass units. For the above values, from Eq. (5.3.2) we expect a residual Doppler width of 33 MHz for the 14_0^1 transition and 72 MHz for the $14_0^1 1_0^1$ transition.

In order to determine the residual Doppler width of rotational transitions in both vibronic bands, the linewidth of the rotational transition $J = 11$, $K = 8$ in the Q-branch of the 14_0^1 band and $J = 12$, $K = 11$ in the Q-branch of the $14_0^1 1_0^1$ band have been measured for different pressures. The resulting Stern–Vollmer plot is shown in Fig. 5.11. From this we extrapolate a linewidth at zero pressure of 57 and 97 MHz for the 14_0^1 and the $14_0^1 1_0^1$ vibronic bands, respectively. This is in good agreement with the calculated residual Doppler width if we take into account a contribution of about 20 MHz from the frequency width of the Ar^+ laser, dye laser, and transient time broadening.

The result demonstrates that it is possible to obtain reasonable spectral resolution below 100 MHz without a Doppler-broadened background in a two-color experiment. This resolution is sufficient to resolve most of the rotational lines within the electronic spectrum of polyatomic molecules (see Section 5.3.3.1) and is comparable to the spectral resolution obtained with the pulsed amplified light of a cw laser.

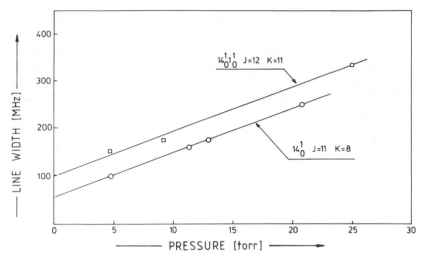

Fig. 5.11 Measured linewidth of two single rotational lines as a function of benzene (C_6H_6) vapor pressure. The two rotational lines are from the vibronic bands 14_0^1 (O) and $14_0^1 1_0^1$ (□), respectively, with different excitation energies. The extrapolated linewidth for zero pressure is mainly given by the residual Doppler broadening due to the use of two photons of different frequency in this experiment.

5.3.3 Experimental Results

5.3.3.1 Rotational Analysis

5.3.3.1.1 General Remarks It is well known that the UV spectrum of polyatomic organic molecules is of band type; i.e., single rotational transitions normally are not resolved and the envelope of these rotational transitions forms the typical rotational contour of a vibronic band. The grouping of many such rotational transitions is the reason for the "microstructure" of a vibronic band that is resolved under an instrumental resolution of about 500,000. A further increase of the instrumental resolution is not expected to increase the effective resolution in the gas-phase spectrum due to the fundamental barrier of the Doppler-broadening. So far, rotational analysis of high-resolution spectra of organic molecules (Ross, 1971; Callomon *et al.*, 1965) has been performed on the basis of the microstructure rather than on the analysis of single rotational lines. Now, Doppler-free two-photon spectroscopy allows us to look behind the Doppler barrier and to increase the spectral resolution by two orders of magnitude. It will be shown in this Section that this increase of resolution reveals new spectroscopic as well as dynamic phenomena that would not have been detected with conventional high-resolution spectroscopic techniques.

The increased resolution possible in Doppler-free two-photon spectroscopy is illustrated in Fig. 5.6, in which part of the two-photon spectrum of benzene (C_6H_6) is shown for different spectral resolutions. The upper trace represents the fundamental vibronic bands of the $S_1 \leftarrow S_0$ two-photon spectrum of C_6H_6 with the inducing vibrations v_{17}, v_{18}, v_{15}, and v_{14} (see Section 5.1). The instrumental resolution of 1 cm^{-1} allows resolution of the rotational branches of the vibronic bands. In the middle trace a magnified portion of the upper spectrum is shown. This is the Q-branch of the 14_0^1 two-photon transition. The resolution here is increased to 0.06 cm^{-1} (Lombardi *et al.*, 1976), the largest resolution possible with Doppler broadening. Here the typical microstructure of the rotational band contour is resolved. Finally, for comparison, in the lowest trace, the result of the Doppler-free two-photon spectroscopy is presented as measured by Riedle *et al.* (1981). Here the blue edge of the Q-branch is magnified by a factor of 47. The vibronic spectrum, which is normally a band-type spectrum, is now completely resolved and displays a sharp line character that is typically observed for much smaller molecules with smaller moments of inertia. The line positions and shapes can be measured very accurately to reveal new structural and dynamic information.

5.3.3.1.2 Quantitative Analysis of the Rotationally Resolved Doppler-Free Two-Photon Spectrum of C_6H_6
In conventional high-resolution spectroscopy there has been no hint of a deviation in the structure of the C_6H_6 molecule from a planar hexagon (D_{6h}) neither in the $^1A_{1g}$ ground state nor in the $^1B_{2u}$ excited state. Hence for the assignment of the spectrum it is assumed initially to be a planar oblate symmetric top structure in the electronically excited state as well as in the ground state. The energy displacement ΔE from the pure electronic transition frequency is then given by (Herzberg, 1966)

$$\Delta E = (B'' - B')[J(J + 1) - \tfrac{1}{2}K^2], \qquad (5.3.3)$$

where $A'' = B'' = 2C''(A' = B' = 2C')$ are the equilibrium rotational constants of the symmetric top in the electronic ground (excited) state. $I_A = I_B = \hbar/4\pi cB$ and $I_C = 2I_A$ are the moments of inertia along the principal axes of the molecule. It is easily seen that the planarity condition

$$I_C - I_A - I_B = 0 \qquad (5.3.4)$$

is fulfilled for the above assumptions.

Strictly speaking, Eq. (5.3.3) is valid only for the equilibrium configuration of a planar symmetric top. This equilibrium structure, however, cannot be observed in rotational spectroscopy since the zero point vibrations are always excited. For conventional high-resolution UV spectroscopy of C_6H_6

this restriction has been neglected (Callomon et al., 1965) due to the limited resolution and accuracy of conventional spectroscopy.

On the basis of Eq. (5.3.3), a computer calculation of the corresponding part of the spectrum was performed. The relative line intensities have been obtained by taking into account the population of the rotational ground state, the M degeneracy, and the statistical weights of the transitions due to nuclear spin statistics (Wilson, 1935). Here the rotational constants $B_0'' = 0.189762$ cm^{-1} (Pliva and Pine, 1982) is taken from high-resolution infrared (IR) measurements, whereas $B_v' = 0.18134$ cm^{-1} (Lombardi, et al., 1976), is the result of Doppler-limited two-photon measurements that are the most accurate excited-state measurements known from the literature. The resulting lines are then convoluted with a Gaussian profile of 30 MHz half-width yielding the best agreement with the measured spectrum. Part of this calculated spectrum is shown in Fig. 5.12a. Figure 5.12b shows our ultrahigh resolution results for the Q-branch of the $14_0^1 1_0^1$ band for comparison. A comparison of both spectra shows rough qualitative agreement in the gross features, such as line positions and strengths. However, there is one fundamental disagreement between this calculation and the Doppler-free measurements.

From Eq. (5.3.3) it can be seen that for a planar symmetric rotor there are particular transitions with J_i, K_j that are accidentally degenerate in energy with other transitions J_k, K_l, independent of the value $B'' - B'$. For example, the transitions with $J = 13$, $K = 13$ and $J = 10$, $K = 5$, with intensity ratio of $1:0.75$, have the same energy according to Eq. (5.3.3).

Actually, in the measured spectrum these two transitions are found to be separated by 0.16 GHz (see Fig. 5.12b). Since a splitting of the accidentally degenerate lines in the spectrum, that is due to centrifugal distortion can be ruled out (Riedle et al., 1981), it is evident that the observed splitting is produced by an inertial defect either of the ground-state rotational constants B_0'' and C_0'' or of the excited-state rotational constants B_v' and C_v' or both.

Unfortunately, there are no precise measurements of C_0'' independent of B_0'' neither from Raman spectroscopy (Jensen and Brodersen, 1979) nor from IR spectroscopy (Pliva and Pine, 1982). Therefore, for a working hypothesis we have to assume that an inertial defect for the ground-state constants B_0'' and C_0'' is small and may be neglected. Then the observed features in the spectrum are due only to an inertial defect

$$\Delta = (\hbar/4\pi c)(1/C_v' - 2/B_v') \neq 0 \qquad (5.3.5)$$

of the excited-state rotational constants.

Here it has been assumed that $A_v' = B_v'$, which has been experimentally confirmed by Riedle et al. (1981) on the basis of Doppler-free two-photon

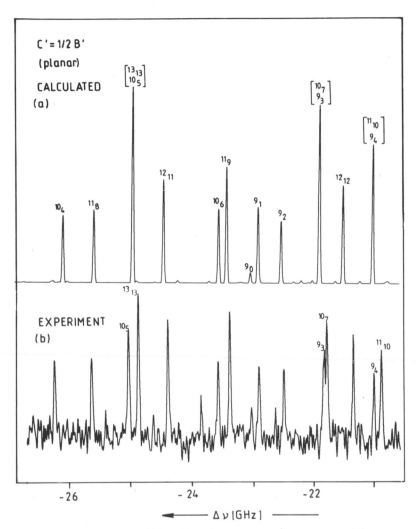

Fig. 5.12 (a) Calculated and (b) measured part of the Q-branch of the $14_0^1 1_0^1$ transition in benzene (C_6H_6). Several lines that are accidentally degenerate in the calculated spectrum under the assumption of $C' = B'/2$ are found to be split in the measured spectrum.

spectroscopy of the 14_0^1 transition. The values for Δ found from Doppler-free two-photon spectra of three vibronic bands under the above-mentioned assumption together with the measured values for the rotational constants B'_v and C'_v are summarized in Table 5.2. For all three bands the value of Δ is negative and, in particular, the value for the 14_0^1 transition is relatively large for an organic ring molecule. The main question that arises is whether

TABLE 5.2

Rotational Constants B'_v and C'_v for Various Vibronic Bands of the First electronically Excited State of Benzene (C_6H_6)

Band	B'_v (cm^{-1})	C'_v (cm^{-1})	$B'_v/2 - C'_v$ (cm^{-1})	Δ (amu A^2)
$14^1_0{}^a$	0.18133 (4.5)^c	0.09074 (8)	-7.5×10^{-5}	-0.154
$14^1_0 1^1_0{}^b$	0.18122 (5)	0.09064 (8)	-3.0×10^{-5}	-0.062
$14^1_0 10^2_0{}^b$	0.18144 (5)	0.09075 (8)	-3.0×10^{-5}	-0.062

^{a} For the calculation, ground-state values of Pliva and Pine (1982) are used. Also given is the difference $B'_v/2-C'_v$ and the inertial defect $\Delta = (\hbar/4\pi c)(1/C'_v - 2/B'_v)$.
^{b} From Riedle *et al.* (1982a).
^{c} The expression 0.18133 (4.5) means 0.18133 ± 0.000045.

this inertial defect may be attributed solely to an anharmonic coupling or to rotation–vibration interaction, or whether there exists already an inertial defect in the excited-state equilibrium configuration. There are several ways to come closer to an answer to this question, both experimentally and theoretically. Experimentally, it is necessary to measure precise rotational constants for other vibronic bands as well as for other benzene isotopes. Theoretically, from force field calculations (Robey and Schlag, 1977) one might obtain a reasonable value for the magnitude of the rotation–vibration interaction and then compare it with the observed values. Certainly, this is an important and advantageous task since precise structural information may offer important hints for the understanding of radiationless processes in the first electronically excited state of benzene and various other polyatomic molecules.

5.3.3.2 Linewidth Measurements

Since the first order Doppler shift increases linearly with the light frequency, the importance of broadening due to the Doppler effect is greater in UV spectroscopy than it is in IR spectroscopy, for example. As a consequence, in conventional UV spectroscopy of polyatomic molecules it is barely possible to observe the homogeneous collisional width of defined single molecular transitions below pressures of 100 Torr. The inhomogeneous Doppler width is larger than 1.5 GHz, and therefore remarkable additional homogeneous broadening can be observed only at these high pressures. In addition, the problem of closely spaced rovibronic transitions arises, which has already been discussed in Section 5.3.3.1.1. To sum up, because of the Doppler broadening in conventional gas-phase spectroscopy of polyatomic molecules, it has not been possible to observe the homogeneous linewidth of a single transition, and therefore dynamical information has not been available from spectral linewidth measurements. As a result, dynamic information in the gas phase was obtained solely from population decay-time measurements in this spectral regime. (There is, of course, the exception of very short-lived excited states, which lead to a broadening of more than the Doppler width due to the uncertainty principle. However, for a polyatomic molecule with closely spaced rovibronic transitions the observation and separation of defined quantum states is no longer possible.)

Doppler-free two-photon spectroscopy eliminates the inhomogeneous broadening produced by molecular groups with different velocities. After elimination of the Doppler broadening in the gas phase, the linewidth of a rovibronic transition is governed by collisional broadening and by the collisionless linewidth due to the finite lifetime of the excited state of the isolated molecule. For benzene, with an excited-state lifetime in the range of some 100 ns (Spears and Rice, 1971), collisional broadening is the dominant effect for pressures above 1 Torr. This is illustrated in Fig. 5.13, in which the same part of the Q-branch of the 14_0^1 two-photon transition is shown for three different gas pressures (20.8, 13.0, and 4.8 Torr). Generally, single rovibronic transitions in benzene can be observed only at pressures less than 4 Torr. But there exists a particular single rotational line ($J = 11$, $K = 8$), with a distance of nearly 1 GHz from neighboring lines, which is marked by an arrow in Fig. 5.13. Its linewidth can be measured over a large pressure range. From similar measurements the collisional line-broadening parameter of this particular eigenstate, due to C_6H_6—C_6H_6 collisions (Riedle and Neusser, 1984), was found to be 15 MHz/Torr, a result which may now be compared with those from fluorescence quantum yield and decay-time measurements.

Next, it would be of interest to investigate the remaining linewidth when the pressure is zero and the collisional broadening is negligible. The corresponding (collisionless) linewidth contains dynamical information about the excited state within the molecule without interaction with neighboring molecules. Then a comparison of population decay-time measurements with linewidth measurements would be possible, as has already been made for the case of atoms (Giacobino and Cagnac, 1980). At present, the signal-to-noise ratio does not allow analysis at pressures less than 1 Torr, for which the collisionless linewidth is expected to be observed for the prototype molecule benzene. However, an increase in laser intensity and narrow-band amplification should bring us closer to this goal, which represents an important step for the understanding of the dynamical behavior of excited states of polyatomic molecules.[2]

---

[2] Recently, Riedle and Neusser (1984) have measured collisionless, homogeneous linewidths of two-photon states in S_1-benzene. They showed that collisionless linewidths strongly varied with the rotational quantum number for constant vibrational excess energy.

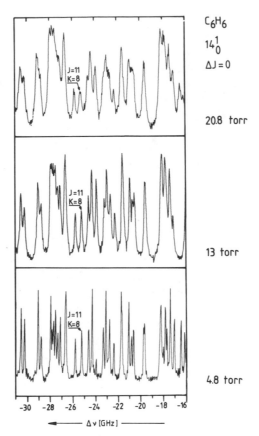

Fig. 5.13 Part of the two-photon spectrum of benzene (C_6H_6) as measured for different gas pressures.

5.3.3.3 *Information about Nonradiative Processes from Doppler-Free Two-Photon Spectra*

As we have discussed in the preceding section, the homogeneous linewidth in general contains dynamical information about the excited molecular states. In this section, we shall show that sometimes it is also possible to obtain information about dynamical behavior from intensity changes and perturbations in the spectrum, even though the natural linewidth is not observed. However, it is necessary to resolve the different rovibronic lines.

Riedle *et al.* (1982b) have succeeded in measuring the high-resolution (sub-Doppler) excitation spectra of $14_0^1 1_0^1$ and $14_0^1 1_0^2$ bands in the $S_1 \leftarrow S_0$ system of C_6H_6 by using the multiphoton technique. These spectra are shown in Fig. 5.14. We can see that many rovibronic lines that are present

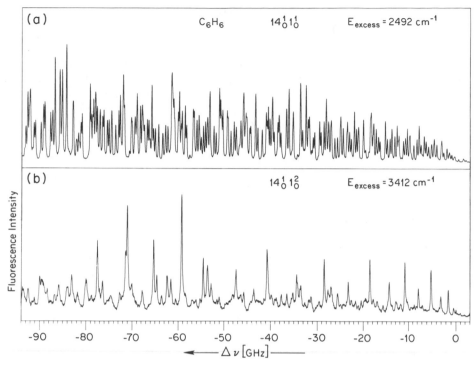

Fig. 5.14 Doppler-free two-photon fluorescence excitation spectra of benzene (C_6H_6). (a) Part of the Q-branch of the $14_0^1 1_0^1$ band at an excess energy of 2492 cm^{-1}; (b) corresponding part of the $14_0^1 1_0^2$ band at 3412 cm^{-1}. Indicated are the $K = 0$ lines. [From Riedle et al. (1982b).]

in the $14_0^1 1_0^1$ spectrum are missing in the $14_0^1 1_0^2$ spectrum. Since the intensity of various rovibronic lines in fluorescence excitation spectra is proportional to the quantum yields of these rovibronic states, they attribute this absence of spectral lines to the opening of a new nonradiative channel, in this case intramolecular vibrational relaxation (IVR). Some spectral lines in the $14_0^1 1_0^2$ spectrum have been assigned as rovibronic transitions. A main feature here is that in the first part of the spectrum the $K = 0$ lines are very pronounced.

There has been considerable interest in the theoretical analysis of systems of coupled anharmonic oscillators (Noid et al., 1981; Heller and Davis, 1982; Hamilton et al., 1982; Kosloff and Rice, 1981; Strott et al., 1980; Shapiro and Child, 1982). As energy is increased, many classical systems are known to undergo a transition from regular to stochastic motion. Considerable effort in this area has been devoted to discovering the phenomenology of the analogous quantum transition. In this connection, the density matrix method has been applied to treat coupled anharmonic systems (Lin, 1982). For this purpose, the adiabatic approximation has been employed to find the basis set (Lin, 1980b; Shapiro and Child, 1982). It has been shown that the energy eigenvalues obtained from the adiabatic approximation are in very good agreement with exact ones and are on the whole better than those obtained from the SCF calculation (Shapiro and Child, 1982; Zhi-Ding *et al.*, 1982). Furthermore, the adiabatic approximation can provide analytical expressions for the wave functions and energy eigenvalues of a coupled anharmonic system. It has been shown by using the Schrödinger equation method that both the dephasing and amplitude of wave functions play important roles in determining the quantum stochasticity of coupled anharmonic oscillators (Shapiro and Child, 1982). It should be noted that the density matrix method can serve this purpose very well; the diagonal elements of the density matrix of the system describe the evolution of the population of the system, while the off-diagonal elements of the density matrix provide information about the phase evolution of the system. With the density matrix method, the time-dependent behavior of the memory function can be used to determine whether a system undergoes regular or stochastic motion.

Next we shall apply the theory of IVR (Rettschnick, 1980; Parmenter, 1982) based on the application of the adiabatic approximation to the IVR of C_6H_6 from the $14_0^1 1_0^2$ spectrum. For convenience of discussion this theory is outlined in Appendix III.B.

Notice that the IVR of a vibrational mode q_i can in general be induced by the Born–Oppenheimer coupling $\hat{H}'_{BO}$ obtained with the anharmonic potential $(1/3!) V_{i\alpha\beta} q_i Q_\alpha Q_\beta$ and by the Coriolis coupling $\hat{H}'_{Cor}$. In other words, the IVR rate from a rotational level (JK) of the $14_0^1 1_0^2$ spectrum of C_6H_6 should consist of that contributed from $\hat{H}'_{BO}$, $W^{(BO)}_{vnJK}$, and of that contributed from $\hat{H}'_{Cor}$, $W^{(Cor)}_{vnJK}$.

We first consider $W^{(BO)}_{vnJK}$:

$$W^{(BO)}_{vnJK} = \frac{2\pi}{\hbar} \sum_{n'J'K'} \left| \sum_{i\alpha\beta} \frac{1}{3!} V_{i\alpha\beta} \langle \Psi_{v'n'J'K'} | q_i Q_\alpha Q_\beta | \Psi_{vnJK} \rangle \right|^2$$

$$\times D(E_{v'n'J'K'} - E_{vnJK}). \tag{5.3.6}$$

Using the relations

$$\theta_{vnJK} = Y_{JK}(x) \prod_j \chi_{vnj}(Q_j) \tag{5.3.7}$$

and

$$\theta_{v'n'J'K'} = Y_{J'K'}(x) \prod_j \chi_{v'nj'}(Q_{j'}), \tag{5.3.8}$$

we obtain

$$W_{vnJK}^{(BO)} = \frac{2}{\hbar^2} \sum_{i\alpha\beta} \left|\frac{1}{3!} V_{i\alpha\beta}\right|^2 \left(\frac{\hbar V_i}{2\omega_i}\right) \mathrm{Re} \int_0^\infty dt\, e^{-it\omega_i} K_{n_\alpha}(t) K_{n_\beta}(t) \prod_j^{j \neq \alpha,\beta} G_{n_j}(t),$$

$$(5.3.9)$$

where

$$K_{n_\alpha}(t) = \sum_{n_\alpha'} |\langle \chi_{v'n_\alpha'} | Q_\alpha | \chi_{vn_\alpha} \rangle|^2$$
$$\times \exp[it(\omega_{n_\alpha'} - \omega_{n_\alpha}) - t\Gamma_{v'n_\alpha',vn_\alpha}] \qquad (5.3.10)$$

and

$$G_{n_j}(t) = \sum_{n_j'} |\langle \chi_{v'n_j'} | \chi_{vn_j} \rangle|^2 \exp[it(\omega_{n_j'} - \omega_{n_j}) - t\Gamma_{v'n_j',vn_j}]. \qquad (5.3.11)$$

As shown in Appendix III.B, for nontotally symmetric modes only frequency shifts can be introduced by anharmonic couplings; in this case, $G_{n_j}(t)$ is given by (see Appendix III.C)

$$G_{n_j}(t) = G_{n_j=0}(t) \exp[-n_j t(i\omega_j + \Gamma_j)]$$
$$\times \sum_{m_j=0}^{n_j} \frac{n_j!}{m_j!\{[(n_j - m_j)/2]!\}^2} \left(\frac{b_j}{2}\right)^{m_j} \left(\frac{a_j}{2}\right)^{n_j - m_j}, \qquad (5.3.12)$$

where $G_{n_j=0}(t)$ represents the function $G_{n_j}(t)$ at $n_j = 0$ and is defined by

$$G_{n_j=0}(t) = \frac{\exp[\frac{1}{2} it(\omega_j' - \omega_j)]}{\eta_j \{1 + \rho_j \exp[-2t(\Gamma_j - i\omega_j')]\}^{1/2}}, \qquad (5.3.13)$$

$$\omega_j = \omega_j(\{v_i\}), \qquad \omega_j' = \omega_j(\{v_i'\}), \qquad (5.3.14)$$

$$\rho_j = -\left(\frac{\omega_j' - \omega_j}{\omega_j' + \omega_j}\right)^2, \qquad (5.3.15)$$

$$\eta_j = \frac{\omega_j + \omega_j'}{(4\omega_j\omega_j')^{1/2}}, \qquad (5.3.16)$$

$$a_j = \frac{(\omega_j'^2 - \omega_j^2)\{1 - \exp[-2t(\Gamma_j - i\omega_j')]\}}{(\omega_j + \omega_j')^2\{1 + \rho_j \exp[-2t(\Gamma_j - i\omega_j')]\}}, \qquad (5.3.17)$$

$$b_j = \frac{2\exp[-t(\Gamma_j - i\omega_j')]}{\eta_j^2\{1 + \rho_j \exp[-2t(\Gamma_j - i\omega_j')]\}}. \qquad (5.3.18)$$

If Q_α is nontotally symmetric, then $K_{n_\alpha}(t)$ can be expressed in terms of $G_{n_\alpha}(t)$.

For totally symmetric modes, both vibrational coordinates and frequencies can be shifted by anharmonic couplings. Vibrational coordinate shifts are usually more important in IVR than vibrational frequency shifts. Ignoring the effect of vibrational frequency shifts, $G_{n_\beta}(t)$ defined by Eq. (5.3.11) can be expressed as

$$G_{n_\beta}(t) = \exp\left[-\tfrac{1}{2}\Delta_\beta^2(1 - e^{t(i\omega_\beta - \Gamma_\beta)})\right]$$

$$\times \sum_{m_\beta = 0}^{n_\beta} \frac{n_\beta!}{m_\beta![(n_\beta - m_\beta)!]^2}\left[\frac{\Delta_\beta^2}{2}(1 - e^{t(i\omega_\beta - \Gamma_\beta)})^2\right]^{n_\beta - m_\beta}$$

$$\times \exp\left[-it\omega_\beta(n_\beta - m_\beta) - t(n_\beta + m_\beta)\Gamma_\beta\right], \qquad (5.3.19)$$

where

$$\Delta_\beta^2 = \frac{\omega_\beta}{\hbar}\left[\sum_i \frac{\hbar V_{ii\beta}}{6\omega_i\omega_\beta^2}(v_i' - v_i)\right]^2. \qquad (5.3.20)$$

To apply the preceding results for $W_{vnJK}^{(BO)}$ to a single rovibrational level (SRVL) IVR rate of $14_0^1 1_0^2$, we should notice that $n_\beta = 2$, $n_\alpha = 0$, $n_j = 0$, and $v_i = 1$ (i.e., $v_i = 1 \rightarrow v_i' = 0$). In this case we have

$$K_{n_\alpha = 0}(t) = \frac{\hbar}{2\omega_\alpha}\exp\left[t(\Gamma_\alpha + i\omega_\alpha)\right]G_{n_\alpha = 1}(t). \qquad (5.3.21)$$

Here $G_{n_\alpha = 1}(t)$ can be obtained from Eq. (5.3.12)

From Eq. (5.3.9) we can see that the type of anharmonic potential $(1/3!)$ $V_{i\alpha\beta}q_iQ_\alpha Q_\beta$ plays the promoting role in $W_{vnJK}^{(BO)}$. Thus if q_i refers to the v_{14} mode (B_{2u}), then the symmetry condition requires that (Q_α, Q_β) can have only (A_{2g}, B_{1u}), (E_{1g}, E_{2u}), and (E_{1u}, E_{2g}). Notice that for C_6H_6 we have

$$v_3(A_{2g}) = 1210 \qquad v_{12}(B_{1u}) = 995 \qquad v_{10}(E_{1g}) = 585,$$
$$v_{16}(E_{2u}) = 237 \qquad v_{17}(E_{2u}) = 712 \qquad v_{18}(E_{1u}) = 923,$$
$$v_{19}(E_{1u}) = 1470 \qquad v_6(E_{2g}) = 522 \qquad v_8(E_{2g}) = 1469,$$
$$v_9(E_{2g}) = 1148 \quad (cm^{-1}).$$

On the other hand, if q_i refers to the v_1 mode (A_{1g}), then as long as Q_α and Q_β belong to the same symmetry, the symmetry condition in $(1/3!)$ $V_{i\alpha\beta}q_iQ_\alpha Q_\beta$ will be satisfied.

In obtaining $W_{vnJK}^{(BO)}$, as given by Eq. (5.3.9), the rotational effect has been ignored. The rotational effect in IVR in $W_{vnJK}^{(BO)}$ can be caused by the rotational–vibrational coupling due to the vibrational coordinate dependence of (I_x, I_y, I_z) and by the effect of the Coriolis coupling on the initial and final states involved in IVR. To this approximation, $W_{vnJK}^{(BO)}$ is independent of JK.

Next we consider $W_{vnJK}^{(Cor)}$. Note that for benzene, $\hat{M}_z$ belongs to A_{2g}, whereas $(\hat{M}_x, \hat{M}_y)$ belongs to E_{1g}. Thus we shall discuss the calculation of $W_{vnJK}^{(Cor)}$ due to the $\hat{M}_z$ component of $\hat{H}'_{Cor}$ and that due to the $(\hat{M}_x, \hat{M}_y)$ components of $\hat{H}'_{Cor}$ separately. For the case of the $\hat{M}_z$ component of $\hat{H}'_{Cor}$, using Eqs. (5.3.7) and (5.3.8) and the relations

$$\hat{H}'_{Cor}(M_z) = -\hat{M}_z \hat{m}_z / I_z \qquad (5.3.22)$$

and

$$\hat{m}_z = \sum_i \sum_\alpha \zeta_{i\alpha}^{(z)} (q_i P_\alpha - p_i Q_\alpha), \qquad (5.3.23)$$

we obtain

$$W_{vnJK}^{(Cor)}(\hat{M}_z) = \frac{2}{\hbar^2} \sum_{i\alpha} \left(\zeta_{i\alpha}^{(z)} \frac{K\hbar}{I_z} \right)^2 \left(\frac{v_i \hbar^2}{4} \right) \left(2 + \frac{\omega_\alpha}{\omega_i} + \frac{\omega_i}{\omega_\alpha} \right)$$

$$\times \operatorname{Re} \int_0^\infty dt \exp[t(-i\omega_i + i\omega_\alpha + \Gamma_\alpha)] G_{n_\alpha = 1}(t) G_{n_\beta = 2}(t)$$

$$\times \prod_j^{j \neq \alpha\beta} G_{n_j = 0}(t), \qquad (5.3.24)$$

where $G_{n_\alpha = 1}(t)$ and $G_{n_\beta = 2}(t)$ should be calculated from Eqs. (5.3.12) and (5.3.19), respectively.

From Eq. (5.3.24) we can see that for the case in which the IVR is induced by the $\hat{M}_z$ component of the Coriolis coupling, the SRVL IVR rate is proportional to K^2.

Now we consider the calculations of $W_{vnJK}^{(Cor)}$ due to the $(\hat{M}_x, \hat{M}_y)$ components of $\hat{H}'_{Cor}$. Noticing that in this case $I_x = I_y$, we obtain

$$W_{vnJK}^{(Cor)}(M_x, M_y) = W_{vnJK}^{(Cor)}(+) + W_{vnJK}^{(Cor)}(-), \qquad (5.3.25)$$

where $W_{vnJK}^{(Cor)}(+)$ represents the IVR rate constant for $(JK) \to (J, K+1)$,

$$W_{vnJK}^{(Cor)}(+) = \frac{2}{\hbar^2} \left(\frac{\hbar}{2I_x} \right)^2 \left(\frac{v_i \hbar^2}{4} \right) (J-K)(J+K+1) \sum_{i\alpha} \left(|\zeta_{i\alpha}^{(x)}|^2 + |\zeta_{i\alpha}^{(y)}|^2 \right)$$

$$\times \left(2 + \frac{\omega_i}{\omega_\alpha} + \frac{\omega_\alpha}{\omega_i} \right) \operatorname{Re} \int_0^\infty dt \exp[t(-i\omega_i + i\omega_\alpha + \Gamma_\alpha)]$$

$$+ it\Delta B(2K+1)] G_{n_\alpha = 1}(t) G_{n_\beta = 2}(t) \prod_j^{j \neq \alpha, \beta} G_{n_j = 0}(t), \qquad (5.3.26)$$

and $W_{vnJK}^{(Cor)}(-)$ represents the IVR rate constant for $(JK) \rightarrow (J,\ K-1)$,

$$W_{vnJK}^{(Cor)}(-) = \frac{2}{\hbar^2} \left(\frac{\hbar}{2I_x}\right)^2 \left(\frac{v_i \hbar^2}{4}\right)(J+K)(J-K+1)\sum_{i\alpha}(|\zeta_{i\alpha}^{(x)}|^2 + |\zeta_{i\alpha}^{(y)}|^2)$$

$$\times \left(2 + \frac{\omega_i}{\omega_\alpha} + \frac{\omega_\alpha}{\omega_i}\right) \mathrm{Re} \int_0^\infty dt \exp[t(-i\omega_i + i\omega_\alpha + \Gamma_\alpha)$$

$$+ it\,\Delta B(-2K+1)]G_{n_\alpha=1}(t)G_{n_\beta=2}(t)\prod_j^{j\neq\alpha,\beta} G_{n_j=0}(t). \quad (5.3.27)$$

Here the following energy expression for a rigid rotator has been used:

$$E_{JK} = \hbar BJ(J+1) + \Delta B\hbar K^2. \quad (5.3.28)$$

From the viewpoint of symmetry, for the IVR of $14_0^1 1_0^1$, if q_i refers to the v_{14} mode, which has a symmetry of B_{2u}, then in the $\hat{M}_z$ component of $\hat{H}'_{Cor}$ the Q_α modes have B_{1u} symmetry; in this case there is only $v_{12} = 936$ cm^{-1}, which has a frequency lower than $v_{14} = 1567$ cm^{-1}. Similarly, in the $(\hat{M}_x, \hat{M}_y)$ components of $\hat{H}'_{Cor}$, the Q_α modes have E_{2u} symmetry; in this case $v_{16} = 237$ cm^{-1} and $v_{17} = 712$ cm^{-1} can be the Q_α modes in $W_{vnJK}^{(Cor)}$. Next, if q_i refers to the v_1 mode, which has a symmetry of A_{1g}, then in the $\hat{M}_z$ component of $\hat{H}'_{Cor}$ the Q_α modes have A_{2g} symmetry; in this case there is only one A_{2g} mode, that is, $v_3 = 1246$ cm^{-1}, which will not be very effective in promoting IVR. Similarly, in the $(\hat{M}_x, \hat{M}_y)$ components of $\hat{H}'_{Cor}$, the Q_α modes have E_{1g} symmetry; in this case the $v_{10} = 585$ cm^{-1} mode can be a promoting mode in $W_{vnJK}^{(Cor)}$.

From Appendix III.B, we see that only totally symmetric modes can have vibrational coordinate and frequency displacements induced by anharmonic couplings; because of these displacements, the totally symmetric modes can accept, in principle any number of quanta of excitation energy within the limits of energy conservation. On the other hand, the vibrational modes of other symmetries can have only vibrational frequency shifts induced by anharmonic couplings, and thus these modes can accept only even numbers of quanta of vibrational energy if these particular modes are initially not excited.

Now we are ready to apply the IVR theory to interpret the spectra shown in Fig 5.14. As mentioned before, a main feature in this spectrum is that the $K = 0$ spectral lines are very pronounced (see Fig. 5.15). From Eq. (5.3.24) we can see that the SRVL IVR rate due to the $\hat{M}_z$ component of $\hat{H}'_{Cor}$ is proportional to K^2. Thus for the $K = 0$ lines the IVR rate due to this particular mechanism is zero, and hence the quantum yields for the $K = 0$ lines are higher in this range.

For the high-resolution spectrum shown in Fig. 5.14, each spectral line is due to a rovibronic transition and the intensity of each such spectral line

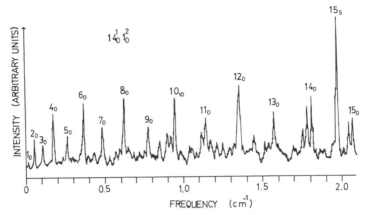

Fig. 5.15 Spectral assignments of the $14_0^1 1_0^2$ spectrum of benzene (C_6H_6) given in Fig. 5.14b.

in this case is proportional to the rotational Boltzmann factor of the ground electronic state and the quantum yield of that particular single rovibronic level. The nonradiative processes involved in the quantum yield can be due to electronic relaxation and IVR. From Eqs. (5.3.9), (5.3.24), and (5.3.25), we can see that the SRVL IVR rate for $K = 0$ can be expressed as

$$W_{vnJK=0} = W_{vn}^{(BO)} + J(J+1)W_{vn}^{(Cor)}, \qquad (5.3.29)$$

where $W_{vn}^{(BO)}$ is given by Eq. (5.3.9) and $W_{vn}^{(Cor)}$ is defined by

$$W_{vn}^{(Cor)} = \frac{4}{\hbar^2}\left(\frac{\hbar}{2I_x}\right)^2\left(\frac{v_i\hbar^2}{4}\right)\sum_{i\alpha}(|\zeta_{i\alpha}^{(x)}|^2 + |\zeta_{i\alpha}^{(y)}|^2)$$

$$\times\left(2 + \frac{\omega_i}{\omega_\alpha} + \frac{\omega_\alpha}{\omega_i}\right)\mathrm{Re}\int_0^\infty dt\,\exp[t(-i\omega_i + i\Delta B + i\omega_\alpha + \Gamma_\alpha)]$$

$$\times G_{n_\alpha=1}(t)G_{n_\beta=2}(t)\prod_j^{j\neq\alpha,\beta} G_{n_j=0}(t). \qquad (5.3.30)$$

The rotational dependence of SRVL electronic relaxation rates and radiative rates in this J value range is small or negligible for C_6H_6 (Henke *et al.*, 1982). Thus we may expect that the plot of the inverse intensity of the $K = 0$ line, after correcting the rotational Boltzmann factor versus $J(J+1)$, will be linear. These plots for odd and even J cases are shown in Fig. 5.16 by using Fig. 5. 14.b (see also Fig. 5.15). From Fig. 5.16 we can see that reasonably good linear relations have been obtained.

Now if one vibrational quantum of the v_1 mode is added to $14_0^1 1_0^2$, then the density of states will increase by at least two orders of magnitude, and both $W_{vnJK}^{(BO)}$ and $W_{vnJK}^{(Cor)}$ will increase accordingly. Whether one of them

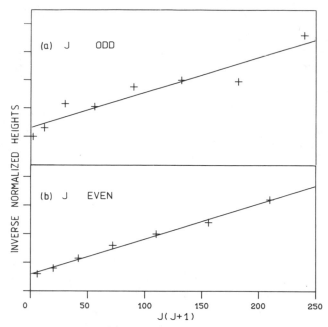

Fig. 5.16 Inverse normalized heights of $K = 0$ peaks plotted as a function of $J(J + 1)$ as found from the measured spectrum in Fig. 5.14b: (a) for J odd, (b) for J even. The solid lines are least-square fits to the data points. [From Riedle *et al.* (1983).]

will become dominant depends on the rate of increase of $W_{vnJK}^{(BO)}$ and $W_{vnJK}^{(Cor)}$ with respect to the increase of vibrational excitation energy.

REFERENCES

Albrecht, A. C. (1960). *J. Chem. Phys.* **33**, 156,169.
Bernstein, R. B. (1982). *J. Phys. Chem.* **86**, 1178.
Biraben, F., Cagnac, B., and Grynberg, G. (1974). *Phys. Rev. Lett.* **32**, 643.
Boesl, U., Neusser, H. J., and Schlag, E. W. (1976). *Chem. Phys.* **15**, 167.
Bray, R. G., Hochstrasser, R. M., and Sung, H. N. (1975). *Chem. Phys. Lett.* **33**, 1.
Callomon, J. H., Dunn, T. M., and Mills, I. M. (1965). *Philos. Trans. R. Soc. London Ser. A* **259**, 99.
Demtröder, W., (1981). "Laser Spectroscopy", p. 525. Springer-Verlag, Berlin.
Esherick, P., Zinsli, P., and El-Sayed, M. A. (1975). *Chem. Phys.* **10**, 415.
Evleth, E. M., and Peticolas, W. L. (1964). *J. Chem. Phys.* **41**, 1400.
Filseth, S. V., Wallenstein, R., and Zacharias, H. (1977). *Opt. Commun.* **23**, 231.
Friedrich, D. M., and McClain, W. M. (1975). *Chem. Phys. Lett.* **32**, 541.
Friedrich, D. M., and McClain, W. M. (1980). *Ann. Rev. Phys. Chem.* **31**, 559.
Fujimura, Y., and Lin, S. H. (1981). *J. Chem. Phys.* **74**, 3726.

Giacobino, E., and Cagnac, B. (1980). In "Progress in Optics," Vol. XVII (E. Wolf, ed.), p. 87. North-Holland Publ., Amsterdam.

Goodman, L., and Rava, R. P. (1982). *Adv. Laser Spectrosc.* **1**, 21.

Goodman, L., and Rava, R. P. (1983). *Adv. Chem. Phys.* **54**, 177.

Hamilton, I., Carter, D., and Brumer, P. (1982). *J. Phys. Chem.* **86**, 2124.

Hampf, W., Neusser, H. J., and Schlag, E. W. (1977). *Chem. Phys. Lett.* **46**, 406.

Hänsch, T. W., Harven, K., Meisel, G., and Schawlow, A. L. (1974). *Opt. Commun.* **11**, 50.

Hawkins, R. T., Hill, W. T., Kowalski, F. V., Schawlow, A. L., and Svanberg, S. (1977). *Phys. Rev. A* **15**, 967.

Heller, E. J., and Davis, M. J. (1982), *J. Phys. Chem.* **86**, 2118.

Henke, W. E., Selzle, H. L., Hays, T. R., Schlag, E. W., and Lin, S. H. (1982). *J. Chem. Phys.* **26**, 1335.

Herzberg, G. (1966). "Molecular Spectra and Molecular Structure," Vol. III, p. 24. Van Nostrand-Reinhold, Princeton, New Jersey.

Hochstrasser, R. M., and Wessel, J. E. (1974). *Chem. Phys. Lett.* **24**, 1.

Hohlneicher, G., and Dick, B. (1979). *J. Chem. Phys.* **70**, 5427.

Honig, B., Jortner, J., and Szöke, A. (1967). *J. Chem. Phys.* **46**, 2714.

Jensen, H. B., and Brodersen, S. (1979). *J. Raman Spectrosc.* **8**, 103.

Johnson, P. M. (1980). *Acc. Chem. Res.* **13**, 21.

Knoth, I., Neusser, H. J., and Schlag, E. W. (1979). *Z. Naturforsch. A* **34**, 979.

Knoth, I., Neusser, H. J., and Schlag, E. W. (1982). *J. Phys. Chem.* **86**, 891.

Kosloff, R., and Rice, S. A. (1981). *J. Chem. Phys.* **74**, 1340.

Letokhov, V. S., and Chebotayev, V. P. (1977). "Nonlinear Laser Spectroscopy." Springer-Verlag, Berlin and New York.

Levenson, M. D., and Bloembergen, N. (1974). *Phys. Rev. Lett.* **32**, 645.

Liao, P. F., and Bjorkholm, J. E. (1976). *Phys. Rev. Lett.* **36**, 1543.

Lin, S. H. (ed.) (1980a). "Radiationless Transitions." Academic Press, New York.

Lin, S. H. (1980b). *Chem. Phys. Lett.* **70**, 492.

Lin, S. H. (1982). *Chem. Phys. Lett.* **86**, 533.

Lin, S. H., Zheng, X. G., Qian, Z. D., Li, X. W., and Eyring, H. (1982). *Proc. Natl. Acad. Sci, U.S.A.* **79**, 1356.

Lombardi, J. R., Wallenstein, R., Hänsch, T. W., and Friedrich, D. M. (1976). *J. Chem. Phys.* **65**, 2357.

McClain, W. A. (1974). *Acc. Chem. Res.* **7**, 131.

Marchese, F. T., Seliskar, C. J., and Jaffé, H. H. (1980). *J. Chem. Phys.* **72**, 4194.

Metz, F. (1975). *Chem. Phys. Lett.* **34**, 109.

Metz, F., Howard, W. E., Wunsch, L., Neusser, H. J., and Schlag, E. W. (1978). *Proc. R. Soc. London Ser. A* **363**, 381.

Mikami, N., and Ito, M. (1977). *Chem. Phys.* **23**, 141.

Monson, P. R., and McClain, W. M. (1970). *J. Chem. Phys.* **53**, 29.

Murakami, J., Kaya, K., and Ito, M. (1980). *J. Chem. Phys.* **72**, 3263.

Noid, D. W., Koszykowski, M. L., and Marcus, R. A. (1981). *Annu. Rev. Phys. Chem.* **31**, 267.

Parmenter, C. S. (1972). *Adv. Chem. Phys.* **22**, 365.

Parmenter, C. S. (1982). *J. Phys. Chem.* **86**, 1735.

Pliva, J., and Pine, A. S. (1982). *J. Mol. Spectrosc.* **93**, 209.

Rava, R. P., Goodman, L., and Jesperson, K. K. (1981). *J. Chem. Phys.* **74**, 273.

Rava, R. P., and Goodman, L. (1982). *J. Am. Chem. Soc.* **104**, 3815.

Rava, R. P., Goodman, L., and Phils, J. G. (1982). *J. Chem. Phys.* **77**, 4912.

Rettschnick, R. P. (1980). *In* "Radiationless Transitions" (S. H. Lin, ed.), pp. 185–218. Academic Press, New York.

Riedle, E. (1983). Doctoral thesis, Technische Universität München, Munich, West Germany.

Riedle, E., and Neusser, H. J. (1984). To be published.

Riedle, E., Neusser, H. J., and Schlag, E. W. (1981). *J. Chem. Phys.* **75**, 4231.

Riedle, E., Moder, R., and Neusser, H. J. (1982a). *Opt. Commun.* **43**, 388.

Riedle, E., Neusser, H. J., and Schlag, E. W. (1982b). *J. Chem. Phys.* **86**, 4847.

Riedle, E., Neusser, H. J., Schlag, E. W. and Lin, S. H. (1983). To be published.

Robey, M. J. and Schlag, E. W. (1977). *J. Chem. Phys.* **67**, 2775.

Robey, M., and Schlag, E. W. (1978). *Chem. Phys.* **30**, 9.

Roche, M., and Jaffe, H. H. (1974). *J. Chem. Phys.* **60**, 1193.

Ross, I. G. (1971). *Adv. Chem. Phys.* **20**, 341.

Salour, M. M. (1977). *Opt. Commun.* **22**, 202.

Shapiro, M., and Child, M. S. (1982). *J. Chem. Phys.* **76**, 6176.

Spears, K. G., and Rice, S. A. (1971). *J. Chem. Phys.* **55**, 5561.

Strott, R. M., Handy, N. C., and Miller, W. H. (1980). *J. Chem. Phys.* **71**, 3311.

Thakur, S. N., and Goodman, L. (1983). *J. Chem. Phys.* **78**, 4356.

Timmermann, A., and Wallenstein, R. (1981). *Opt. Commun.* **39**, 239.

Vasilenko, L. S., Chebotayev, V. P., and Shishaev, A. V. (1970). *JETP Lett.* **12**, 113.

Vasudev, R., and Brand, J. C. D. (1979). *Chem, Phys.* **37**, 211.

Wilson, E. B., (1935). *J. Chem. Phys.* **3**, 276.

Woerdman, J. P., (1976). *Chem. Phys. Lett.* **43**, 279.

Wunsch, L., Neusser, H. J., and Schlag, E. W. (1975a). *Chem. Phys. Lett.* **31**, 433.

Wunsch, L., Neusser, H. J., and Schlag, E. W. (1975b). *Chem. Phys. Lett.* **32**, 210.

Wunsch, L., Metz, F., Neusser, H. J., and Schlag, E. W. (1977). *J. Chem. Phys.* **66**, 386.

Zhi-Ding, Q., Xing-Guo, Z., Xing-Wen, L., Kono, H., and Lin, S. H. (1982). *Mol. Phys.* **47**, 713.

Ziegler, L., and Albrecht, A. C. (1974). *J. Chem. Phys.* **60**, 3558.

CHAPTER

6

Spectroscopic Results

In the past few years a great effort has been made to observe new electronic states that are forbidden in one-photon transitions by using visible and UV multiphoton spectroscopy. Here we are concerned with the spectroscopic results of excited states that were discovered or confirmed by applying the multiphoton technique. One of the merits of multiphoton spectroscopy is that by varying the laser frequency one can detect various electronically excited states associated with Rydberg transitions as well as valence-type transitions of molecules. The multiphoton ionization (MPI) technique developed by Johnson *et al.* (1975), Johnson (1975, 1976), Petty *et al.* (1975), and Dalby *et al.* (1977) has especially made it possible to observe excited states in the wide range from the lower excited states to the ionization threshold. Advanced techniques combined with MPI, such as the use of two or more tunable dye lasers, the detection of the kinetic energy of photoelectrons, or the development of multiphoton ionization mass spectrometry (Bernstein, 1982), have now been applied to investigate the static and dynamic properties of electronically excited states.

Vast numbers of articles on the spectroscopic results obtained by using multiphoton excitation techniques have been written [for a review, see Johnson and Otis (1981)]. We shall focus our attention on the very restricted spectroscopic results of current interest. In Section 6.1, spectroscopic results for Rydberg transitions of molecules are described. In Section 6.2, we shall survey spectroscopic results concerning transitions of the valence electron, the 2^1A_g states of linear polyenes, I_2 molecules, and so on. One of the great successes of multiphoton spectroscopy has been the observation of the forbidden 2^1A_g excited states of linear polyenes. It should be noted that the classification of molecular electronic transitions into Rydberg or valence transitions is not always justified, although the difference between them can sometimes be identified, for example, by the pressure effect (Robin, 1974, 1975). The classification presented here is therefore conventional, and in fact mixed Rydberg or valence characters can be expected when the energy of the Rydberg orbital is nearly equal to that of the valence orbital with the same symmetry (Ashfold *et al.*, 1979).

6.1 RYDBERG TRANSITIONS

A Rydberg orbital is defined as an orbital that lies outside the valence shells of atoms in molecules. Rydberg transitions can be ordered into series converging to an ionization potential I_p that is related to an electronic state of the positive ion (see Duncan, 1971; Sandorfy, 1979; Ashfold *et al.*, 1979). The transition frequency $\tilde{v}$ is given by the simple expression

$$\tilde{v}_n = I_p - R/(n - \delta)^2, \tag{6.1.1}$$

where R is the Rydberg constant, n the principal quantum number of the excited electron, δ the quantum defect, and $R/(n - \delta)^2$ the term values. The magnitude of the quantum defect depends on the degree to which the Rydberg orbital penetrates the molecular ion core. The term value of a Rydberg transition depends not only on the principal quantum number n but also on the azimuthal quantum number l. The values of δ are typically $\delta(ns) \sim 1$, $\delta(np) \sim 0.6$, and $\delta(nd) \sim 0.1$ for Rydberg states when $l = 0$, 1, and 2, respectively. When n is small, as in the Born–Oppenheimer approximation, the total wave function Φ can be expressed as

$$\Psi = (A\Phi_{core}\Phi_R)\chi\eta, \tag{6.1.2}$$

where Φ_{core} is the wave function of the core electrons, Φ_R that of the Rydberg electron, χ the vibrational wave function, and η the rotational wave function, respectively. On the other hand, for high values of n

$$\Phi = (\Phi_{core}\chi\eta)\Phi_R. \tag{6.1.3}$$

Equation (6.1.3) shows that the Rydberg electron moves in the field of a rotating, vibrating core. The potential energy curves of the Rydberg states are similar in shape because their potential energies are determined by the molecular ion core, except for cases in which the principal quantum number is small. The characteristic feature of the Rydberg states is their sensitivity to external perturbation, for example, collisional broadening under the influence of high pressures of dilute inert gas (Robin, 1975).

6.1.1 Acetaldehyde

Heath *et al.* (1980) have applied the MPI technique to study the molecular geometry and hindered internal rotation of acetaldehyde in its lowest Rydberg state in vapor. The MPI spectrum observed by them is shown in Fig. 6.1. The Rydberg nature of the resonant state due to the $n \rightarrow 3s$ transition from a nonbonding orbital to the 3s orbital has been verified by the extreme broadening of the two-photon absorption when pressurized by about 80 atm of argon. Though both one- and two-photon transitions are allowed because of the low C_s symmetry, the one-photon spectra of acetaldehyde were recorded in the past with poorer (~ 10 cm^{-1}) resolution than that of the MPI experiment at 1 cm^{-1} resolution. The MPI spectrum consists mainly of the transitions associated with ν_{15} and ν_{10} modes, which are, respectively, nontotally symmetric CH_3 torsion of frequency 150 cm^{-1} and CCO deformation of 509 cm^{-1} in the ground state. The interval of 58 cm^{-1} in the spectrum on both sides of the origin (55,039 cm^{-1}) is assigned as the $\nu'_{15}-\nu''_{15}$ sequence. The sequences were analyzed with the help of the ground state torsional data of Fateley and Miller (1961): Each level of the ν_{15} mode with $\nu_{15} \geq 2$ is split into a and e components in the local C_{3v} symmetry of the CH_3—C group. It is interesting to discuss the potential energy difference of the internal rotation of the methyl group between the ground state and the resonant Rydberg state. Its potential is generally expressed as a function of rotation α by

$$V = (V_3/2)(1 + \cos 3\alpha), \qquad (6.1.4)$$

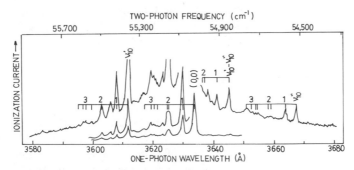

Fig. 6.1 Multiphoton ionization spectrum of acetaldehyde vapor resonant at the second photon. The horizontal ladders mark the positions of the $\nu\nu'_{15}-\nu\nu''_{15}$ sequences having $\nu = 0$–3. [From Heath *et al.* (1980).]

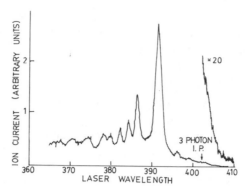

Fig. 6.2 Corrected spectrum of the three-photon ionization region of benzene showing the two-photon resonance with a state at 51,085 cm^{-1}. [From Johnson (1976).]

where V_3 is the torsional barrier height (Herschbach, 1959). Based on an analysis of MPI, Heath *et al.* (1980) estimated $V_3 = 755 \pm 10$ cm^{-1} in the resonant state and on the other hand, $V_3 = 413$ cm^{-1} in the ground state.

6.1.2 Benzene

Benzene has a broad one-photon absorption band attributed to the $^1E_{1u}$ valence state, with much sharper Rydberg bands than those assigned to the $^1A_{2u}3p$ Rydberg transition. Johnson (1975, 1976) has observed a new state of benzene at 391.4 nm in the vicinity of the $^1E_{1u}$ state by using the multiphoton ionization (MPI) technique. The MPI spectrum in the 400–365 nm region shown in Fig. 6.2 originates from the (2 + 1) ionization via the new two-photon allowed state. From the similarity of the vibrational structure of the new state to the Rydberg states observed in one-photon absorption spectra, Johnson has tentatively assigned it as the $^1E_{1g}$ Rydberg state formed by excitation from the highest occupied $\pi(e_{1g})$ to 3s orbitals, although the possibility of the $^1E_{2g}$ valence state cannot be neglected.

6.1.3 Methyl-Substituted Benzenes

Uneberg *et al.* (1980) have reported two of the 3p Rydberg transitions of methyl-substituted benzenes (toluene and *p*-xylene) in MPI spectra. They have assigned these to states that include the $3p_z$ orbitals with symmetries $^1A_1$ and $^1B_2$ in toluene and $^1B_{1u}$ and $^1B_{2u}$ in *p*-xylene. The origins of these states appear at 52,839 and 54,734 cm^{-1} in toluene, as indicated in Fig. 6.3. The Rydberg nature of these states has been confirmed by observing the pressure broadening and shifts of the lines (Fig. 6.3). Furthermore, using the

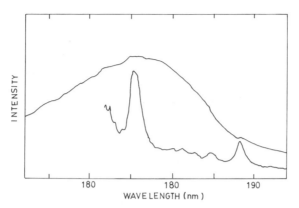

Fig. 6.3 Vacuum ultraviolet and MPI spectra of toluene. The upper curve is the one-photon spectrum, which primarily represents the valence absorptions, the lower curve is the MPI spectrum and displays two of the Rydberg states in this energy region. The actual wavelength of the laser used in the MPI spectrum was double that shown here. [From Uneberg *et al.* (1980).]

formula for the p-type Rydberg series,

$$\tilde{\nu}_n = 71180 - R/(n - 0.5)^2, \tag{6.1.5}$$

which has been adopted by Price and Walsh (1947) for $n \geq 4$, Uneberg *et al.* (1980) have estimated the transition frequency to be close to those observed for $n = 3$.

6.1.4 *trans*-1,3-Butadiene

Johnson (1976) has measured the multiphoton ionization spectrum of *trans*-1,3-butadiene by using dye lasers at 365–468 nm and he (Johnson, 1980) has reviewed the (2 + 1) and (3 + 1) ionization mechanisms of this molecule. In Fig. 6.4, the observed multiphoton ionization spectrum is shown together with the energy level diagram. The spectrum is separated into two regions, one to the blue side of 410 nm and one to the red side. In the former region, the structure in the multiphoton ionization spectrum is characteristic of an allowed two-photon resonance with the $\tilde{B}$ state designated by Herzberg, and the two-photon spectrum confirms the assignment of the vacuum UV absorption bands, which have been analyzed by McDiarmid (1976). Johnson has argued that the $\tilde{B}$ state has ^1B_g symmetry and is formed by the removal of a π electron to an s-type Rydberg orbital.

Many three-photon resonances with Rydberg states have been measured, from the (3 + 1) four-photon ionization region to below 410 nm. The observed multiphoton ionization and the vacuum UV spectra analyzed by McDiarmid (1976) are very similar to each other, because three-photon

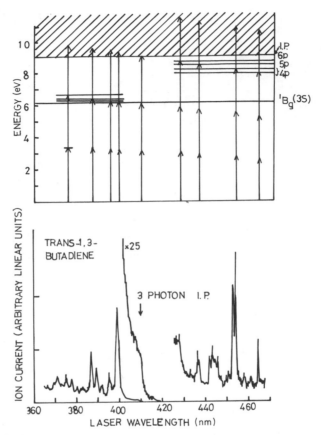

Fig. 6.4. Multiphoton ionization spectrum of *trans*-1,3-butadiene and an energy level diagram. [Reprinted with permission from Johnson (1980). Copyright 1982 American Chemical Society.]

transitions have the same selection rules as one-photon transitions in a C_{2h} molecule. Four Rydberg series with quantum defects of 0.67, 0.42, 0.21, and 0.04 (0.08 by McDiarmid) have been identified. McDiarmid has claimed that the discrepancy of the quantum defect with the lowest value between 0.04 and 0.08 originates from the fact that the $n = 3$ member has a quantum defect very different from those of the higher members because of a strong mixing between the Rydberg and valence antibonding orbitals for the $n = 3$ member. The McDiarmid value should be interpreted as the average quantum defect.

An excited valence state ^1A_g, which was expected to be observed by using multiphoton techniques, was not found below 48,765 cm^{-1} (see Fig. 6.4). An attempt to observe the low-lying excited g state of *trans*-butadiene in both the gas and liquid phases has been reported by Vaida *et al.* (1978), who used multiphoton ionization and thermal lensing techniques.

Berg *et al.* (1978) have measured the polarization ratio Ω for the state assigned as a B_{1g} state by Johnson: Ω, the ratio of absorption for two corotating circularly polarized photons to that for two parallel linearly polarized photons, as discussed in Chapter 4, should be $\frac{3}{2}$ for nonrotating molecular states having less than full molecular symmetry and be less than $\frac{3}{2}$ for totally symmetric states. The observed value is found to be 1.4–1.6, which supports Johnson's assignment.

6.1.5 Iodine

Petty *et al.* (1975) first reported the multiphoton ionization spectrum of the I_2 molecule. Dalby *et al.* (1977) have analyzed the vibronic bands appearing in the ionization spectrum (Fig. 6.5) that arises from a two-photon coherent excitation to a Rydberg level followed by one-photon ionization such as

$$I_2 \xrightarrow{2\nu} I_2^*(R) \xrightarrow{\nu} I_2^+ + e.$$

A least-squares fit of the term value (in cm^{-1}) is expressed as

$$T_{v'} = 53{,}562.75 \pm 1.0 \pm (241.41 \pm 0.04)(v' + \tfrac{1}{2})$$
$$- (0.58 \pm 0.36)(v' + \tfrac{1}{2})^2.$$

The vibrational frequency of the ground state is $214\ cm^{-1}$. By applying the Franck–Condon principle to the vibronic band intensity distribution, Dalby *et al.* found that the bond length in the Rydberg state is smaller than that in the ground state ($r_e'' = 2.667$ Å), that is, its difference is $\Delta = 0.099$ Å. The closeness of the values of the molecular constants between the new state and the ground state in fact reflect the Rydberg character of the new state. With

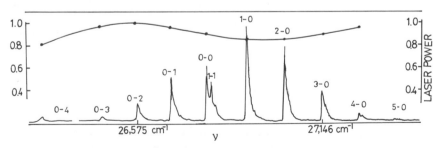

Fig. 6.5 Photoionization spectrum of I_2 at room temperature. The relative peak heights are uncorrected for variations in laser power. [From Dalby *et al.* (1977). Reproduced by permission of the National Research Council of Canada from the *Canadian Journal of Physics*, Volume 55, 1977.]

the help of the photoelectron spectra of I_2 observed by Frost *et al.* (1967), Dalby *et al.* have concluded that the new state is a low-lying Rydberg state. The lowest lying I_2 Rydberg states of g symmetry arise from the configuration $(2\Pi_{3/2g})6s$ and $(^2\Pi_{1/2g})6s$ (Mulliken, 1934). A further investigation of the assignment of the Rydberg states of I_2 has been made by Lehman *et al.* (1978).

6.1.6 Nitric oxide

The first application of the MPI method to nitric oxide was done by Johnson *et al.* (1975). Nitric oxide is one of the heteronuclear diatomic molecules whose electronic structures have been extensively studied experimentally and theoretically (Dressler and Miescher, 1965; Jungen, 1970). Because the ground state configuration expressed by $KK(3\sigma)^2(4\sigma)^2(5\sigma^2)(1\pi)^4(2\pi)$ is sufficiently approximated by that of an alkali atom, most of the excited states below the first ionization potential can be regarded as of the Rydberg type.

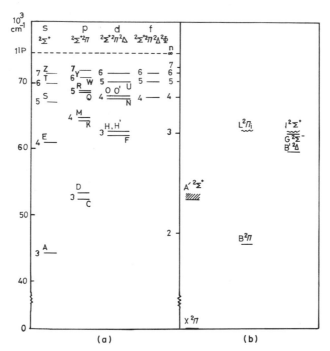

Fig. 6.6 Doublet electronic states of nitric oxide: (a) resonant states $(5\sigma)^2(1\pi)^4(nl\lambda)$; (b) nonresonant states $(5\sigma)^2(1\pi)^4(\sigma 2p)(2\pi)$, $(5\sigma)^2(1\pi)^3(2\pi)^2$, and $(5\sigma)(1\pi)^4(2\pi)^2$. [From Jungen (1970).]

The energy diagram of the doublet Rydberg states together with the non-Rydberg states is shown in Fig. 6.6, in which Rydberg states having a $^1\Sigma^+$ molecular core with a closed-shell configuration $(5\sigma)^2(1\pi)^4$ are shown.

After the work by Johnson *et al.* (1975), MPI studies on the electronic, vibrational, and rotational states of NO were performed by using various detection methods such as those involving fluorescence, ionization, third-harmonic generation, and photoelectron kinetic energy. Only the rotational levels of the low-lying excited $A^2\Sigma^+$ state have been resolved in two-photon excitation spectra by Bray *et al.* (1974) and Zacharias *et al.* (1976). Zakheim and Johnson (1978) have recorded the two- and three-photon resonance ionization spectra of nitric oxide cooled by supersonic expansion in the 400–490 nm region. They have identified 38 different vibronic states by comparison with results from one-photon spectroscopy. Multiphoton ionization processes involved in the transitions consist of the (2 + 2) and (3 + 1) mechanisms, depending on the laser frequency applied.

It is interesting to compare the intensities of the two processes with each other in order to investigate the mechanism of the MPI. The intensity measurement clearly shows that for the (2 + 2) process via resonant intermediate vibronic states $A^2\Sigma^+$, $3s\sigma$ (2, 0), (1, 0), and (0, 0), where v' and v'' in (v', v'') denote the vibrational quantum number in the resonant and initial states, respectively, the band peaks have higher intensities than those associated with the (3 + 1) process via a resonant intermediate Rydberg state ($F, H, M, \ldots$). This has been supported by calculating the cross sections (Cremaschi *et al.*, 1978; Cremaschi, 1981). The (2 + 2) process is schematically represented as

$$X^2\Pi \to (C^2\Pi, D^2\Sigma^+) \to A^2\Sigma^+ \to (C^2\Pi, D^2\Sigma^+) \to C,$$

while the (3 + 1) process is

$$X^2\Pi \to (C^2\Pi, D^2\Sigma^+) \to (A^2\Sigma^+) \to R \to C,$$

where the parentheses contain the virtual intermediate states and R is a resonant Rydberg state. The calculation of the dipole matrix elements yields

$$|\langle C^2\Pi|\mu|A^2\Sigma^+\rangle| \approx |\langle D^2\Sigma^+|\mu|A^2\Sigma^+\rangle| > |\langle A^2\Sigma^+|\mu|R\rangle|$$

$$\text{for}\quad R = F, H, M, \ldots.$$

This difference between the transition moments explains why the intensity of the MPI for the (2 + 2) process is strong compared with that for the (3 + 1) process.

An application of the third-order nonlinear susceptibility method using resonantly enhanced sum-frequency mixing to the NO transition has been

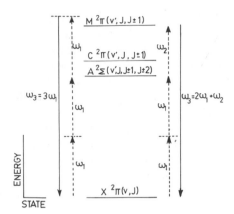

Fig. 6.7 Energy level diagram for nitric oxide showing two- and three-photon resonant enhancement effects. [From Wallace and Innes (1981).]

reported by Wallace and Innes (1980). The spectrum that is obtained is proportional to $|\chi^{(3)}|^2$, where $\chi^{(3)}$ is the third-order nonlinear susceptibility. Two ways to obtain the resonant-enhanced sum-frequency mixing are illustrated in Fig. 6.7. For the case in which $\omega_3 = 3\omega_1$, as shown on the left-hand side of Fig. 6.7, the coincidence of the laser frequency and the molecular energy level is with $A^2\Sigma^+$. In the case in which $C'2\Pi$ is an intermediate state,

$$\chi^{(3)}_{3\omega_1}$$

$$= \frac{\langle X^2\Pi; v=0|\mu|C^2\Pi\rangle\langle C^2\Pi|\mu|A^2\Sigma^+\rangle\langle A^2\Sigma^+|\mu|C^2\Pi\rangle\langle C^2\Pi|\mu|X^2\Pi; v=0\rangle}{4h^3(\omega_C - 3\omega_1)(\omega_A - 2\omega_1)(\omega_C - \omega_1)}.$$

For the double-resonant experiment using two lasers with frequencies ω_1 and ω_2, where the sum frequency $\omega_3 = 2\omega_1 + \omega_2$, with $2\omega_1$ set equal to the energy of a rovibronic energy level of the $A^2\Sigma^+$ state and ω_2 probing the $M^2\Sigma$ state of NO, shown on the right-hand side of Fig. 6.7, the expression for the third-order nonlinear susceptibility is approximately written in terms of a single term as

$$\chi^{(3)}_{2\omega_1 + \omega_2}$$

$$= \frac{3}{4} \frac{\langle \chi^2\Pi|\mu|M^2\Sigma^+\rangle\langle M^2\Sigma^+|\mu|A^2\Sigma^+\rangle\langle A^2\Sigma^+|\mu|C^2\Pi\rangle\langle C^2\Pi|\mu|\chi^2\Pi\rangle}{h^3[\omega_M - (2\omega_1 + \omega_2)](\omega_A - 2\omega_1)(\omega_C - \omega_1)}.$$

Wallace and Innes (1980) recognized that relative intensities of the recorded rotational structure agree with a theoretical calculation, except for cases in which there is accidental double-resonance enhancement of $\chi^{(3)}$ occurring through both two- and three-photon transition probabilities. The anomalous

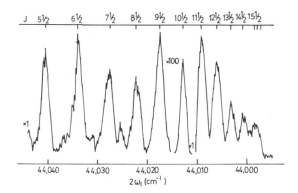

Fig. 6.8   Two-photon spectrum of nitric oxide showing the O_{12} branch of the (0,0) band. [From Wallace and Innes (1980).]

intensity enhancement takes place around $J = 10\frac{1}{2}$ for the O_{12} band (see Fig. 6.8).

Ebata *et al.* (1982) measured the two-photon fluorescence (TPF) excitation and the two-photon resonant four-photon ionization spectra of the $A^2\Sigma^+$ state at the same time. They found a large intensity difference at several rovibronic bands between the two spectra indicated in Fig. 6.9: The intensity of the two-photon fluorescence spectrum is abnormally weak whereas that of the MPI is strong. This can be explained by taking into account the resonant $(2 + 1 + 1)$ MPI process. This mechanism, which involves two resonant states—$A^2\Sigma^+$ and the Rydberg state—enhances the ionization process compared with the case of the $(2 + 2)$ process, as observed by Wallace and Innes (1980), and, on the other hand, depletes the fluorescence

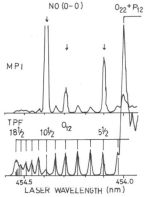

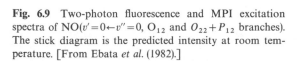

Fig. 6.9   Two-photon fluorescence and MPI excitation spectra of NO($v'=0\leftarrow v''=0$, O_{12} and $O_{22}+P_{12}$ branches). The stick diagram is the predicted intensity at room temperature. [From Ebata *et al.* (1982).]

intensity from the $A^2\Sigma^+$ state because the intensity is proportional to the population of the $A^2\Sigma^+$ state in the steady state.

Motivated by the Wallace–Innes study, Dressler and Miescher (1981) analyzed the rovibronic fine structure appearing in the $A^2\Sigma^+$–$X^2\Pi$ transition of NO. They noticed that the valence states and an np Rydberg complex $(C^2\Pi + D^2\Sigma^+, v = 6)$ are strongly mixed with each other (see the energy level diagram of NO) and emphasized the necessity of multistate interactions in interpreting detailed experimental information on the energy levels observed in highly resolved multiphoton spectra.

Miller and Compton (1981) have reported the photoelectron spectrum of NO following the four-photon absorption via resonant $A^2\Sigma^+$ $(v = 0, 1, 2, 3)$ states. The ionization process is expressed as

$$\text{NO: } X^2\Pi(v = 0) \xrightarrow{2h\nu} \text{NO: } A^2\Sigma^+(v = 0, 1, 2, 3)$$
$$\xrightarrow{2h\nu} \text{NO}^+\text{: } X^1\Sigma^+(v = 0, 1, 2, 3) + \text{e}^- \quad (\text{KE}),$$

where KE means the kinetic energy of the electron. Multiphoton ionization photoelectron spectroscopy is expected to be a useful tool for investigating the ionization mechanisms of molecules, such as direct ionization, vibrational or electronic autoionization, and so on. Miller and Compton have observed two electron kinetic energy peaks (Fig. 6.10) when the laser was tuned to each resonant vibrational state; the high-energy peak corresponds to direct ionization. This can be explained by the Franck–Condon principle. Because the $A^2\Sigma^+$ state and the ground ion $X^2\Sigma^+$ state have the same electronic configuration, the potential curves of the two states are quite similar: $\omega_e =$ 2374.31 and 2376.42 cm^{-1}, and $r_e = 1.0634$ and 1.0632 Å for the $A^2\Sigma^+$ and $X^2\Sigma^+$ states, respectively. Considering the Franck–Condon factor, we see

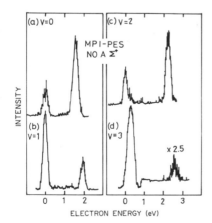

Fig. 6.10 Electron kinetic energy distribution following four-photon ionization resonant with four vibrational levels of the $A^2\Sigma^+$ state of nitric oxide. [From Miller and Compton (1981).]

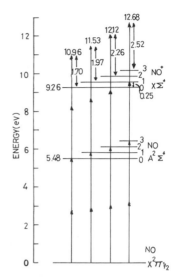

Fig. 6.11 Energy level diagram showing the two-photon resonant, four-photon ionization of nitric oxide via various vibrations of the $A^2\Sigma^+$ state. The final four-photon energy in the continuum and the expected electron energy resulting from direct ionization are given in electron volts (eV). [From Miller and Compton (1981).]

that only those transitions with the vibrational quantum number restriction of $\Delta v = 0$ can take place, and the outgoing electron has taken all the excess energy as kinetic energy (see Fig. 6.11). This explains the appearance of the high-energy peak. The origin of the low kinetic energy peak, on the other hand, is still unknown.

Kimman *et al.* (1982) repeated the experiment of Miller and Compton (1981) by using a new type of spectrometer accepting 2Π sr at a resolution of 15 meV. They observed vibrational distribution peaks not only at $\Delta v = 0$ (Franck–Condon for direct ionization) but also at the highest energetically allowed Δv. In order to explain the ejection of electrons from all energetically allowed vibrational states of the ion, Kimman *et al.* (1982) claimed the presence of a dissociative channel above the ionization continuum. Further experimental and theoretical work is necessary to confirm the experimental result. [For theoretical work on the autoionization of a diatomic molecule, see Jungen and Dill (1980) and Takagi and Nakamura (1981).]

To investigate the multiphoton ionization mechanism and the relaxations taking place in the resonant states, two-color double-resonance experiments for the four-photon ionization of nitric oxide were reported by Ebata *et al.* (1982b, 1983). They simultaneously measured the double-resonance enhanced four-photon ionization spectra and fluorescence dip spectra using two tunable dye lasers. By varying the time delay between the pulses from the first and second dye lasers, collision-induced rotational relaxation rates, which are found to be comparable to the electronic quenching rate, have been observed.

Achiba *et al.* (1983) measured the energy spectra and angular distribution of photoelectrons emitted by the (3 + 1) processes

$$NO(X^2\Sigma) \xrightarrow{3h\nu} NO^* \begin{Bmatrix} F^2\Delta \\ H^2\Sigma^+ \\ H'^2\Pi \end{Bmatrix} \xrightarrow{h\nu} NO^+(X^1\Sigma^+) + e^- \quad (KE).$$

Achiba *et al.* (1983) found that the photoelectron angular distributions may well be interpreted in terms of cosine-square distributions. This is the same cosine-dependence as that found in vacuum UV photoelectron spectroscopy associated with one-photon ionization. This indicates that the ionization step takes place by one-photon direct ionization from the three-photon resonant states.

6.1.7   Pyrrole, N-Methylpyrrole, and Furan

Cooper *et al.* (1980) have reported the resonant multiphoton ionization of the five-membered rings pyrrole (C_4H_5N), N-methylpyrrole (C_5H_7N), and furan (C_4H_4O) in the wavelength region from 365 to 680 nm. Vibrational constants of newly and previously observed Rydberg states obtained by excitation from the highest occupied a_2 orbital or the next highest occupied b_1 orbital to the ns, np, or nd orbitals have been reported. In Fig. 6.12, the three-photon resonant MPI spectrum of furan is shown. This spectrum

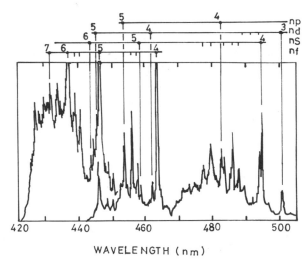

Fig. 6.12 Three-photon resonant MPI spectrum of furan showing $R(ns)$, $R(np)$, $R(nd)$, and $R(nf)$ series with their vibronic structure. The peak heights do not represent relative intensities. [From Cooper *et al.* (1980).]

contains two new Rydberg series converging on the first ionization limit of 71,649 cm^{-1}, which were not identified in the analysis of the one-photon and photoelectron spectra. These Rydberg series have been assigned to the $1a_2 \rightarrow ns$ and $1a_2 \rightarrow nd$ transitions.

6.1.8 Saturated Three-Membered Rings, Cyclopropane-h_6, Cyclopropane-d_6, and Ethylene Imine

Resonant multiphoton ionization spectra of the saturated three-membered rings cyclopropane-h_6, cyclopropane-d_6, and ethylene imine have been recorded by Robin and Kuebler (1978). In the cyclopropanes, a complex band system originating from the $3e' \rightarrow 3s$ Rydberg transition has been observed in the two-photon 51,000–56,000 cm^{-1} region. The MPI spectra are shown in Fig. 6.13. Progressions of 1420 and 1230 cm^{-1} for h_6- and d_6-cyclopropanes are identified. It has been suggested that this mode is the totally symmetric CH$_2$ deformation v'_2 with ground state frequencies of 1479 and 1275 cm^{-1} in the -h_6 and -d_6 cyclopropanes. The appearance of this totally symmetric vibration starting from the origin supports the notion that the transition is electronically allowed, that is, the resonant state symmetry is of the $^1E'$ type. The MPI vibronic structures are complicated by the combination of the C_3 ring deformation $v'_{10}(e')$ of 870 ± 10 cm^{-1} (-h_6) or

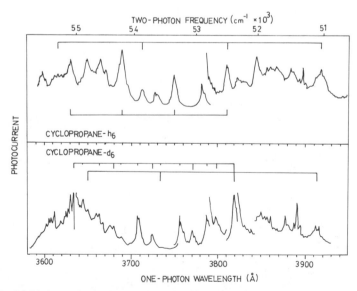

Fig. 6.13 Multiphoton ionization spectra of cyclopropane-h_6 (upper) and cyclopropane-d_6 (lower), each at 50 Torr in the gas phase. [From Robin and Kuebler (1978).]

$660 \pm 20 \text{ cm}^{-1}(\text{-d}_6)$. Robin and Kuebler (1978) have suggested that the resonant Rydberg $^1E'$ state is subject to the Jahn–Teller distortion by the $v'_{10}(e')$ vibration, although a detailed analysis was not given.

6.2 VALENCE STATES

6.2.1 Linear Polyenes

Much experimental and theoretical attention has been given to elucidating low-lying "hidden" electronic excited states (^1A_g) of linear polyenes because of their photochemical and biochemical potentials (Schulten and Karplus, 1972; Hudson and Kohler, 1973, 1974; Gavin *et al.*, 1973). One of the fruitful applications of multiphoton spectroscopy is the direct observation of the excited states of linear polyenes. The $^1A_g \leftarrow {}^1A_g$ transitions that are forbidden for the one-photon process have recently been observed by using multiphoton spectroscopic methods.

In order to understand qualitatively the low-lying excited states of polyenes, let us review a molecular orbital consideration of the simplest polyene, *trans*-butadiene. The molecular orbitals and the electronic configurations are presented in Fig. 6.14. From the singly excited configurations, four excited singlet states are obtained: 1^1B_u, ϕ_1; $^1A_g^+$, $\phi_2 + \phi_3$; $2^1A_g^-$, $\phi_2 - \phi_3$; 2^1B_u, ϕ_4. Here the 1^1B_{1u} state is the lowest excited state and the $1^1A_g^+$ and $2^1A_g^-$ states are located above the 1^1B_{1u} state in the singly excited configuration interaction approximation. Inclusion of the double-excitation configurations, for example, ϕ_5, is known to drastically lower the $2^1A_g^-$ state and to be close to the $1^1B_{1u}^+$ state (Honig *et al.*, 1975). There have been many discussions about the ordering of the $2^1A_g^-$ and $1^1B_u^+$ states (Tavan and Schulten, 1979, and references therein); increasing the π electrons will lower the $2^1A_g^-$ state below the $1^1B_u^+$ state. The ordering of the electronic energy

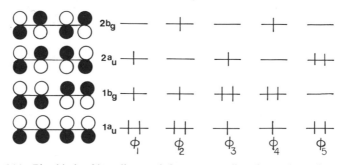

Fig. 6.14 Pi orbitals of butadiene and the corresponding electronic configurations.

TABLE 6.1

Observed $2^1A_g^-$ and $1^1B_u^+$ States in Polyenes

Molecule	E^{00}/(eV)		Gap (eV)	Temperature (K)	Solvent
	$^1A_g^-$	$^1B_u^+$			
Octatetraene[a]	3.55	3.98	0.43	77	Hexane
Octatetraene[b]	3.57	4.01	0.44	4.2	Octane
2,4,6,8-Decatetraene[c]	3.56	3.88	0.32	4.2	Undecane
2,10-Dimethyldecapentaene[d]	3.04	3.43	0.39	4.2	Nonane
2,12-Dimethyltridecahexaene[e]	2.70	3.20	0.50	4.2	Decane
Diphenylbutadiene[f]	3.49	3.50	0.01	77	EPA
Diphenylhexatriene[g]	3.12	3.22	0.10	4.2	Hexane
Diphenylhexatriene[h]	3.11	3.23	0.12	77	EPA
Diphenyloctatetraene[i]	2.80	3.01	0.21	4.2	Pentadecane
Diphenyloctatetraene[i]	2.79	2.90	0.16	1.8	Bibenzyl
Diphenyloctatetraene[g]	2.77	2.96	0.19	4.2	Octane
Diphenyloctatetraene[j]	2.77	3.02	0.25	77	EPA

[a] One-photon absorption (A) and fluorescence (F); Gavin *et al.* (1978).
[b] Two-photon excitation (TPE); Granville *et al.* (1979, 1980).
[c] F; Andrews and Hudson (1978).
[d] A and F; Christensen and Kohler (1975).
[e] A and F; Christensen and Kohler (1976).
[f] TPE; Bennett and Birge (1980).
[g] A and F; Nikitina *et al.* (1976).
[h] TPE; Fang *et al.* (1978).
[i] A and F; Hudson and Kohler (1972, 1973).
[j] TPE; Fang *et al.* (1977).

levels depends, of course, on the substitution groups (D'Amico *et al.*, 1980), as well as on experimental conditions, such as the solvent used, the temperature, and so on.

Let us now mention experimental results concerning the 2^1A_g states of the linear polyenes. Table 6.1 shows the observed energy levels of the $2^1A_g^-$ and $1^1B_u^+$ states of linear polyenes.

6.2.2 *trans*-1,3-Butadiene

The band that exhibits a sharp structure at 6.25 eV was assigned to the $2^1A_g \leftarrow 1^1A_g$ transition (hereafter Pariser alternant symmetry signs $\pm$ are omitted) (Mosher *et al.*, 1973). This assignment, however, has been revised by Johnson (1976) and McDiarmid (1976). They concluded that a Rydberg state is responsible for the transition near 6.25 eV (see the Rydberg cases).

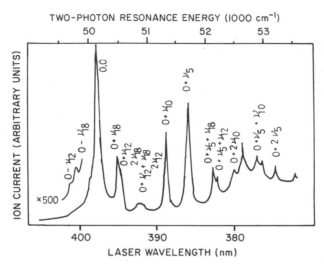

Fig. 6.15 The MPI spectrum of *trans*-1,3,5-hexatriene. [From Parker *et al.* (1976).]

6.2.3 *trans*-1,3,5-Hexatriene

Parker *et al.* (1976) have recorded the multiphoton ionization spectrum of *trans*-hexatriene gas at room temperature. In Fig. 6.15 is shown the spectrum they observed, in which the strong 0–0 band located at 50,280 cm^{-1} supports a two-photon allowed transition. The separations of the main bands from the origin are 350, 420, 1200, and 1560 cm^{-1}, which can be assigned to the ground state vibrations ν_{18}, ν_{12}, ν_{10}, and ν_5, respectively; the remaining bands appear as combinations and overtones of these vibrations. Although the electronic state associated with the transition has never been definitely assigned (either the 2^1A_g or gerade Rydberg state), Parker *et al.* (1976) recognized that the latter assignment agrees with those made for absorptions in the 0, 0 transition energy region in *trans*-1,3-butadiene by McDiarmid (1976) and Johnson (1976).

6.2.4 *trans-trans*-1,3,5,7-Octatetraene

Granville *et al.* (1979) have reported the two-photon excitation spectrum of this molecule in an *n*-octane matrix at liquid helium temperatures, as well as the fluorescence and the one-photon fluorescence excitation spectra (Fig. 6.16). They have also reported the one- and two-photon excitation spectra. The one-photon spectrum is found to be vibronically induced, whereas the two-photon excitation spectrum is electronically allowed. The origin of the strongly allowed state is located at 28,561 cm^{-1}, which is 3540 cm^{-1} below

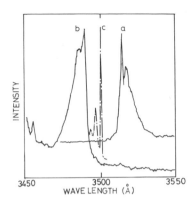

Fig. 6.16   The $2^1 A_g$ origin region of all-*trans*-1,3,5,7-octatetraene in polycrystalline *n*-octane at 4.2 K: (a) the shortest wavelength feature seen in fluorescence, (b) the longest wavelength feature seen in one-photon excitation, and (c) the longest wavelength feature seen in two-photon excitation. The actual excitation wavelength of the two-photon spectrum is twice the abscissa. [From Granville *et al.* (1979).]

the $^1B_u \leftarrow {}^1A_g$ transition. This indicates a $2^1 A_g$ assignment for the two-photon allowed state. Granville *et al.* (1980) carried out the vibrational analysis and noted that most of the measured peaks can be assigned to simple combinations of 11 fundamental frequencies, which are given by five torsional vibrations (220, 320, 340, 400, and 527 cm^{-1}), C–C stretching vibrations (1219 and 1270 cm^{-1}), the symmetric olefin stretching vibration (1754 cm^{-1}), carbon–hydrogen stretching vibration (2965 cm^{-1}), lattice or low-frequency molecular vibration (30 cm^{-1}), and the second harmonic of a b_u-symmetry promoting mode (538 cm^{-1}). From an analysis of the vibronic structure of the one-photon excitation spectrum, Granville *et al.* (1980) assigned the four vibrations (93, 463, 538, and 1054 cm^{-1}) to b_u-symmetry promoting modes, because the progressions are built on four false origins of those frequencies and do not appear in the two-photon excitation spectrum. It is interesting to analyze the relative intensities found in the two-photon spectrum: Granville *et al.* (1980) noticed that both the fundamental and overtone bands of the 1754-cm^{-1} mode have greater intensity than the origin, which suggests that there are large double-bond length changes between the $2^1 A_g$ and $1^1 A_g$ states on excitation. Next, we shall focus our attention on phenyl-substituted linear polyenes.

6.2.5   All-*trans*-1,4-Diphenyl-1,3-Butadiene

Figure 6.17 shows the two-photon excitation spectrum of all-*trans*-1,4-diphenyl-1,3-butadiene in EPA at 77 K, which was observed by Bennett and Birge (1980), together with the one-photon absorption and fluorescence spectra. Bennett and Birge assigned the origin of the lowest excited ^1A_g state at 27,900 cm^{-1}, which is 130 cm^{-1} below the origin of the strongly one-photon allowed ^1B_u state. Compared with the case of the unsubstituted *trans*-butadiene, the fact that the electronic energy gap between the $2^1 A_g$ and

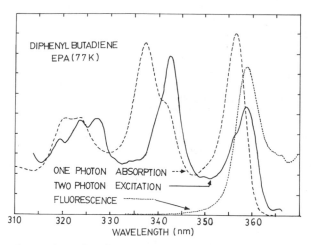

Fig. 6.17 One-photon absorption, fluorescence, and two-photon excitation spectra of diphenylbutadiene in EPA at 77 K. The two-photon excitation spectrum is plotted versus $\lambda_{ex}/2$, where λ_{ex} is the wavelength of the laser excitation. [From Bennett and Birge (1980).]

$1B_u$ states is very close strongly suggests the contribution of the π electrons in phenyl groups to lowering the 2^1A_g state. To investigate the role of π electrons in phenyl groups, Bennett and Birge (1980) have also reported a molecular orbital (MO) configurational analysis using a semiempirical Pariser, Parr, and Pople (PPP) SCF MO CI procedure within singly and doubly excited configurations. Their calculation shows that the optical electron density on the phenyl groups is highly delocalized, which is interesting to compare with the role of the π electrons in the phenyl groups of other linear

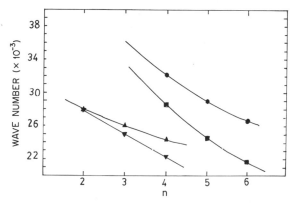

Fig. 6.18 Low-lying excited singlet-state system origin energies in the diphenyl polyenes (EPA at 77 K; ▼, $^1A_g^{*-}$; ▲, $^1B_u^{*+}$) and the linear polyenes (3MP at 77 K; ■, $^1A_g^{*-}$; ●, $^1B_u^{*+}$) plotted as a function of the number n of polyene double bonds. [From Bennett and Birge (1980).]

polyenes. In Fig. 6.18, the 2^1A_g and 1^1B_u electronic state origins of three diphenyl and three linear polyenes in low-temperature (77K) glass solvents are shown (Bennett and Birge, 1980). The solid lines connecting the two excited electronic state origins were calculated by using a quadratic least-squares fit, $\tilde{v} = a + bn + cn^2$. Figure 6.18 indicates that the phenyl groups provide increasing stabilization of the electronic system with decreasing length of the polyene chain.

6.2.6 All-*trans*-Diphenylhexatriene and All-*trans*-Diphenyloctatetraene

The lowest excited singlet states (^1A_g) of all-*trans*-diphenylhexatriene (DPH) and all-*trans*-diphenyloctatetraene (DPO) have been recognized by analysis of the one-photon absorption and emission spectra. Measurements of the two-photon excitation spectra have confirmed this discovery (Holton and McClain, 1976; Fang *et al.*, 1977, 1978). The two-photon excitation spectra of DPO and DPH in a rigid EPA matrix at 77 K observed by Fang *et al.* (1977, 1978) are shown in Figs. 6.19 and 6.20, respectively. The origins of the lowest excited A_g states of DPO and DPH are located, respectively, at 22,360 and 25,050 cm^{-1}, which are at about 2000 and 900 cm^{-1} below each

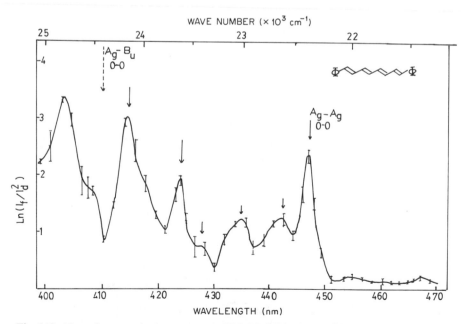

Fig. 6.19 Two-photon excitation spectrum of DPO in EPA at 77 K. [From Fang *et al.* (1978).]

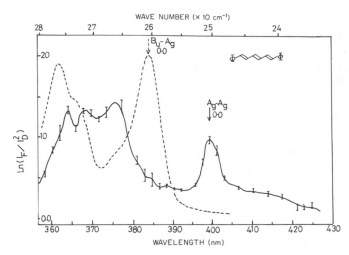

Fig. 6.20 Two-photon excitation spectrum of DPH in EPA at 77 K. The one-photon absorption spectrum is shown by the dashed curve. [From Fang *et al.* (1978).]

origin of the first allowed $^1B_u \leftarrow \, ^1A_g$ transition. Vibrational mode analysis was not given.

To summarize, that the lowest excited singlet states of the linear polyenes are of $^1A_g^-$ symmetry, except for butadiene and presumably hexatriene, has been confirmed by multiphoton excitation techniques together with conventional one-photon absorption and fluorescence spectroscopy. The spectroscopic results can serve as a useful guide for investigating the dynamics that take place in the low-lying excited states of polyenes (Heimbrook *et al.*, 1981).

6.2.7 Iodine

As is well known (Mulliken, 1971), iodine has a number of valence-shell electronic states constructed from electronic configurations $\sigma_g^2 \sigma_u^2 \sigma_g^m \pi_u^p \pi_g^q \sigma_u^n$, with $m + n + p + q = 10$, where σ_g and σ_u are constructed from the 5s ± 5s

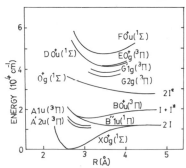

Fig. 6.21 Potential curves for the valence-shell states of iodine. [From Kasatani *et al.* (1981).]

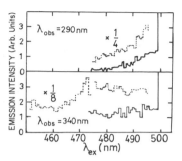

Fig. 6.22 Ultraviolet fluorescence excitation spectra in the 450–500 nm region. The solid line was obtained at 20°C and the dotted one at 70°C. [From Kasatani *et al.* (1981).]

and $5p\sigma \pm 5p\sigma$ states and π_u and π_g from the $5p\pi \pm 5p\pi$ states in the simple LCAO MO description. Potential curves of the valence states are shown in Fig. 6.21. Not only have the electronic states of this molecule been elucidated, but the photophysical properties such as predissociation and magnetic quenching that take place in these excited states have also been extensively studied by many researchers [for example, see Steinfeld and Houston (1978) and Lin and Fujimura (1979)].

We focus our attention on the spectroscopic result associated with the multiphoton excitation mechanism of the I_2 molecule. The mechanism of two- and three-photon absorption of the iodine molecule via dissociative states has been discussed by Kasatani *et al.* (1981). They observed the fluorescence spectra from the excited $D0_u^+(^1\Sigma)$, $F0_u^+(^1\Sigma)$, and $G2_g(^3\Pi)$ states to the lower excited and ground states by varying the laser excitation frequencies in the range 450–610 nm. When the energy of laser light was above the dissociation limit (498.9 nm) of the B state, the emission intensity decreased drastically but was still observed. The fluorescence excitation spectra in the 450–500 nm range are shown in Fig. 6.22. From the measurement of the polarization ratio $\langle W^{(2)} \rangle^{\curvearrowright}/\langle W^{(2)} \rangle^{\uparrow\uparrow} \simeq 1$, which is reflected in the two-photon fluorescence excitation spectrum, the mechanism of the excitation can be described in terms of the sequential process. If the excitation is, on the other hand, described in terms of the simultaneous process, the polarization ratio should be 1.5 for a $\Pi(xz, yz$ polarized$) \leftarrow \Sigma^+$ two-photon transition in D_∞ symmetry (see Section 4.2). The observed polarization data support $G2g(^3\Pi) \leftarrow B''1u(^1\Pi) \leftarrow X0_g^+(^1\Sigma)$, where the repulsive $B''1_u(^1\Pi)$ state is mixed with the $B0_u^+(^3\Pi)$ state.

6.2.8 Naphthalene

Vibronic two-photon transitions to the first electronically excited state of naphthalene in the condensed phase have been reported by several authors (Bergman and Jortner, 1974; Mikami and Ito, 1975; Hochstrasser *et al.*,

TABLE 6.2

Newly Discovered Vibrational Excited State ($^1B_{3u}$) Frequencies of Naphthalene[a]

Assignment	Ground state frequency (cm^{-1})	IR frequency (cm^{-1})	Excited state frequency (cm^{-1})
$C_{10}H_8$			
$4(b_{3u})$	1375	1361	1559
$5(b_{3u})$	1214	1209	1206
$6(b_{2u})$	1142	1125	940
$1(b_{1u})$	1014	955	865
$8(b_{3u})$	641	618	592
$8(b_{2u})$	355	362	355
$C_{10}D_8$			
$4(b_{3u})$	1312	1318	1552
$5(b_{3u})$	1089	1082	1038
$6(b_{2u})$	837	885	796
$1(b_{1u})$	799	790	716
$8(b_{3u})$	614	593	518
$8(b_{2u})$	331	328	331

[a] The corresponding group state frequencies found by the hot-band analysis are compared with the IR data of Scully and Whiffen (1960).

1974). From the one-photon absorption spectra the band origin of naphthalene is found to be at 32,020 cm^{-1}, and the symmetry of the first electronic state is known to be $^1B_{3u}$ in point group D_{2h}. Here the short and long molecular axes are taken as the x and y axes, respectively. In the condensed phase the spectra are subject to frequency shifts, splittings, changes in intensities, and even changes in selection rules. To avoid the problems of perturbations in the condensed phase, Boesl et al. (1976) measured two-photon excitation spectra from 32,900 to nearly 34,750 cm^{-1} in the vapor phase. In Table 6.2 newly discovered vibrational frequencies in the excited state $^1B_{3u}$ are shown. These frequencies were obtained by analyzing the rotational structure, polarization behavior, and hot band information. These methods have been used successfully to analyze the vibrational structure in the benzene molecule, as was shown in Chapter 5.

6.2.9 Pyrazine

From the one-electron excitation of the orbitals constructed by the plus and minus linear combinations of the two lone-pair orbitals to the lowest vacant π orbital, one can construct two excited $n\pi^*$ states of symmetry $^1B_{3u}$

TABLE 6.3

The Vibrational Frequencies of Pyrazine-h_4 and (-d_4) as Determined
from the Gas Phase Two-Photon Spectrum[a]

Vibration	Vibrational symmetry	Frequency (cm^{-1}) in ^1A_g ground state	Frequency (cm^{-1}) in $^1B_{3u}$ excited state	Normalized intensities
16b	b_{3u}	416(398)	235(213)	100(112)
11	b_{3u}	786(597)	639(533)	25(26)
18a or 12	b_{1u}		647(616)	80(82)

[a] All band intensities are normalized to the strongest band induced by v_{16b}
in pyrazine-h$_4$.

and $^1B_{2g}$. Here the N–N direction is taken as the Z axis, the direction
perpendicular to this axis in molecular plane being taken as the y axis.
Transitions to the $^1B_{3u}$ state are one-photon electric dipole allowed, whereas
transitions to the $^1B_{2g}$ state are one-photon forbidden, and it is interesting
to find the $^1B_{2g}$ state by applying two-photon spectroscopy. Esherick *et al.*
(1975) have tried to find the $^1B_{2g}$ state by using the two-photon phosphorescence excitation method. They have observed and analyzed the spectrum of
neat pyrazine crystal at 1.6 K. Their results suggest that the $^1B_{2g}$ state is
located at a position slightly higher than that of the first excited $\pi\pi^{*1}B_{2u}$
state.

Knoth *et al.* (1976) have reported spectroscopic results concerning the
vibrational structure of the $^1B_{3u}$ state of pyrazine-h$_4$ and -d$_4$ in the gas phase
by using the two-photon excitation method. A pure electronic two-photon
transition to the $^1B_{3u}$ state is not possible because of the parity selection
rule. Thus transitions that are vibronically induced by vibrations of u-parity
can be expected in point group D_{2h}. The vibrational frequencies of pyrazine-
h$_4$ and -d$_4$, determined from the gas phase two-photon spectrum, are listed
in Table 6.3.

6.2.10   Pyrimidine

Due to the fact that the pyrimidine molecule belongs to the low symmetry
point group C_{2v}, the transition symmetries of one- and two-photon excitations are identical. Therefore, one would expect that the spectrum obtained
by two-photon excitation should resemble the one-photon spectrum. As
already discussed in Chapter 4, however, one of the advantages of two-
photon spectroscopy over one-photon spectroscopy is that it enables one to
measure the polarization dependence of the rotational bands in order to

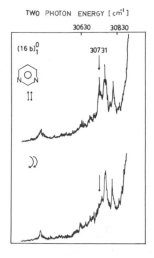

Fig. 6.23 Part of the hot-band region of the two-photon spectrum of pyrimidine measured in linearly (upper trace) and circularly (lower trace) polarized light. [From Knoth *et al.* (1982).]

assign the vibrations. Knoth *et al.* (1982) have reported two-photon excitation spectra of pyrimidine gas in the region of the $S_1({}^1B_1)-S_0({}^1A_1)$ transition. They have shown that the band structures are different from those in the one-photon absorption given by Innes *et al.* (1969) and Knight *et al.* (1975), this being due to the additionally allowed rotational branches in two-photon absorption. In Fig. 6.23, the spectrum of the hot-band region in linearly (upper part) and circularly polarized (lower part) excitation light is shown. The prominent feature is the complete disappearance of the peak at 30,731 cm^{-1}, marked by an arrow in the circularly polarized spectrum. Therefore, the polarization behavior of this peak is that of a totally symmetric transition, as has been discussed in Chapter 4. Since the symmetry of the S_1 state of pyrimidine is ${}^1B_1$, it follows that a totally symmetric two-photon transition has to be vibrationally induced by the vibrational mode of symmetry b_1. Knoth *et al.* (1982) assigned this vibrational frequency of 342 cm^{-1} as ν_{16} from the comparison with the frequency 344 cm^{-1} found for ν_{16b} in liquid-phase IR measurements given by Lord *et al.* (1957). In a similar way, Knoth *et al.* measured the cold-band spectrum and assigned the vibrational frequency of 360 cm^{-1} in the ${}^1B_{1u}$ state as ν_{6b}.

REFERENCES

Achiba, Y., Sato, K., Shobatake, K., and Kimura, K. (1983). *J. Chem. Phys.* **78**, 5474.
Andrews, J. R., and Hudson, B. S. (1978). *Chem. Phys. Lett.* **57**, 600.
Ashfold, M. N. R., Macpherson, M. T., and Simons, J. P. (1979). *Top. Curr. Chem.* **86**, 3.
Bennett, J. A., and Birge, R. R. (1980). *J. Chem. Phys.* **73**, 4234.
Berg, J. O., Parker, D. H., and El-Sayed, M. A. (1978). *J. Chem. Phys.* **68**, 5661.

Bergman, A., and Jortner, J. (1974). *Chem. Phys. Lett.* **26**, 323.

Bernstein, R. B. (1982). *J. Phys. Chem.* **86**, 1178.

Birge, R. R. (1982). *Methods Enzymol.* **88**, 522.

Birge, R. R., Bennett, J. A., Hubbard, L. M., Fang, H. L., Pierce, B. M., Kliger, D. S., and Leroi, G. E. (1982). *J. Am. Chem. Soc.* **104**, 2519.

Bray, R. G., Hochstrasser, R. M., and Wessel, J. E. (1974). *Chem. Phys. Lett.* **27**, 167.

Christensen, R. L., and Kohler, B. E. (1975). *J. Chem. Phys.* **63**, 1837.

Christensen, R. L., and Kohler, B. E. (1976). *J. Phys. Chem.* **80**, 2197.

Cooper, C. D., Williamson, A. D., Miller, J. C., and Compton, R. N. (1980). *J. Chem. Phys.* **73**, 1527.

Cremaschi, P. (1981). *J. Chem. Phys.* **75**, 3944.

Cremaschi, P., Johnson, P. M., and Whitten, J. L. (1978). *J. Chem. Phys.* **69**, 4341.

Dalby, F. W., Petty-Sil, G., Price, M. H., and Tai, C. (1977). *Can. J. Phys.* **55**, 1033.

D'Amico, K. L., Manos, C., and Christensen, R. L. (1980). *J. Am. Chem. Soc.* **102**, 1777.

Dressler, K., and Miescher, E. (1965). *Astrophys. J.* **141**, 1266.

Dressler, K., and Miescher, E. (1981). *J. Chem. Phys.* **75**, 4310.

Duncan, A. B. F. (1971). "Rydberg Series in Atoms and Molecules." Academic Press, New York.

Ebata, T., Abe, H., Mikami, N., and Ito, M. (1982). *Chem. Phys. Lett.* **86**, 445.

Ebata, T., Imajo, T., Mikami, N., and Ito, M. (1982). *Chem. Phys. Lett.* **89**, 45.

Ebata, T., Mikami, N., and Ito, M. (1983). *J. Chem. Phys.* **78**, 1132.

Esherick, P., Zinsli, P., and El-Sayed, M. A. (1975). *Chem. Phys.* **10**, 415.

Fang, H. L. B., Thrash, R. J., and Leroi, G. E. (1977). *J. Chem. Phys.* **67**, 3389.

Fang, H. L. B., Thrash, R. J., and Leroi, G. E. (1978). *Chem. Phys.* **57**, 59.

Fateley, G. W., and Miller, F. A. (1961). *Spectrochim. Acta* **17**, 857.

Frost, D. C., McDowell, C. A., and Vroom, D. A. (1967). *J. Chem. Phys.* **46**, 4255.

Gavin, R. M., Risemberg, S., and Rice, S. A. (1973). *J. Chem. Phys.* **58**, 3160.

Gavin, R. M., Jr., Weisman, C., McVey, J. K., and Rice, S. A. (1978). *J. Chem. Phys.* **68**, 522.

Granville, M. F., Holton, G. R., Kohler, B. E., Christensen, R. L., and D'Amico, K. L. (1979). *J. Chem. Phys.* **70**, 593.

Granville, M. F., Holton, G. R., and Kohler, B. E. (1980). *J. Chem. Phys.* **72**, 4671.

Heath, B. A., Robin, M. B., Kuebler, N. A., Fisanick, G. J., and Eichelberger, T. S., IV (1980). *J. Chem. Phys.* **72**, 5565.

Heimbrook, L. A., Kenny, J. E., Kohler, B. E., and Scott, G. W. (1981). *J. Chem. Phys.* **75**, 4338.

Herschbach, D. R. (1959). *J. Chem. Phys.* **31**, 91.

Hochstrasser, R. M., Sung, H. N., and Wessel, J. E. (1974). *Chem. Phys. Lett.* **24**, 168.

Holton, G. R., and McClain, W. M. (1976). *Chem. Phys. Lett.* **44**, 436.

Honig, B., Warshel, A., and Karplus, M. (1975). *Acc. Chem. Res.* **8**, 92.

Hudson, B. S., and Kohler, B. E. (1972). *Chem. Phys. Lett.* **14**, 299.

Hudson, B. S., Kohler, B. E. (1973). *J. Chem. Phys.* **59**, 4984.

Hudson, B. S., and Kohler, B. E. (1974). *Annu. Rev. Phys. Chem.* **25**, 437.

Innes, K. K., McSwiney, H. D., Jr., Simmons, J. D., and Tilford, S. G. (1969). *J. Mol. Spectrosc.* **31**, 76.

Johnson, P. M. (1975). *J. Chem. Phys.* **62**, 4562.

Johnson, P. M. (1976). *J. Chem. Phys.* **64**, 4638.

Johnson, P. M. (1980). *Acc. Chem. Res.* **13**, 20.

Johnson, P. M., and Otis, C. E. (1981). *Annu. Rev. Phys. Chem.* **32**, 139.

Johnson, P. M., Berman, M. R., and Zakheim, D. (1975). *J. Chem. Phys.* **62**, 2500.

Jungen, C. (1970) *J. Chem. Phys.* **53**, 4168.

Jungen, C., and Dill, D. (1980). *J. Chem. Phys.* **73**, 3338.

Kasatani, K., Tanaka, Y., Shibuya, K., Kawasaki, M., Obi, K., Sato, H., and Tanaka, I. (1981). *J. Chem. Phys.* **74**, 895.

Kimman, J., Kruit, P., and Wiel, M. J. (1982). *Chem. Phys. Lett.* **88**, 576.

Knight, A. E. W. W., Lawburgh, C. M., and Parmenter, C. S. (1975). *J. Chem. Phys.* **63**, 4336.

Knoth, I., Neusser, H. J., and Schlag, E. W. (1978). *Z. Naturforsch.* **34a**, 979.

Knoth, I., Neusser, H. J., and Schlag, E. W. (1982). *J. Phys. Chem.* **86**, 891.

Lehman, K. K., Smolarek, J., and Goodman, L. (1978). *J. Chem. Phys.* **69**, 1569.

Lin, S. H., and Fujimura, Y. (1979). *Excited States* **4**, 237.

Lord, R. C., Marston, A. L., and Miller, F. A. (1957). *Spectrochim. Acta* **9**, 113.

McDiarmid, R. (1976). *J. Chem. Phys.* **64**, 514.

Mikami, N., and Ito, M. (1975). *Chem. Phys. Lett.* **31**, 472.

Miller, J. C., and Compton, R. N. (1981). *J. Chem. Phys.* **75**, 22.

Mosher, O. A., Flicker, W. M., and Kuppermann, A. (1973). *J. Chem. Phys.* **59**, 6502.

Mulliken, R. S. (1934). *Phys. Rev.* **46**, 549.

Mulliken, R. S. (1971). *J. Chem. Phys.* **55**, 288.

Nikitina, A. N., Ponomareva, N. A., Yanovskaya, L. A., Dombrovskii, V. A., and Kucherov, V. F. (1976). *Opt. Spectrosc. (Engl. Transl.)* **40**, 144.

Parker, D. H., Sheng, S. J., and El-Sayed, M. A. (1976). *J. Chem. Phys.* **65**, 5534.

Petty, G., Tai, C., and Dalby, F. W. (1975). *Phys. Rev. Lett.* **34**, 1207 (1975).

Pierce, B. M., Bennett, J. A., Birge, R. R. (1982). *J. Chem. Phys.* **77**, 6343.

Pierce, B. M., and Birge, R. R. (1982). *J. Phys. Chem.* **86**, 2651.

Price, W. C., and Walsh, W. D. (1947). *Proc. R. Soc. London, Ser. A* **191**, 22.

Robin, M. B. (1974). "Higher Excited States of Polyatomic Molecules," Vol. 1. Academic Press, New York.

Robin, M. B. (1975). "Higher Excited States of Polyatomic Molecules," Vol. 2. Academic Press, New York.

Robin, M. B., and Kuebler, N. A. (1978). *J. Chem. Phys.* **69**, 806.

Sandorfy, C. (1979). *Top. Curr. Chem.* **86**, 91.

Schulten, K., and Karplus, M. (1972). *Chem. Phys. Lett.* **14**, 305.

Scully, D. B., and Whiffen, D. H. (1960). *Spectrochim. Acta* **16**, 1409.

Steinfeld, J. I., and Houston, P. (1978). *In* "Laser and Coherence Spectroscopy" (J. I. Steinfeld, ed.), p. 1. Plenum, New York.

Takagi, H., and Nakamura, H. (1981). *J. Chem. Phys.* **74**, 5808.

Tavan, P., and Schulten, K. (1979). *J. Chem. Phys.* **70**, 5407.

Uneberg, G., Campo, P., and Johnson, P. M. (1980). *J. Chem. Phys.* **73**, 1110.

Vaida, V., Turner, R. E., Casey, J. L., and Colson, S. D. (1978). *Chem. Phys. Lett.* **54**, 25.

Wallace, S. C., and Innes, K. K. (1980). *J. Chem. Phys.* **72**, 4805.

Zacharias, H., Halpern, J. B., and Welge, K. H. (1976). *Chem. Phys. Lett.* **43**, 41.

Zakheim, D., and Johnson, P. M. (1978). *J. Chem. Phys.* **68**, 3644.

7

New Developments in Multiphoton Spectroscopy

In this chapter we shall present some of the new developments in experiment and theory related to multiphoton spectroscopy of molecules. In Section 7.1, multiphoton ionization mass spectroscopy, whose setup was mentioned in Chapter 3, will be described by focusing our attention on theoretical approaches for the explanation of the fragmentation patterns of polyatomic molecules. In Section 7.2, a theory about multiphoton circular dichroism will be discussed by using the time-dependent perturbation method. Ion dip spectroscopy, which was developed by Cooper *et al.* (1981), will be presented in Section 7.3. In this spectroscopy the reduction of ion current is measured. The reduction originates from the depopulation of the pumped level due to stimulated emission induced by a second probe laser. In Section 7.4, we shall describe a theory of two-photon magnetic circular dichroism (MCD). We shall show that the expression for the transition probability of resonant two-photon MCD consists of an R term (called the resonance term), and A, B, and C terms that contribute to the spectrum of a one-photon MCD.

7.1   MULTIPHOTON IONIZATION MASS SPECTROSCOPY

The multiphoton ionization method combined with mass detection, as discussed in Chapter 3, allows us to identify molecules by the optical spectrum related to a resonant intermediate state as well as by their mass. For example, the isotopic species ^{13}C-benzene can be preferentially ionized in a natural isotopic mixture by shifting the wavelength by 1.6 cm^{-1} from the absorption band of light benzene (Boesl et al., 1981). This demonstrates that trace components in a mixture can be ionized without ionizing the major components if the intermediate state spectrum shows sharp features at a resolution of 1 cm^{-1}. Different compounds in a gas mixture can be selectively ionized at different wavelengths to yield the mass spectra of each constituent of the mixture. The fragmentation patterns in multiphoton ionization mass spectroscopy (MPIMS) depend on the laser intensity, and they are different from those obtained by electron impact excitation, charge exchange excitation, and other methods. In MPIMS, ions with small mass weights are produced as compared with those obtained by other methods; for example, C^{+} ions are observed for benzene as shown in Fig. 3.8 (Reilly and Kompa, 1980; Boesl et al., 1980; Bernstein, 1982). Thus the intensity studies add a useful third dimension to MPIMS, which is already two-dimensional in mass and wavelength. For a resonant two-photon ionization process the light intensity allows direct quantitative control of ionization and fragmentation, ranging from soft ionization that produces pure parent-ion mass spectra to hard fragmentation that produces small fragments.

In this section, general features of multiphoton absorption and ionization of molecules will be considered in a qualitative manner. Next, we shall semi-quantitatively explain the fragmentation pattern of polyatomic molecules [for reviews see Bernstein (1982) and Schlag and Neusser (1983)].

7.1.1   General Features

It must be recognized that to be efficient the first one or two absorption steps of multiphoton ionization must involve a transition into a real intermediate level of the S_1 state. A second (or additional) photon(s) then transports the molecule into the ionization continuum. From photoionization spectroscopy it is well known that in general there exist two contributions to the ionization cross section:

(1)   direct ionization to a vibronic level of the molecular ion, and
(2)   autoionization out of Rydberg states.

For the case of multiphoton ionization we expect similar behavior as long as the absorption step that crosses the ionization potential is also the final step of the total multiphoton absorption process. When an additional photon is added after having crossed the ionization potential, the absorption can proceed either via the neutral molecule or the parent ion. In the case of benzene, with an absorption coefficient of about 10^{-17} cm^2 and pumped with a 10^8 W cm^{-2} N$_2$ laser-pumped dye laser system, photon absorption occurs on a time scale of typically 100 ps within the absorption ladder. All of the various other possible processes must be considered in relation to this time scale. Lifetimes of S_1 are usually longer than this. In these cases optical pumping will outrun decay and hence preserve the electronically excited molecule until ionization. Above the ionization threshold, ionization occurs typically within 0.1 to 1 ps and will outrun the optical pumping process. Therefore, optical pumping of the neutral molecule will not be interrupted, and the molecule will cross to a new ladder of the molecular ions. This is referred to as the ladder switching model (Dietz *et al.*, 1982). This ladder switching will be typical for most molecules. However, exceptions may be expected in the case of higher power lasers (> 1 MW) and laser pulses in the picocosecond range, since absorption is much faster for the higher intensity of ultrashort laser pulses. As another exception, small molecules with predissociating states having lifetimes in the 10 ps range would not produce intact parent molecular ions. In this case, ladder switching to neutral fragments occurs even below ionization in the neutral molecules.

When the molecular ions in turn acquire 2–3 photons of energy they will dissociate into fragment ions. These fragment ions again will absorb 2–3 photons, which leads to dissociation. This process of ladder switching continues until atomic fragment ions (C$^+$, etc.) are finally produced.

Three theoretical approaches have been proposed to explain the fragmentation patterns of polyatomic molecules:

(i)   the information-theoretical approach,
(ii) the statistical approach, and
(iii) the rate equation approach.

In Sections 7.1.2–7.1.4 the three approaches will be outlined briefly.

7.1.2   The Information-Theoretical Approach

Let us first outline the information-theoretical approach proposed by Silberstein and Levine (1980, 1981). This method belongs to the extreme statistical limit in which fragmentation is assumed to be independent of the laser intensity, resonance condition, dynamics of the molecule, etc. Fragmentation patterns that exhibit maximum entropy are obtained under the constraints of absorbed energy, elements, and charge.

The number of molecules X_j of species j is given by (Silberstein and Levine, 1980, 1981)

$$X_j = XQ_j \exp\left(-\sum_{k=1}^{K+1} \gamma_k a_{kj}\right),$$ (7.1.1)

where $X = \sum_j X_j$ is the total number of molecules including the parent molecule and fragments, and Q_j is the partition function defined in terms of the energies ε_{ij} of the quantum state i of species j,

$$Q_j = \sum_i \exp(-\beta \varepsilon_{ij}),$$ (7.1.2)

with all the species assumed to be in a heat bath of temperature T ($\beta = 1/kT$). The partition function $\bar{Q}_j$, defined by its internal energy measured from the ground state of the species, is given by

$$Q_j = \exp(-\beta \Delta E_j)\bar{Q}_j,$$ (7.1.3)

where ΔE_j is the heat of formation of species j at 0K. In Eq. (7.11) a_{kj} is the number of atoms of type k (or the charge) of species j. The Lagrangian parameters β and γ_k are determined by taking into account the equations for the constraints of energy, elements, and charge. The mean energy $\langle E \rangle$ absorbed is defined as

$$\langle E \rangle = \sum_j X_j \sum_i \varepsilon_{ij} x_{ij},$$ (7.1.4)

where x_{ij} is the fraction of molecules of species j in the quantum state i,

$$x_{ij} = \exp(-\beta \varepsilon_{ij})/Q_j.$$ (7.1.5)

The total number of atoms of element k or of elementary charge is denoted by C_k, which is given by

$$C_k = \sum_j a_{kj} X_j, \qquad k = 1, 2, \ldots, K + 1,$$ (7.1.6)

where k is the number of different elements in the parent molecule. The $\langle E \rangle$ and C_k are given, and Eqs. (7.1.4) and (7.1.6) are to be solved for β and the γ_k.

By varying $\langle E \rangle$, the breakdown curve that shows the total fraction of positive ions containing a given number of atoms can be found. Silberstein and Levine (1980, 1981) have applied the information-theoretical approach to benzene, toluene, t-butylbenzene, triethylenediamine (1,4-diazobicyclo-[2.2.2]octane, or DABCO) and iodomethane.

7.1.3 The Statistical Approach

We shall outline the statistical approach proposed by Rebentrost *et al.* (1981) and Rebentrost and Ben-Shaul (1981). Rebentrost *et al.* (1981) used an absorption and multiple fragmentation model; that is, only the parent molecular ion absorbs the laser photon, but none of the daughter fragments receive energy from the laser. The fragmentation patterns were calculated by using quasi-equilibrium (see Appendix IV.A), RRKM, or phase-space theory and assuming that the transition states are identical to the products.

Following the treatment by Rebentrost *et al.* (1981), let $P_{ij}(E_i, E_j)\,dE_j$ represent the probability that an ion i with internal energy E_i will decompose into a smaller ion j, with its internal energy in the range of E_j and $E_j + dE_j$, and into a complementary neutral fragment N_{ij}. The branching ratios of the channel $i \to j$ are given by

$$F_{ij}(E_i) = \int dE_j\, P_{ij}(E_i, E_j), \qquad (7.1.7)$$

with

$$\sum_j F_{ij}(E_i) = 1.$$

Energy distributions in the ionic fragments $f_{ij}(E_i, E_j)$ are defined by

$$f_{ij}(E_i, E_j) = P_{ij}(E_i, E_j)/F_{ij}(E_i). \qquad (7.1.8)$$

If the initial energy distribution $P_1(E_1)$ of the parent ion is given, then the energy distributions $P_j(E_j)$ prior to fragmentation can be obtained as

$$P_j(E_j) = \sum_i \int dE_i\, P_i(E_i)P_{ij}(E_i, E_j), \qquad (7.1.9)$$

where the summation is taken over all ions immediately preceding the jth ion. The final probability of observing the j ion is given by

$$P_j = \int_0^{E_j^0} dE_j\, P_j(E_j), \qquad (7.1.10)$$

where E_j^0 is the smallest threshold for any further fragmentation $I_j^+ \to I_k^+ + N_{jk}$.

The probability function $P_{ij}(E_i, E_j)$ in the product phase space model is given by

$$P_{ij}(E_i, E_j) = \alpha_{ij}\rho_j(E_j) \int d\varepsilon_t\, \frac{\rho_n(E_n - \varepsilon_t)\rho_t(\varepsilon_t)}{\Omega_i(E_i)}, \qquad (7.1.11)$$

where α_{ij} is the reaction path degeneracy, ρ_j and ρ_n represent the internal rovibronic state densities of the product ion I_j^+ and its complementary neutral fragment, and Ω_i is the total phase space density at energy E_i for all products formed from I_i^+; that is,

$$\Omega_i(E_i) = \sum_j \alpha_{ij} \int dE_j\, \rho_j(E_j) \int d\varepsilon_t\, \rho_n(E_n - \varepsilon_t)\rho_t(\varepsilon_t). \qquad (7.1.12)$$

These densities can be calculated by assuming harmonic oscillators, classical rotators, and the usual translational motion.

7.1.4 The Rate Equation Approach

In the two approaches previously described fragmentation and subsequent absorption of energy by fragment ions during the period of a laser pulse were ignored. Fragmentations and energy relaxations that compete with photon absorption during a light pulse of duration on the order of 1 ns may take place. This may be called the multiple absorption–fragmentation (MAF) model. In this model, the fast unimolecular reactions produce switching of the excitation ladder to products. By the rate equation approach we can take into account the ladder-switching effect (Fisanick *et al.*, 1980; Dietz *et al.*, 1982).

Let us deal with the rate equation approach proposed by Dietz *et al.* (1982). The rate equation approach for optical processes without fragmentation has been discussed in Sections 4.1.3 and 5.2. To apply the rate equation approach to the MAF model, Dietz *et al.* (1982) assumed that photon absorption is dominated by the stepwise process, that relaxations are expressed in terms of the direct relaxation of electronic excited states to the vibrationally excited electronic ground state (Freiser and Beauchamp, 1975; Orlowski *et al.*, 1976), and that the rovibronic energy E is uniformly distributed among the available molecular states between E and $E + dE$ with a given probability. The fragmentation occurs from the electronic ground state via a bottleneck of the transition states.

Let $C_i(E_i, t)$ denote the population per unit energy (energy distribution function) of ion i with rovibronic energy of the electronic ground state E_i at time t. Taking into account the unimolecular loss to the product ion and unimolecular gain from the precursor ion, and the population loss and gain due to the excitations, the rate equation can be expressed as

$$\frac{dC_i(E_i, t)}{dt} = -k_i(E_i)C_i(E_i, t) + \int_{E_{i-1,i}^u}^{\infty} k_{i-1,i}(E_{i-1}, E_i)$$
$$\times\, C_{i-1}(E_{i-1}, t)dE_{i-1} - k^{op}(E_i)C_i(E_i, t)$$
$$+ k_i^{op}(E_i - \Delta)C_i(E_i - \Delta, t), \qquad (7.1.13)$$

where the first term represents the unimolecular loss with the total rate constant $k_i(E_i)$ for precursor ion populations with energy E_i. The second term in Eq. (7.1.13) is the unimolecular gain from the precursor ion $i - 1$ with the constraint $E_{i,j}^u = \max\{E_{i,j}^0, E_j + E_{i,j}^{00}\}$, where $E_{i,j}^0$ is the threshold for $i \rightarrow j$ fragmentation and $E_{i,j}^{00}$ denotes the difference between the heat of formation of the product ion j plus the corresponding neutral product and that of the precursor ion i. The third and fourth terms represent the loss and gain, respectively, by laser excitation with energy $\Delta = h\nu$. Finally, $k^{op}(E)$ is the average absorption rate constant,

$$k^{op}(E) = \sum_{E \leq E_i \leq E + \Delta E} \sum_{E' \leq E_f \leq E' + \Delta E} \frac{2\pi |V_{if}|^2}{\hbar} \frac{1}{\Delta E^2 \, N(E)}, \quad (7.1.14)$$

with $V_{if} = \langle i | \boldsymbol{\mu} \cdot \mathbf{E}_0 | f \rangle$ and $E' = E + \Delta$. As an initial condition, the precursor ion A_1^+ with an initial energy distribution $C_1(E_1, 0)$ is present at the beginning ($t = 0$) of the rectangular laser pulse.

To take into account the switching of the excitation ladder, Dietz et al. (1982) defined a switching point by the equation

$$k_i^{op}(E_i^M) = k_i(E_i^M), \quad (7.1.15)$$

where E_i^M denotes the energy at the switching point. The unimolecular rate $k_i(E_i)$ is zero below E_i^M, and the absorption rate $k_i^{op}(E_i)$ is zero above that point. Let $C_i^>(E_i, t)$ and $C_i^<(E_i, t)$ represent the energy distributions of ions with energies above and below the switching point, respectively; they can be expressed as

$$C_i^>(E_i, t) = k_i^{-1}(E_i)\left[k_i^{op}(E_i - \Delta)C_i^<(E_i - \Delta, t)\right.$$
$$\left. + \int_{E_{i-1}^M}^{E_{i-1}^M + \Delta} k_{i-1,i}(E_{i-1}, E_i)C_{i-1}^>(E_{i-1}, t)\,dE_{i-1}\right] \quad (7.1.16)$$

and

$$C_i^<(E_i, t) = \int_0^t \exp\left[-k_i^{op}(E_i)(t - \tau)\right]\left[k_i^{op}(E_i - \Delta)C_i^<(E_i - \Delta, \tau)\right.$$
$$\left. + \int_{E_{i-1}^M}^{E_{i-1}^M + \Delta} k_{i-1,i}(E_{i-1}, E_i)C_{i-1}^>(E_{i-1}, \tau)\,dE_{i-1}\right]d\tau.$$
$$(7.1.17)$$

In deriving Eq. (7.1.16), $k_i(E_i)\tau_p \gg 1$ for $E_i > E_i^M$ has been used. Here τ_p is the pulse duration.

Applying Eq. (7.1.16) several times in Eq. (7.1.17) yields a connection between the stable energy distributions:

$$C_i^<(E_i, t) = \hat{T}_i(E_i, t)\left[C_i^<(E_i - \Delta, \tau) + \sum_{s=1}^{i-1} \hat{J}_s^i(E_i) \frac{k_s^{op}(E_s - \Delta)}{k_i^{op}(E_i - \Delta)} C_s^<(E_s - \Delta, \tau)\right]$$
$$(7.1.18)$$

for $i > 1$ and

$$C_1^<(E_1, t) = \hat{T}_1(E_1, t)C_1(E_1 - \Delta, \tau) + \exp[-k_1(E_1)t]C_1(E_1, 0) \quad (7.1.19)$$

for $i = 1$. Here the operators $\hat{T}_i$ and $\hat{J}_s^i$ are defined as

$$\hat{T}_i(E_i, t) = k_i^{op}(E_i - \Delta)\int_0^t \exp[-k_i^{op}(E_i)(t - \tau)]f(\tau)\,d\tau \quad (7.1.20)$$

and

$$\hat{J}_s^i(E_i)f(E_s) = \int_{E_{i-1}^M}^{E_{i-1}^M + \Delta} q_{i-1}(E_{i-1}, E_i)\int_{E_{i-2}^M}^{E_{i-2}^M + \Delta} q_{i-2}(E_{i-2}, E_{i-1})$$

$$\cdots \int_{E_s}^{E_s^M + \Delta} q_s(E_s, E_{s+1})f(E_s)dE_s \cdots dE_{i-2}\,dE_{i-1}, \quad (7.1.21)$$

where q_i is the probability function for obtaining ion $i + 1$ with internal energy E_{i+1} from ion i with E_i and is defined by

$$q_i(E_i, E_{i+1}) = k_{i,i+1}(E_i, E_{i+1})/k_i(E_i). \quad (7.1.22)$$

The operator $\hat{J}_s^i(E_i)$ is a generating operator that produces energy distribution $f_i(E_i)$ in ion i from any energy distribution $f_s(E_s)$ in ion s after photon absorption in the ions and $i - s$ consecutive unimolecular processes.

If the absorption rates $k_i(E_i)$ are independent of the energy E_i and, further, if the k_i for the different ions i are replaced by a mean absorption rate $\bar{k}$, Eq. (7.1.19) can be expressed as

$$C_i(E_i, t) = \sum_\alpha \rho_\alpha^i(E_i)H_\alpha(t), \quad (7.1.23)$$

where $\rho_\alpha^i(E_i)$, the abundance of ion i with energy E_i after the absorption of α photons, is given by

$$\rho_\alpha^i(E_i) = \rho_{\alpha-1}^i(E_i - \Delta) + \sum_{s=1}^{i-1} \hat{J}_s^i(E_i)\rho_{\alpha-1}^s(E_s - \Delta). \quad (7.1.24)$$

In Eq. (7.1.23), $H_\alpha(t)$, the probability for the absorption of α photons, has the recursion formula

$$H_\alpha(t) = \hat{T}(t)H_{\alpha-1}(\tau) = \bar{k}\int_0^t \exp[-\bar{k}(t - \tau)]H_{\alpha-1}(\tau)\,d\tau. \quad (7.1.25)$$

The abundance $C_i(t)$ of ion i can finally be obtained by integrating Eq. (7.1.23) over E_i:

$$C_i(t) = \int_0^{E_i^M} C_i(E_i, t)\,dE_i = \sum_\alpha \rho_\alpha^i H_\alpha(t). \quad (7.1.26)$$

So far multiphoton ionization and fragmentation of a single reaction chain have been considered. There may be many fragmentation paths that lead from the parent molecule to the smallest atomic ions. A generalization of the preceding treatment to all possible branching reactions $s \to j$ with $j > s$ can be also explained as follows:

(1) In Eq. (7.1.13), $k_i(E_i)$ is replaced by the sum $\sum_{j>i} k_{i,j}(E_i)$.
(2) The unimolecular gain is replaced by the sum $\sum_{j<i} \int_{E_{j,i}^u}^{\infty} k_{j,i}(E_j, E_i)$ $C_j(E_j, t) dE_j$ over all precursor ions j.

By substituting these expressions for the rate constants into Eq. (7.1.13) and performing the same operations as those described in the preceding treatment, we again obtain Eq. (7.1.21) with a new meaning for $\hat{J}_s^i(E_i)$:

$$\hat{J}_s^i(E_i) = \sum_{r=1}^{i-s} \hat{A}_{r,s}^i(E_i), \qquad (7.1.27)$$

where

$$\hat{A}_{r,s}^i(E_i) f(E_s) = \sum_{j'<i} \int_{E_{j'}^M}^{E_{j'}^M+\Delta} q_{j'}(E_{j'}, E_i) \sum_{j''<j'} \int_{E_{j''}^M}^{E_{j''}^M+\Delta} q_{j''}(E_{j''}, E_{j'})$$

$$\cdots \sum_{j(r-1)<j(r-2)} \int_{E_{j(r-1)}^M}^{E_{j(r-1)}^M+\Delta} q_{j(r-1)}(E_{j(r-1)}, E_{j(r-2)})$$

$$\times \int_{E_s}^{E^M+\Delta} q_s(E_s, E_{j(r-1)}) f(E_s) \, dE_s \, dE_{j(r-1)} \cdots dE_{j'}. \qquad (7.1.28)$$

This rate equation approach has been applied to the benzene molecule (Dietz et al., 1982). It is recognized (Boesl et al., 1980) that the initial process of multiphoton absorption and fragmentation of benzene, which is excited by using a laser with $\lambda = 2590$ Å, is a switching of neutral to ionic benzene as soon as the first ionization potential is reached. The contribution of highly excited autoionization states can be neglected in this case. The condition at $t = 0$ is

$$\rho_3^1(E_1) = \begin{cases} 1 & \text{for } E_1 = 4.9 \text{ eV}, \\ 0, & \text{otherwise;} \end{cases}$$

$$A_1^+ = C_6H_6^+;$$

$$\rho_i^i(E_i) = 0 \qquad \text{for } i = 1, 2, \ldots, 20.$$

The parameter $q_i(E_i, E_k)$ for each unimolecular reaction step is calculated by using the quasi-equilibrium theory described in Appendix IV.A.

In Fig. 7.1 the breakdown curve is shown as a function of laser intensity. We note that there is a slow onset of C^+ abundance from about 8.0 to 8.4 W cm^{-2}, followed by a sharp onset from 1 to 30% C^+ over a half order

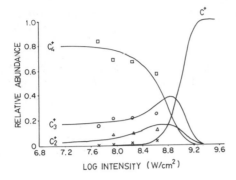

Fig. 7.1 Relative C_i^+ abundance versus laser intensity. Here C_i stands for all ions with different numbers of H atoms but a constant number of C atoms: calculated (—) and measured ($\bigcirc, \square, \blacktriangle, X$) values. [From Dietz *et al.* (1982).]

of magnitude. In Fig. 7.2, the calculated and experimental mass spectra of benzene are shown. The appearance of the C groups is satisfactorily reproduced, although there are some discrepancies of the peak heights in other C groups between the experimental and calculated spectra. To explain quantitatively the fragmentation patterns, it is necessary to determine the potential surface of each unimolecular reaction. It seems that the vari-

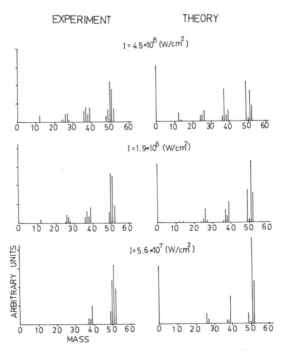

Fig. 7.2 Experimental versus theoretical multiphoton mass spectra of benzene for three different light intensities. Fragments with $m > 76$ are not shown. [From Dietz *et al.* (1982).]

ous approximations adopted here should be examined with further refined treatments.

7.2 MULTIPHOTON CIRCULAR DICHROISM

Measurements of one-photon optical activity have played a very important role in determining the molecular structures of organic and inorganic molecules (Caldwell and Eyring, 1971). In view of the increasing interest in multiphoton absorption, one can expect that multiphoton optical activity will soon become an important tool in determining molecular structures.

Circular dichroism is defined as the difference in the absorption coefficients for left and right circularly polarized light (Caldwell and Eyring, 1971). Any molecule that lacks a point or plane of symmetry can demonstrate circular dichroism; the molecule is said to be optically active or chiral. Circular dichroism is related to optical rotation by the Kramers–Kronig relation.

In this section we shall show how to treat multiphoton optical activity. For this purpose, we choose to study two-photon circular dichroism (Tinoco, 1975; Power, 1975; Andrews, 1976), and the time-dependent perturbation method will be employed. Notice that for the time-dependent Schrödinger equation

$$\hat{H}\Psi = i\hbar \, \partial\Psi/\partial t \tag{7.2.1}$$

and

$$\hat{H} = \hat{H}_0 + \hat{H}', \tag{7.2.2}$$

where $\hat{H}'$ is time dependent. Equation (7.2.1) can be solved by letting (see Chapter 2)

$$\Psi = \sum_n C_n \Psi_n^0, \tag{7.2.3}$$

where

$$\Psi_n^0 = \psi_n \exp(-itE_n/\hbar).$$

Substituting Eq. (7.2.3) into Eq. (7.2.1) yields

$$i\hbar \frac{dC_m}{dt} = \sum_k C_k \langle \Psi_m^0 | \hat{H}' | \Psi_k^0 \rangle. \tag{7.2.4}$$

To apply the perturbation method to Eq. (7.2.4), we first introduce the perturbation parameter λ into Eq. (7.2.4),

$$i\hbar \frac{dC_m}{dt} = \lambda \sum_k C_k \langle \Psi_m^0 | \hat{H}' | \Psi_k^0 \rangle, \tag{7.2.5}$$

and then use the perturbation expansion

$$C_m = C_m^{(0)} + \lambda C_m^{(1)} + \lambda^2 C_m^{(2)} + \cdots. \qquad (7.2.6)$$

If the system is initially in the nth state, then from Eqs. (7.2.5) and (7.2.6) we obtain

$$C_m^{(1)} = \frac{1}{i\hbar} \int_0^t d\tau \langle \Psi_m^{(0)}(\tau) | \hat{H}'(\tau) | \Psi_n^{(0)}(\tau) \rangle, \qquad (7.2.7)$$

$$C_m^{(2)} = \frac{1}{(i\hbar)^2} \sum_k \int_0^t d\tau_1 \langle \Psi_m^0(\tau_1) | \hat{H}'(\tau_1) | \Psi_k^0(\tau_1) \rangle$$

$$\times \int_0^{\tau_1} d\tau_2 \langle \Psi_k^0(\tau_2) | \hat{H}'(\tau_2) | \Psi_n^0(\tau_2) \rangle, \qquad (7.2.8)$$

and so on.

In the semiclassical theory of radiation, $\hat{H}'$ takes the form

$$\hat{H}' = \hat{V} \exp(it\omega) + \hat{V}^* \exp(-it\omega), \qquad (7.2.9)$$

where ω is the optical frequency. In other words, we shall treat only the two-photon circular dichroism (TPCD) of identical frequencies. Substituting Eq. (7.2.9) into Eq. (7.2.8) yields

$$C_m^{(2)} = \sum_k \frac{V_{mk}^* V_{kn}^*}{\hbar^2 (\omega_{kn} - \omega)} \left(\frac{e^{it(\omega_{mn} - 2\omega)} - 1}{\omega_{mn} - 2\omega} - \frac{e^{it(\omega_{mk} - \omega)} - 1}{\omega_{mk} - \omega} \right), \qquad (7.2.10)$$

where, for example, $\omega_{kn} = (E_k - E_n)/\hbar$, and so forth. Near the resonance condition, that is, $2\omega \sim \omega_{mn}$, the second term in Eq. (7.2.10) is negligible and thus in the long time limit we obtain

$$\frac{|C_m^{(2)}|^2}{t} = \frac{2\pi}{\hbar^4} \left| \sum_k \frac{V_{mk}^* V_{kn}^*}{\omega_{kn} - \omega} \right|^2 \delta(\omega_{mn} - 2\omega). \qquad (7.2.11)$$

As shown in Chapter 2, when the interaction between the system and the reservoir is taken into account, the delta function $\delta(\omega_{mn} - 2\omega)$ is replaced by the Lorentzian, and a level shift and a damping constant (or dephasing constant) should be introduced into the term $\omega_{kn} - \omega$.

To determine $\hat{V}$ or $\hat{V}^*$ for optical activity, we notice that

$$\hat{H}' = -\frac{e}{mc} \sum_j \mathbf{A}_j \cdot \mathbf{P}_j + \cdots, \qquad (7.2.12)$$

with

$$\mathbf{A}_j = \tfrac{1}{2} [\mathbf{A}(\mathbf{r}_j) e^{it\omega} + \mathbf{A}^*(\mathbf{r}_j) e^{-it\omega}], \qquad (7.2.13)$$

and so

$$\hat{H}' = -\frac{e}{2mc} \sum_j \mathbf{A}(\mathbf{r}_j) \cdot \mathbf{P}_j e^{it\omega} - \frac{e}{2mc} \sum_j \mathbf{A}^*(\mathbf{r}_j) \cdot \mathbf{P}_j e^{-it\omega}. \qquad (7.2.14)$$

It follows that

$$\hat{V}^* = -\frac{e}{2mc} \sum_j \mathbf{A}^*(\mathbf{r}_j) \cdot \mathbf{P}_j, \qquad (7.2.15)$$

where

$$\mathbf{A}(\mathbf{r}_j) = \mathbf{A} \exp(-2\pi i \mathbf{k} \cdot \mathbf{r}_j). \qquad (7.2.16)$$

Here $\mathbf{k}$ denotes the wave vector that determines the direction of propagation of incident radiation. In most cases, $2\pi |\mathbf{k} \cdot \mathbf{r}_j| < 1$. Thus we have

$$\mathbf{A}^*(\mathbf{r}_j) = \mathbf{A}^* \exp(2\pi i \mathbf{k}^* \cdot \mathbf{r}_j) = \mathbf{A}^*(1 + 2\pi i \mathbf{k}^* \cdot \mathbf{r}_j + \cdots). \qquad (7.2.17)$$

The vector potential $\mathbf{A}$ will depend on the polarization of light, that is,

$$\hat{e}_\pm = (1/\sqrt{2})(\hat{i} \pm i\hat{j}) \qquad (7.2.18)$$

for circularly polarized light (see Chapter 4). The corresponding vector potentials can be expressed as

$$\hat{A}_\pm = \hat{e}_\pm A_0. \qquad (7.2.19)$$

With Eq. (7.2.17), $\hat{V}^*$ can be written as

$$\hat{V}^* = -\frac{e}{2mc} [(\mathbf{A}^* \cdot \mathbf{P}) + 2\pi i \sum_j (\mathbf{k}^* \cdot \mathbf{r}_j)(\mathbf{A}^* \cdot \mathbf{P}_j)], \qquad (7.2.20)$$

where $\mathbf{P} = \sum_j \mathbf{P}_j$.
  Notice that

$$2 \sum_j \mathbf{r}_j \mathbf{P}_j = \sum_j \{[(\mathbf{r}_j \mathbf{P}_j) - (\mathbf{P}_j \mathbf{r}_j)] + [(\mathbf{r}_j \mathbf{P}_j) + (\mathbf{P}_j \mathbf{r}_j)]\} \qquad (7.2.21)$$

and

$$\mathbf{k}^* \cdot \sum_j [(\mathbf{r}_j \mathbf{P}_j) - (\mathbf{P}_j \mathbf{r}_j)] \cdot \mathbf{A}^* = (\mathbf{k}^* \times \mathbf{A}^*) \cdot \mathbf{M}, \qquad (7.2.22)$$

where $\mathbf{M}$ represents the angular momentum operator

$$\mathbf{M} = \sum_j \mathbf{r}_j \times \mathbf{P}_j. \qquad (7.2.23)$$

Thus Eq. (7.2.20) becomes (Lin, 1971)

$$\hat{V}^* = -\frac{e}{2mc} [(\mathbf{A}^* \cdot \mathbf{P}) + \pi i (\mathbf{k}^* \times \mathbf{A}^*) \cdot \mathbf{M} + \pi i (\mathbf{k}^* \cdot \bar{\bar{Q}} \cdot \mathbf{A}^*)], \quad (7.2.24)$$

where $\bar{\bar{Q}}$ is defined by

$$\bar{\bar{Q}} = \sum_j (\mathbf{r}_j \mathbf{P}_j + \mathbf{P}_j \mathbf{r}_j) \qquad (7.2.25)$$

and is proportional to the quadrupole moment operator.

From Eqs. (7.2.11) and (7.2.15) we can see that the transition probability for two-photon absorption depends on the polarization of the light. For TPCD, we are concerned with the difference in transition probabilities due to the absorption of $\hat{e}_+$ light and $\hat{e}_-$ light, that is,

$$R_{mn} = |C_m^{(2)}|_+^2/t - |C_m^{(2)}|_-^2/t$$
$$= \frac{2\pi}{\hbar^4} g(\omega_{mn} - 2\omega)\left(\left|\sum_k \frac{V_{mk}^*(+)V_{kn}^*(+)}{\omega_{kn} - \omega}\right|^2 - \left|\sum_k \frac{V_{mk}^*(-)V_{kn}^*(-)}{\omega_{kn} - \omega}\right|^2\right),$$
$$(7.2.26)$$

where $g(\omega_{mn} - 2\omega)$ represents the line shape function (it may be a delta function or a Lorentzian) and, for example,

$$V_{mk}^*(+) = -\frac{e}{2mc}\left[(\mathbf{A}_+^* \cdot \mathbf{P}_{mk}) + \pi i(\mathbf{k}^* \cdot \mathbf{A}_+^*) \cdot \mathbf{M}_{mk} + \pi i(\mathbf{k}^* \cdot \bar{\bar{Q}}_{mk} \cdot \mathbf{A}_+^*)\right].$$
$$(7.2.27)$$

For randomly oriented molecules, a spatial average must be carried out over the expression for R_{mn} given by Eq. (7.2.26). This has been accomplished by Monson and McClain (1970). In terms of their notation, we obtain

$$R_{mn} = \frac{\pi}{15\hbar^4}\left(\frac{e}{2mc}|A_0|\right)^4 [F\,\Delta(S_{\alpha\alpha}\bar{S}_{\beta\beta}) + G\,\Delta(S_{\alpha\beta}\bar{S}_{\alpha\beta}) + H\,\Delta(S_{\alpha\beta}\bar{S}_{\beta\alpha})],$$
$$(7.2.28)$$

where $F = -2$, $G = H = 3$,

$$\Delta(S_{\alpha\alpha}\bar{S}_{\beta\beta}) = -4\pi|\mathbf{k}|\,\text{Re}\{\alpha_{mn}^{\text{PP}}(\omega)[\alpha_{mn}^{\text{PM}}(\omega)^* + \alpha_{mn}^{\text{MP}}(\omega)^*]\}, \qquad (7.2.29)$$

$$\Delta(S_{\alpha\beta}\bar{S}_{\alpha\beta}) = -4\pi|\mathbf{k}|\,\text{Re}[\bar{\bar{\alpha}}_{mn}^{\text{PM}}(\omega):\bar{\bar{\alpha}}_{nm}^{\text{PP}}(\omega) + \bar{\bar{\alpha}}_{mn}^{\text{PP}}(\omega):\bar{\bar{\alpha}}_{nm}^{\text{PM}}(\omega)], \qquad (7.2.30)$$

and

$$\Delta(S_{\alpha\beta}\bar{S}_{\beta\alpha}) = -4\pi|\mathbf{k}|\,\text{Re}[\bar{\bar{\alpha}}_{mn}^{\text{PP}}(\omega):\bar{\bar{\alpha}}_{mn}^{\text{MP}}(\omega)^* + \bar{\bar{\alpha}}_{mn}^{\text{PP}}(\omega):\bar{\bar{\alpha}}_{mn}^{\text{PM}}(\omega)^*]. \quad (7.2.31)$$

In Eqs. (7.2.29)–(7.2.31) the following notations have been used:

$$\alpha_{mn}^{\text{PP}}(\omega) = \sum_k \frac{\mathbf{P}_{mk} \cdot \mathbf{P}_{kn}}{\omega_{kn} - \omega}, \qquad (7.2.32)$$

$$\alpha_{mn}^{\text{PM}}(\omega) = \sum_k \frac{\mathbf{P}_{mk} \cdot \mathbf{M}_{kn}}{\omega_{kn} - \omega}, \qquad (7.2.33)$$

$$\bar{\alpha}_{mn}^{PP}(\omega) = \sum_k \frac{\mathbf{P}_{mk}\mathbf{P}_{kn}}{\omega_{kn} - \omega}, \tag{7.2.34}$$

$$\bar{\alpha}_{mn}^{PM}(\omega) = \sum_k \frac{\mathbf{P}_{mk}\mathbf{M}_{kn}}{\omega_{kn} - \omega}, \tag{7.2.35}$$

and so on.

Note that in this section we have treated the TPCD of two photons with identical frequency and parallel propagation; other cases have been discussed by Tinoco (1975). Furthermore, only the expression of TPCD for a particular transition $n \to m$ has been given; for practical application, this expression has to be modified by including the Boltzmann average, the medium effect, if it exists, and the summation over all possible transitions consistent with energy conservation. The adiabatic approximation can then be introduced into the TPCD expression to interpret the experimental results.

7.3 ION DIP SPECTROSCOPY

Ion dip spectroscopy (IDS), a new high-resolution multiphoton spectroscopy based on competition between ionization and stimulated emission (or stimulated absorption), has been developed by Cooper *et al.* (1981). Its principle for the two-photon ionization (TPI) case can be described by a four-level system with ground state $|a\rangle$, an excited vibrational state $|v\rangle$, an excited electronic state $|m\rangle$, and an ionized state $|n\rangle$ (see Fig. 7.3). Two photons at ω_1 induce efficient ionization because resonance enhancement occurs through state $|m\rangle$. A high-intensity laser probe at frequency ω_2 is introduced. When $\omega_1 - \omega_2$ matches suitable ground-state vibrational frequencies, stimulated emission competes with ionization; this decreases the effective population in $|m\rangle$ and reduces the ionization signal (i.e., ion dip). Ion dips may also take place if $|v\rangle$ is at a higher energy than $|m\rangle$ (i.e., in this case, it is due to stimulated absorption). Applying the master equation approach to

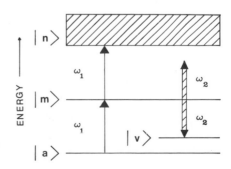

Fig. 7.3 Simplified energy level for ionization dip spectroscopy.

this four-level model, we find

$$dp_{mm}/dt = k^{(1)}_{a \to m}\rho_{aa} - (k^{(0)}_{mm} + k^{(1)}_{m \to v} + k^{(1)}_{m \to n})\rho_{mm} \qquad (7.3.1)$$

and

$$dp_{nn}/dt = k_{a \to n}\rho_{aa} + k^{(1)}_{m \to n}\rho_{mm}, \qquad (7.3.2)$$

where $k^{(1)}_{m \to v}$ denotes the stimulated radiative rate constant for $m \to v$. Here, for simplicity, the total decay rate of $|n\rangle$ (i.e., the molecular ion) has been ignored. The four-level model of IDS can be solved exactly. In the following discussion we present some preliminary results for the case

$$\rho_{aa} \gg \rho_{nn}, \qquad \rho_{aa} \gg \rho_{mm}, \qquad \rho_{aa} \gg \rho_{vv}.$$

If we let $I(k^{(1)}_{m \to v} \neq 0)$ and $I(k^{(1)}_{m \to v} = 0)$ represent the ion currents in the presence and absence of $k^{(1)}_{m \to v}$, respectively, then the observed ion dip signal can be expressed as

$$\Delta I = I(k^{(1)}_{m \to v} \neq 0) - I(k^{(1)}_{m \to v} = 0)$$

$$= -K \frac{\rho_{aa} k^{(1)}_{a \to m} k^{(1)}_{m \to n} k^{(1)}_{m \to n}}{(k^{(0)}_{mm} + k^{(1)}_{m \to n})(k^{(0)}_{mm} + k^{(1)}_{m \to n} + k^{(1)}_{m \to n})}. \qquad (7.3.3)$$

Here it has been assumed that $t(k^{(0)}_{mm} + k^{(1)}_{m \to n}) \gg 1$ and K is a proportional constant. From the preceding equation we can see that if the saturation effect due to ω_2 is negligible, then the ion dip signal ΔI is indeed proportional to $k^{(1)}_{m \to v}$, the stimulated radiative rate constant for $m \to v$, and that the preceding four-level model can also explain the saturation effect due to ω_2. It is apparent that the theoretical model for IDS can easily be extended to

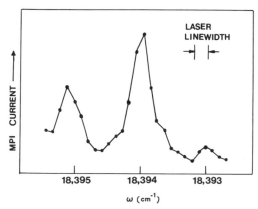

Fig. 7.4 Multiphoton ionization spectrum of I_2 in the region $B(v' = 26) \leftarrow X(v'' = 0)$.

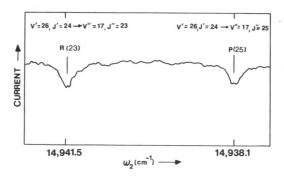

Fig. 7.5 Ionization dip spectrum of I_2 recorded with $\omega_1 = 18,393.9$ cm^{-1}.

multiphoton ionization (MPI) processes that require more than two ω_1 photons for ionization. More realistic multilevel models of IDS should be developed to study the effect of vibrational relaxation in each electronic manifold on dip spectra.

To demonstrate the use IDS, Cooper *et al.* (1981) applied this technique to I_2. In the presence of ω_1 set at the MPI peak at 18,393.9 cm^{-1} (Fig. 7.4) a second laser induces spectrally sharp dips ($\Delta\omega < 0.2$ cm^{-1}) in the ionization current at $\omega_2 = 14,941.5$ cm^{-1} and $\omega_2 = 14,938.1$ cm^{-1}. These ω_2 transitions are readily assigned to $R(23)$ and $P(25)$ of $B(v' = 26) \rightarrow X(v'' = 17)$ (see Fig. 7.5). Similar dips were observed for transitions $R(23)$ and $P(25)$ of $v' = 26 \rightarrow v'' = 15$ and $P(25)$ of $v' = 26 \rightarrow v'' = 14$. Failure to observe dips for $v' = 26 \rightarrow v'' = 16$ is consistent with the small Franck–Condon factor for this transition. A second set of ion dips corresponding to $R(15)$ and $P(17)$ was observed for $v' = 26 \rightarrow v'' = 15, 14$ ($v'' = 17$ was not investigated) when the ω_1 laser was tuned into resonance with the second strong MPI feature at 18,395.1 cm^{-1} (Fig. 7.4). For all transitions the measured positions are within ± 0.3 cm^{-1} of the accurately known $B \rightarrow X$ transition energies, and P to R branch intervals are within ± 0.2 cm^{-1} of calculated values. Thus the ion dip technique allows assignment of specific intermediate states in an MPI spectrum. The simplicity of the ion dip spectra indicates that each MPI peak involves a single rovibronic state in the B electronic stage.

Cooper *et al.* (1981) also studied the ion dip transition amplitude on I_2 with $\omega_1 = 18,395.1$ cm^{-1} as a function of the intensity of the ω_2 beam for $R(15)$ and $P(17)$ of $v' = 26 \rightarrow v'' = 15, 14$. Observed data points for three separate transitions are well represented by the expected functional form of optical saturation (Fig. 7.6) of a two-level system ($|m\rangle - |a\rangle$). This saturation behavior is indeed predicted by Eq. (7.3.3).

Although I_2 has been studied extensively (Mulliken, 1971), relatively few of the 43 valence shell states have been assigned. The $B \leftrightarrow X$ system in the

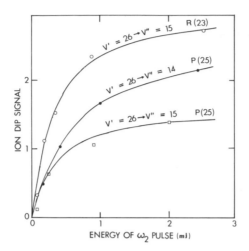

Fig. 7.6 Optical saturation curves of I_2 obtained by varying the neutral density in the ω_2 beam.

region of $15{,}000–20{,}000 \text{ cm}^{-1}$ is now well characterized and the lower energy A and A' states are firmly established (Ashby and Johnson, 1980; Tellinghuisen, 1977; Simmons and Hougen, 1977). Several Rydberg states have been studied by one-wavelength MPI spectroscopy (Lehman *et al.*, 1978; Tai and Dalby, 1978). The study of ion pair states is currently a very active field, and the electronic term energy and assignment of several states are uncertain (Chen *et al.*, 1978; Guy *et al.*, 1980; Williamson, 1979; King *et al.*, 1981). These states have characteristically low vibrational frequencies and large internuclear separations.

Cooper and Wessel (1982) have recently employed the IDS technique to detect a new state in I_2. As described earlier, the first step is excitation to the $v' = 26$ level of the B electronic state by using $\omega_1 = 18{,}395.1 \text{ cm}^{-1}$ (see Fig. 7.4). The optical field generated by a second laser (ω_2) interacts with the intermediate state (perhaps the F state that is $3\omega_1$ above the ground state) and causes transitions upward to a newly discovered state that is arbitrarily labeled the K state.

The new state is characterized by an extensive 37-cm^{-1} vibrational progression, an internuclear separation of about $5.3 \pm 0.2 \text{ Å}$, and a term energy around $70{,}000 \text{ cm}^{-1}$, which is near the ionization limit.

The preceding results show that IDS can become an effective method for studying transitions in polyatomic molecules that are not easily accessible by conventional techniques. An application of IDS to polyatomic molecules was reported by Murakami *et al.* (1982), who observed the ion dip intensities of the vibronic bands of toluene and aniline. They used a supersonic free jet in order to avoid an obscuring of the ion dip intensity by simultaneous multiphoton ionization from thermally populated vibrational levels.

7.4   MULTIPHOTON MAGNETIC CIRCULAR DICHROISM

When a magnetic field is applied along a propagation direction of circularly polarized light, circular dichroism in a medium can be observed; this is called magnetic circular dichroism (MCD). While natural circular dichroism is a relatively rare phenomenon, MCD is a property of all matter. One-photon MCD has been widely used by chemists and physicists to study the electronic structure of molecules by measuring MCD in regions of electron absorption (Buckingham and Stephens, 1966; Stephens, 1976; Caldwell and Eyring, 1971). However, to our knowledge, the observation of multiphoton MCD spectra has not been reported. In this section we shall present a theoretical consideration of the multiphoton MCD in hope that this will help experimenters to observe the MCD spectra of molecules. For this purpose, we restrict ourselves to two-photon MCD.

Wagnière (1979) has theoretically analyzed the polarization effects on two-photon MCD and derived the transition probability of two-photon MCD for nondegenerate initial and final states. The temperature effect was not included in his treatment. A general theory of two-photon MCD caused by the absorption of identical photons has been developed for nonresonant and resonant intermediate state cases by Lin and co-workers (1983). This theory is applicable to both degenerate and nondegenerate transitions, and both temperature and resonance effects are considered in the theory.

By using the semiclassical theory of radiation, the interaction Hamiltonian between the system and the radiation field in the dipole approximation can be expressed as

$$\hat{H}' = -(e/2mc)[\mathbf{A}_0 \exp(it\omega_R) + \mathbf{A}_0^* \exp(-it\omega_R)] \cdot \mathbf{P}, \qquad (7.4.1)$$

where $\mathbf{A}_0$ denotes the vector potential of the radiation field and $\mathbf{P}$ represents the total linear momentum operator of the system. For convenience, Eq. (7.4.1) will be rewritten as

$$\hat{H}' = \hat{V} \exp(it\omega_R) + \hat{V}^* \exp(-it\omega_R), \qquad (7.4.2)$$

where

$$\hat{V} = -(e/2mc)\mathbf{A}_0 \cdot \mathbf{P} \qquad (7.4.3)$$

and

$$\hat{V}^* = -(e/2mc)\mathbf{A}_0^* \cdot \mathbf{P}. \qquad (7.4.4)$$

By taking into account dephasing (or damping) effects, the transition probability for two-photon absorption is given by

$$W(\omega_R) = \frac{2\pi}{\hbar^4} \sum_a \sum_b P_a \left| \sum_k \frac{V_{bk}^* V_{ka}^*}{\omega_{ka} - \omega_R + i\gamma_{ka}} \right|^2 f_{ab}(\omega_R), \qquad (7.4.5)$$

where P_a represents the Boltzmann factor, γ_{ka} is the dephasing (or damping) constant between states k and a, and $f_{ab}(\omega_R)$ the line shape function for $a \rightarrow b$, i.e.,

$$f_{ab}(\omega_R) = \frac{1}{\pi} \frac{\gamma_{ba}}{(\omega_{ba} - 2\omega_R)^2 + \gamma_{ba}^2} \qquad (7.4.6)$$

for the Lorentzian line shape. Note that as $\gamma_{ba} \rightarrow 0$, $f_{ab}(\omega_R)$ reduces to the delta function.

We first consider two-photon MCD in the case of the nonresonant intermediate state. Using the relation $\mathbf{P}_{ba} = im\omega_{ba}\mathbf{R}_{ba}/e$, where $\mathbf{R}_{ba}$ is the matrix element of the transition moment, and $\mathbf{A}_0 = A_0\hat{\mathbf{e}}$, where $\hat{\mathbf{e}}$ represents the unit polarization vector, Eq. (7.4.5) can be rewritten as

$$W(\omega_R) = K\omega_R^4 \sum_a \sum_b P_a M_{ab}(\omega_R) f_{ab}(\omega_R), \qquad (7.4.7)$$

where $K = A_0^4/8\pi^3 c^4$ and

$$M_{ab}(\omega_R) = \left| \sum_k \frac{(\hat{\mathbf{e}} \cdot \mathbf{R}_{ak})(\hat{\mathbf{e}} \cdot \mathbf{R}_{kb})}{\omega_{ka} - \omega_R} \right|^2. \qquad (7.4.8)$$

In the case of two identical photons, MCD can be induced if the magnetic field is applied along the direction of light propagation. In this case, the transition probability of two-photon MCD is given by

$$\Delta W(\omega_R) = K\omega_R^4 \sum_a \sum_b P_a \Delta M_{ab}(\omega_R) f_{ab}(\omega_R), \qquad (7.4.9)$$

where

$$\Delta M_{ab}(\omega_R) = \left| \sum_k \frac{(\hat{\mathbf{e}}_l \cdot \mathbf{R}_{ak})(\hat{\mathbf{e}}_l \cdot \mathbf{R}_{kb})}{\omega_{ka} - \omega_R} \right|^2 - \left| \sum_k \frac{(\hat{\mathbf{e}}_r \cdot \mathbf{R}_{ak})(\hat{\mathbf{e}}_r \cdot \mathbf{R}_{kb})}{\omega_{ka} - \omega_R} \right|^2 \qquad (7.4.10)$$

and

$$\hat{\mathbf{e}}_l = (\hat{i} - i\hat{j})/\sqrt{2} \quad \text{and} \quad \hat{\mathbf{e}}_r = (\hat{i} + i\hat{j})/\sqrt{2} \qquad (7.4.11)$$

Here $\hat{i}$ and $\hat{j}$ denote the unit vectors along the space-fixed X and Y directions, respectively. If the applied magnetic field H is not strong, $\Delta W(\omega_R)$ can be expanded in power series of H as

$$\Delta W(\omega_R) = \Delta W^{(0)}(\omega_R) + H \Delta W^{(1)}(\omega_R) + \cdots. \qquad (7.4.12)$$

For a randomly oriented system, $\Delta W^{(0)}(\omega_R) = 0$. As in the case of one-photon MCD, we can separate $\Delta W^{(1)}(\omega_R)$ as

$$\Delta W^{(1)}(\omega_R) = \Delta W^{(1)}(\omega_R)_A + \Delta W^{(1)}(\omega_R)_B + \Delta W^{(1)}(\omega_R)_C, \qquad (7.4.13)$$

where

$$\Delta W^{(1)}(\omega_R)_A = K\omega_R^4 \sum_a \sum_b P_a^{(0)} \langle \Delta M_{ab}^{(0)}(\omega_R) f_{ab}^{(1)}(\omega_R) \rangle, \qquad (7.4.14)$$

$$\Delta W^{(1)}(\omega_R)_B = K\omega_R^4 \sum_a \sum_b P_a^{(0)} \langle \Delta M_{ab}^{(1)}(\omega_R) \rangle f_{ab}^{(0)}(\omega_R), \qquad (7.4.15)$$

and

$$\Delta W^{(1)}(\omega_R)_C = K\omega_r^4 \sum_a \sum_b \langle P_a^{(1)} \Delta M_{ab}^{(0)}(\omega_R) \rangle f_{ab}^{(0)}(\omega_R), \qquad (7.4.16)$$

where $\langle \ \rangle$ denotes the spatial averaging over the orientations of the system.

First we consider $\Delta W^{(1)}(\omega_R)_A$. By using the perturbation method, we obtain

$$\omega_{ba} = \omega_{ba}^{(0)} - [(\hat{\mu}_Z)_{bb} - (\mu_Z^\dagger)_{aa}]H/\hbar + \cdots, \qquad (7.4.17)$$

and

$$f_{ab}^{(1)}(\omega_R) = [(\hat{\mu}_Z)_{bb} - (\hat{\mu}_Z)_{aa}]f_{ab}'(\omega_R)/\hbar, \qquad (7.4.18)$$

where $\hat{\mu}_Z$ denotes the Z-component of the total magnetic moment (i.e., including the contributions from both orbital and spin). If the line-shape function is of the delta function, then

$$f_{ab}'(\omega_R) = -\tfrac{1}{2}(d/d\omega_R)\,\delta(2\omega_R - \omega_{ba}^{(0)}). \qquad (7.4.19)$$

Substitution of Eqs. (7.4.17) and (7.4.18) into Eq. (7.4.14) and averaging over the orientation of molecules yield

$$\Delta W^{(1)}(\omega_R)_A = -(1/15\hbar)\,K\,\omega_R^4 \sum_a \sum_b P_a^{(0)} A_{ab}^{(2)}(\omega_R) f_{ab}'(\omega_R), \qquad (7.4.20)$$

where

$$A_{ab}^{(2)}(\omega_R) = I_m\Big\{ \sum_k \sum_{k'} (\omega_{ka}^{(0)} - \omega_R)^{-1}\,(\omega_{k'a}^{(0)} - \omega_R)^{-1}$$

$$\times \{(\mathbf{R}_{ak}^0 \cdot \mathbf{R}_{bk'}^0)[\mathbf{R}_{kb}^0 \cdot (\mathbf{R}_{k'a}^0 \times \Delta\boldsymbol{\mu}_{ba})] + (\mathbf{R}_{kb}^0 \cdot \mathbf{R}_{bk'}^0)[\mathbf{R}_{ak}^0 \cdot (\mathbf{R}_{k'a}^0$$

$$\times \Delta\boldsymbol{\mu}_{ba})] + (\mathbf{R}_{ak}^0 \cdot \mathbf{R}_{k'a}^0)[\mathbf{R}_{kb}^0 \cdot (\mathbf{R}_{bk'}^0 \times \Delta\boldsymbol{\mu}_{ba})] + (\mathbf{R}_{kb}^0 \cdot \mathbf{R}_{k'a}^0)$$

$$\times [\mathbf{R}_{ak}^0 \cdot (\mathbf{R}_{bk'}^0 \times \Delta\boldsymbol{\mu}_{ba})]\}$$

$$(7.4.21)$$

and

$$\Delta\boldsymbol{\mu}_{ba} = \boldsymbol{\mu}_{bb} - \boldsymbol{\mu}_{aa}. \qquad (7.4.22)$$

Next we consider $\Delta W^{(1)}(\omega_R)_C$. Using the relation

$$P_a^{(1)} = \beta(\hat{\mu}_Z)_{aa}P_a^{(0)}, \qquad (7.4.23)$$

where $\beta = (kT)^{-1}$, Eq. (7.4.16) becomes

$$\Delta W^{(1)}(\omega_R)_C = -\frac{1}{15}\beta K \omega_R^4 \sum_a \sum_b P_a^{(0)} C_{ab}^{(2)}(\omega_R) f_{ab}^{(0)}(\omega_R), \qquad (7.4.24)$$

where

$$\begin{aligned}
C_{ab}^{(2)}(\omega_R) = I_m\bigg(&\sum_k \sum_{k'}(\omega_{ka}^{(0)} - \omega_R)^{-1}(\omega_{k'a}^{(0)} - \omega_R)^{-1} \\
&\times \{(\mathbf{R}_{ak}^0 \cdot \mathbf{R}_{bk'}^0)[\mathbf{R}_{kb}^0 \cdot \mathbf{R}_{k'a}^0 \times \boldsymbol{\mu}_{aa}] \\
&+ (\mathbf{R}_{kb}^0 \cdot \mathbf{R}_{bk'}^0)[\mathbf{R}_{ak}^0 \cdot (\mathbf{R}_{k'a}^0 \times \boldsymbol{\mu}_{aa})] \\
&+ (\mathbf{R}_{ak}^0 \cdot \mathbf{R}_{k'a}^0)[\mathbf{R}_{kb}^0 \cdot \mathbf{R}_{bk'}^0 \times \boldsymbol{\mu}_{aa}] \\
&+ (\mathbf{R}_{kb}^0 \cdot \mathbf{R}_{k'a}^0)[\mathbf{R}_{ak}^0 \cdot (\mathbf{R}_{bk'}^0 \times \boldsymbol{\mu}_{aa})]\}\bigg).
\end{aligned} \qquad (7.4.25)$$

Finally we consider $\Delta W^{(1)}(\omega_R)_B$. Using the relations such as

$$X_{ak} = X_{ak}^0 + HX_{ak}^{(1)} + \cdots, \qquad (7.4.26)$$

and

$$X_{ak}^{(1)} = \sum_l^{l \neq k} X_a^0(\hat{\mu}_Z)_{lk}^{(1)} - \sum_l^{l \neq a}(\hat{\mu}_Z)_{al}^{(1)} X_{lk}^0, \qquad (7.4.27)$$

where

$$(\hat{\mu}_Z)_{lk}^{(1)} = (\hat{\mu}_Z)_{lk}/(E_l^0 - E_k^0), \qquad (7.4.28)$$

we obtain

$$\Delta W^{(1)}(\omega_R)_B = -\frac{1}{15} K \omega_R^4 \sum_a \sum_b P_a^{(0)} B_{ab}^{(2)}(\omega_R) f_{ab}^{(0)}(\omega_R), \qquad (7.4.29)$$

where

$$B_{ab}^{(2)}(\omega_R) = \sum_{i=1}^4 B_i^{(2)}(\omega_R)_{ab}. \qquad (7.4.30)$$

The expressions for $B_i^{(2)}(\omega_R)_{ab}$ are given by Lin *et al.* (1983).

By using Eqs. (7.4.12), (7.4.20), (7.4.24), and (7.4.29), we obtain the transition probability of nonresonant two-photon MCD as

$$\Delta W(\omega_R) = -\frac{1}{15} K H \omega_R^4 \sum_a \sum_b P_a^{(0)}$$

$$\times \left\{ A_{ab}^{(2)}(\omega_R)\frac{f_{ab}'(\omega_R)}{\hbar} + [\beta C_{ab}^{(2)}(\omega_R) + B_{ab}^{(2)}(\omega_R)] f_{ab}^{(0)}(\omega_R) \right\}.$$

$$(7.4.31)$$

If the initial electronic state is nondegenerate, then $C_{ab}^{(2)}(\omega_R) = 0$ because $\boldsymbol{\mu}_{aa} = 0$, and if both initial and final electronic states are nondegenerate, then $A_{ab}^{(2)}(\omega_R) = 0$ and $C_{ab}^{(2)}(\omega_R) = 0$ because in this case $\boldsymbol{\mu}_{bb}$ is also zero. Note that just as in the case of one-photon MCD, from Eq. (7.4.31) we can see that if all three terms are present, the relative contributions from the A, B and C terms are approximately in the ratio $(\hbar\gamma_{ba})^{-1}:(\Delta E_{ba}^0)^{-1}:(kT)^{-1}$, where ΔE_{ka}^0 denotes the order of magnitude of zero-field state separations between the initial and final states. In other words, unless $A_{ab}^{(2)}(\omega_R)$ and $C_{ab}^{(2)}(\omega_R)$ vanish their contributions are more important than that of $B_{ab}^{(2)}(\omega_R)$.

We shall consider two-photon MCD in the resonant intermediate state case. In this case $\omega_{ka} \simeq \omega_R$, the H-dependence both in the energy denominator $\omega_{ka}(H) - \omega_R$ and in the damping constant $\gamma_{ka}(H)$ in Eq. (7.4.5), has to be taken properly into account. By neglecting the H-dependence of $\gamma_{ka}(H)$ and utilizing the expansion

$$[\omega_{ka}(H) - \omega_R - i\gamma_{ka}]^{-1} \doteqdot (\omega_{ka}^{(0)} - \omega_R - i\gamma_{ka})^{-1}$$
$$\times \{1 - [(\hat{\mu}_Z)_{kk} - (\hat{\mu}_Z)_{aa}]H/\hbar(\omega_{ka}^{(0)} - \omega_R - i\gamma_{ka})\},$$

(7.3.32)

we can derive the expression for the transition probability of two-photon MCD in the resonant case. In addition to $\Delta W^{(1)}(\omega_R)_A, \Delta W^{(1)}(\omega_R)_B$, and $\Delta W^{(1)}(\omega_R)_C$ with the damping constant γ_{ka}, the resonance (R) term $\Delta W^{(1)}(\omega_R)_R$ makes a contribution to resonant two-photon MCD:

$$\Delta W(\omega_R) = -\frac{1}{15} KH\omega_R^4 \sum_u \sum_b P_a^{(0)} \left\{ A_{ab}^{(2)}(\omega_R) \frac{f_{ab}'(\omega_R)}{\hbar} \right.$$

$$\left. + \left[\beta C_{ab}^{(2)}(\omega_R) + B_{ab}^{(2)}(\omega_R) - \frac{R_{ab}^{(2)}(\omega_R)}{\hbar} \right] f_{ab}^{(0)}(\omega_R) \right\}, \quad (7.4.33)$$

where the R term is given as

$$R_{ab}^{(2)}(\omega_R) = 2I_m \left(\sum_k \sum_{k'} (\omega_{ka}^{(0)} - \omega_R + i\gamma_{Ra})^{-2}(\omega_{k'a}^{(0)} - \omega_R - i\gamma_{k'a})^{-1} \right.$$

$$\times \{(\mathbf{R}_{ak}^0 \cdot \mathbf{R}_{k'a}^0)[\mathbf{R}_{kb}^0 \cdot (\mathbf{R}_{bk'}^0 \times \Delta\boldsymbol{\mu}_{ka})]$$

$$+ (\mathbf{R}_{kb}^0 \cdot \mathbf{R}_{k'a}^0)[\mathbf{R}_{ak}^0 \cdot (\mathbf{R}_{bk'}^0 \times \Delta\boldsymbol{\mu}_{ka})]$$

$$+ (\mathbf{R}_{ak}^0 \cdot \mathbf{R}_{bk'}^0)[\mathbf{R}_{kb}^0 \cdot (\mathbf{R}_{k'a}^0 \times \Delta\boldsymbol{\mu}_{ka})]$$

$$\left. + (\mathbf{R}_{kb}^0 \cdot \mathbf{R}_{bk'}^0)[\mathbf{R}_{ak}^0 \cdot (\mathbf{R}_{k'a}^0 \times \Delta\boldsymbol{\mu}_{ka})]\} \right) \quad (7.4.34)$$

with $\Delta\mu_{ka} = (\mu_{kk}) - (\mu)_{aa}$ and where

$$A_{ab}^{(2)}(\omega_R) = I_m\left(\sum_k \sum_{k'} (\omega_{ka}^{(0)} - \omega_R + i\gamma_{ka})^{-1}(\omega_{k'a}^{(0)} - \omega_R - i\gamma_{k'a})^{-1}\right.$$

$$\times \{(\mathbf{R}_{ak}^0 \cdot \mathbf{R}_{bk'}^0)[\mathbf{R}_{k'b}^0 \cdot (\mathbf{R}_{k'a}^0 \times \Delta\mu_{ba})]$$

$$+ (\mathbf{R}_{kb}^0 \cdot \mathbf{R}_{bk'}^0)[\mathbf{R}_{ak}^0 \cdot (\mathbf{R}_{k'a}^0 \times \Delta\mu_{ba})]$$

$$+ (\mathbf{R}_{ak}^0 \cdot \mathbf{R}_{k'a}^0)[\mathbf{R}_{kb}^0 \cdot (\mathbf{R}_{bk'}^0 \times \Delta\mu_{ba})]$$

$$\left.+ (\mathbf{R}_{kb}^0 \cdot \mathbf{R}_{k'a}^0)[\mathbf{R}_{ak}^0 \cdot (\mathbf{R}_{bk'}^0 \times \Delta\mu_{ba})]\} \right). \qquad (7.4.35)$$

and $B_{ab}^{(2)}(\omega_R)$ and $C_{ab}^{(2)}$ can be expressed in a way similar to that for the nonresonant case.

In the above formulation, it has been shown that the expression for the transition probability of two-photon MCD consists of the A, B, C, and R terms, depending on the nature of the electronic state manifolds relevant to the transition. In the nonresonant intermediate state case, compared with those of the A, B, and C terms, the contribution of the R term is considerably smaller, except for the case in which the intermediate electronic state is degenerate and at the same time both initial and final electronic states are nondegenerate. In this case, the A and C terms vanish and the B and R terms are in the same order, being approximately in the ratio of $(\Delta E_{ba}^0)^{-1}:(\Delta E_{ka}^0)^{-1}$. In the near-resonant or the resonant case, on the other hand, the R term may make a significant contribution to the two-photon MCD spectrum; the relative contributions from the A, B, C, and R terms are approximately in the ratio of $(\hbar\gamma_{ba})^{-1}:(\Delta E_{ba}^0)^{-1}:(kT)^{-1}:(\hbar\gamma_{ka})^{-1}$. In order to calculate two-photon MCD of molecules, we can use the adiabatic approximation (Lin *et al.*, 1983). The transition probability can be easily derived as in the case of the two-photon absorption described in Chapter 5.

REFERENCES

Andrews, D. L. (1976). *Chem. Phys.*, **16**, 419.
Ashby, R. A., and Johnson, C. W. (1980). *J. Mol. Spectrosc.* **84**, 41.
Bernstein, R. B. (1982). *J. Phys. Chem.* **86**, 1178.
Boesl, U., Neusser, H. J., and Schlag, E. W. (1980). *J. Chem. Phys.* **72**, 4327.
Boesl, U., Neusser, H. J., and Schlag, E. W. (1981). *J. Am. Chem. Soc.* **103**, 5058.
Buckingham, A. D., and Stephens, P. J. (1966). *Ann. Rev. Phys. Chem.* **17**, 390.
Caldwell, D. J., and Eyring, H. (1971). "The Theory of Optical Activity." Wiley (Interscience), New York.
Chen, K., Steenhoek, L. E., and Yeung, E. S. (1978). *Chem. Phys. Lett.* **59**, 222.
Cooper, D. E., and Wessel, J. E. (1982). *J. Chem. Phys.* **76**, 2155.

Cooper, D. E., Klimcak, C. M., and Wessel, J. E. (1981). *Phys. Rev. Lett.* **46**, 324.

Dietz, W., Neusser, H. J., Boesl, U., Schlag, E. W., and Lin, S. H. (1982). *Chem. Phys.* **66**, 105.

Fisanick, G. J., Eichelberger, T. S., Heath, B. A., and Robin, M. B. (1980). *J. Chem. Phys.* **72**, 5571.

Freiser, B. S., and Beauchamp, J. L. (1975). *Chem. Phys. Lett.* **35**, 35.

Guy, A. L., Viswanathan, K. S., Sur, A. and Tellinghuisen, J. (1980). *Chem. Phys. Lett.* **73**, 582.

King, G. W., Littlewood, I. M., and Robins, J. R. (1981). *Chem. Phys.* **56**, 145.

Lehman, K. K., Smolarek, J., and Goodman, L. (1978). *J. Chem. Phys.* **69**, 1569.

Lin S. H. (1971). *J. Chem. Phys.* **55**, 3546.

Lin, S. H., Fujimura, Y., Saito, M., and Nakajima, T. (1983). *J. Phys. Chem.* **87**, 2895.

Monson P. R., and McClain, W. M. (1970). *J. Chem. Phys.* **53**, 29.

Mulliken, R. S. (1971). *J. Chem. Phys.* **55**, 288.

Murakami, J., Kaya, K., and Ito, M. (1982). *Chem. Phys. Lett.* **91**, 401.

Orlowski, T. E., Freiser, B. S., and Beauchamp, J. L. (1976). *Chem. Phys.* **16**, 439.

Power, E. A. (1975). *J. Chem. Phys.* **63**, 1348.

Rebentrost, F. and Ben-Shaul, A. (1981). *J. Chem. Phys.* **74**, 3255.

Rebentrost, F., Kompa, K. L., and Ben-Shaul, A. (1981). *Chem. Phys. Lett.* **77**, 394.

Reilly, J. P., and Kompa, K. L. (1980). *J. Chem. Phys.* **73**, 5468.

Schlag, E. W., and Neusser, H. J. (1983). *Acc. Chem. Res.* **16**, 453.

Silberstein, J., and Levine, R. D. (1980). *Chem. Phys. Lett.* **74**, 6.

Silberstein, J., and Levine, R. D. (1981). *J. Chem. Phys.* **75**, 5735.

Simmons, J. D., and Hougen, J. T. (1977). *J. Res. Natl. Bur. Stand., Sect. A* **81**, 25.

Stephens, P. J. (1976). *Adv. Chem. Phys.* **35**, 197.

Tai, C., and Dalby, F. W. (1978). *Can. J. Phys.* **56**, 183.

Tellinghuisen, J. (1977). *Chem. Phys. Lett.* **49**, 485.

Tinoco, I., Jr. (1975). *J. Chem. Phys.* **62**, 1006.

Wagniére, G. (1979). *Chem. Phys.* **40**, 119.

Williamson, A. (1979). *Chem. Phys. Lett.* **60**, 451.

APPENDIX

I

In this appendix to Chapter 2 we shall review the calculation of mode density, discuss an application of the coherent state method, and derive the susceptibility method.

I.A CALCULATION OF MODE DENSITY

In this section we are concerned with the calculation of the mode density (i.e., the number of normal modes in a given frequency range contained in a cavity of volume V_L in the radiation field). If $\hat{i}, \hat{j}$, and $\hat{k}$ are three unit vectors along the cube edges of the cavity, the position vector is

$$\mathbf{r} = x\hat{\mathbf{i}} + y\hat{\mathbf{j}} + z\hat{\mathbf{k}}, \tag{I.A.1}$$

and the propagation vector is

$$\mathbf{k}_l = k_{lx}\hat{i} + k_{ly}\hat{j} + k_{lz}\hat{k}. \tag{I.A.2}$$

The periodic boundary conditions of the vector potential require that

$$\mathbf{A}(\mathbf{r} + L\hat{i}, t) = \mathbf{A}(\mathbf{r} + L\hat{j}, t) = \mathbf{A}(\mathbf{r} + L\hat{k}, t) = \mathbf{A}(\mathbf{r}, t), \tag{I.A.3}$$

where L represents the length of the cavity cube.

Since the vector potential can be written as a linear superposition of plane waves (i.e., $\mathbf{A}_l \alpha \exp[\pm i(\mathbf{k}_l \cdot \mathbf{r} - \omega_l t)]$), the periodic boundary conditions given by Eq. (I.A.3) are satisfied if

$$\mathbf{k}_l = (2\pi/L)(l_1\hat{i} + l_2\hat{j} + l_3\hat{k}), \tag{I.A.4}$$

where l_1, l_2, and l_3 are integers from $-\infty$ to ∞. That is, the propagation constants are restricted to a discrete set of values by virtue of the boundary conditions (Louisell, 1973; Heitler, 1954). For each set of integers (l_1, l_2, l_3) there are two traveling modes (one for each polarization). If we set $(l_1, l_2, l_3) \to (-l_1, -l_2, -l_3)$ (which is designated simply by $l \to -l$), from Eq. (I.A.4) we find that

$$\mathbf{k}_{-l} = -\mathbf{k}_l, \tag{I.A.5}$$

and from $k_l^2 = \omega_l^2/c^2$, we find that

$$\omega_{-l} = \omega_l. \tag{I.A.6}$$

In other words, if we change l to $-l$, the two corresponding plane waves travel in opposite directions.

Each set of integers $(l_1\ l_2\ l_3)$ corresponds to two traveling wave modes, since there are two polarizations. We may represent each mode of a given polarization by a point in a three-dimensional space generated by $(l_1\ l_2\ l_3)$, i.e.,

$$l_1^2 + l_2^2 + l_3^2 = (L/2\pi)^2 k_l^2. \tag{I.A.7}$$

In a small element of volume $dl_1\,dl_2\,dl_3$ the number of normal modes for a single polarization is

$$d_N = dl_1\,dl_2\,dl_3. \tag{I.A.8}$$

By using Eq. (I.A.4), Eq. (I.A.8) becomes

$$dN = (L/2\pi)^3\,dk_x\,dk_y\,dk_z. \tag{I.A.9}$$

Notice that in spherical coordinates

$$dk_x\,dk_y\,dk_z = k^2\,dk\,\sin\theta\,d\theta\,d\phi \equiv k^2\,dk\,d\Omega. \tag{I.A.10}$$

It follows that (Heitler, 1954; Lin, 1967; Louisell, 1973)

$$dN = (L/2\pi)^3 k^2\,dk\,d\Omega. \tag{I.A.11}$$

By using the relation $\omega^2 = c^2 k^2$, we obtain

$$dN = (L^3/8\pi^3 c^3)\omega^2\,d\omega\,d\Omega = g(\omega)\,d\omega\,d\Omega, \tag{I.A.12}$$

where $g(\omega)$ is the mode density

$$g(\omega) = (L^3/8\pi^3 c^3)\omega^2. \tag{I.A.13}$$

In summations of discrete values of l (l_1, l_2, l_3), when we let $L \to \infty$, l_1/L, l_2/L, and l_3/L become continuous variables and we may replace sums by integrals as

$$\sum_{l \,\text{or}\, \mathbf{k}} \rightarrow \frac{L^3}{8\pi^3} \int_{-\infty}^{\infty} \int\int dk_x \, dk_y \, dk_z = \frac{L^3}{8\pi^3 c^3} \int_0^{\infty} d\omega \, \omega^2 \int_{\Omega} d\Omega \qquad \text{(I.A.14)}$$

I.B   THE COHERENT STATE METHOD

In this section we shall discuss the application of the coherent state method to optical problems. If $|n\rangle$ is the n-photon state and $\hat{a}$ is the usual lowering (or annihilation) operator for the coherent electromagnetic mode under consideration, then the Glauber coherent state $|\alpha\rangle$ is defined as the eigenfunction of $\hat{a}$ with eigenvalue α (Glauber, 1963; Louisell, 1973):

$$\hat{a}|\alpha\rangle = \alpha|\alpha\rangle. \qquad \text{(I.B.1)}$$

The coherent state $|\alpha\rangle$ can be expressed in terms of the photon number (or Fock) state $|n\rangle$ as

$$|\alpha\rangle = \sum_{n=0}^{\infty} C_n |n\rangle. \qquad \text{(I.B.2)}$$

To find C_n we apply $\hat{a}$ to both sides of Eq. (I.B.2):

$$\alpha|\alpha\rangle = \sum_{n=0}^{\infty} C_n \sqrt{n}|n-1\rangle = \alpha \sum_n C_n |n\rangle. \qquad \text{(I.B.3)}$$

We obtain

$$C_n = (\alpha/\sqrt{n})C_{n-1}. \qquad \text{(I.B.4)}$$

It follows that

$$C_n = (\alpha^n/\sqrt{n!})C_0 \qquad \text{(I.B.5)}$$

and that

$$|\alpha\rangle = C_0 \sum_{n=0}^{\infty} \frac{\alpha^n}{\sqrt{n!}} |n\rangle. \qquad \text{(I.B.6)}$$

Here C_0 can be determined by the normalization condition of $|\alpha\rangle$, i.e.,

$$\langle\alpha|\alpha\rangle = 1 = |C_0|^2 \sum_{n=0}^{\infty} \frac{(|\alpha|^2)^n}{n!} \qquad \text{(I.B.7)}$$

or

$$|C_0|^2 = \exp(-|\alpha|^2). \qquad \text{(I.B.8)}$$

Substituting Eq. (I.B.8) into Eq. (I.B.6) yields

$$|\alpha\rangle = \exp\left(-\frac{1}{2}|\alpha|^2\right)\sum_{n=0}^{\infty}\frac{\alpha^n}{\sqrt{n!}}|n\rangle. \tag{I.B.9}$$

Furthermore, because of the relation

$$\hat{a}^\dagger|n\rangle = \sqrt{n+1}\,|n+1\rangle, \tag{I.B.10}$$

where $\hat{a}^\dagger$ denotes the raising (or creation) operator, we obtain

$$|n\rangle = \frac{\hat{a}^\dagger}{\sqrt{n}}|n-1\rangle = \frac{(\hat{a}^\dagger)^n}{\sqrt{n!}}|0\rangle \tag{I.B.11}$$

and

$$|\alpha\rangle = \exp\left(-\frac{1}{2}|\alpha|^2\right)\sum_{n=0}^{\infty}\frac{(\alpha\hat{a}^\dagger)^n}{n!}|0\rangle. \tag{I.B.12}$$

By using the expression

$$|\alpha'\rangle = \exp\left(-\frac{1}{2}|\alpha'|^2\right)\sum_{n=0}^{\infty}\frac{\alpha'^n}{\sqrt{n!}}|n\rangle \tag{I.B.13}$$

and Eq. (I.B.9), we find that

$$\langle\alpha|\alpha'\rangle = \exp\left[-\frac{1}{2}(|\alpha|^2+|\alpha'|^2)\right]\sum_{n=0}^{\infty}\frac{(\alpha^*\alpha')^n}{n!}$$
$$= \exp\left[\alpha^*\alpha' - \frac{1}{2}(|\alpha|^2+|\alpha'|^2)\right]. \tag{I.B.14}$$

Equation (I.B.14) shows that the coherent states are nonorthogonal.

To show the application of the coherent state method, we consider a molecule to be a two-level system with a ground state $|g\rangle$ and an excited level $|e\rangle$. Without any loss in generality, the former is taken to have zero energy, while the latter has energy E_1. The incident light field may be represented as a statistical distribution of wave packets constructed from Eq. (I.B.12). Since the requisite averages may be performed after evaluating expectation values, the simple form of Eq. (I.B.12) is retained for now because it poses the central technical problem to be surmounted for this subject. In the rotating wave approximation, the Hamiltonian for this two-level system in the single-mode coherent field is (Freed and Villaeys, 1979; Klauder and Sudarshan, 1968)

$$\hat{H} = |e\rangle E_1\langle e| + \hbar\omega\hat{a}^\dagger\hat{a} + |e\rangle\mu_{eg}\langle g|\hat{a} + |g\rangle\mu_{ge}\langle e|\hat{a}^\dagger, \tag{I.B.15}$$

where ω is the frequency of the coherent mode and μ_{eg} is the dipole transition matrix element for the $g \to e$ absorption that would arise from a one-photon state of the radiation field.

The central problem in the Green's function method (see Chapter 2) is the evaluation of the matrix elements of the evolution operator $\exp[(-it/\hbar)\hat{H}]$, i.e.,

$$\left\langle \alpha j \left| \exp\left(-\frac{it}{\hbar}\hat{H} \right) \right| \alpha' k \right\rangle = \frac{i}{2\pi} \int_c dE \, \exp\left(-\frac{itE}{\hbar} \right) \langle \alpha j | \hat{G}(E) | \alpha' k \rangle, \quad \text{(I.B.16)}$$

where $\hat{G}(E)$ is given by

$$\hat{G}(E) = 1/(E - \hat{H}) \quad \text{(I.B.17)}$$

and j and k belong to (g, e). Notice that

$$(E - \hat{H})\hat{G}(E) = 1. \quad \text{(I.B.18)}$$

It follows that

$$E\langle \alpha j | \hat{G}(E) | \alpha' k \rangle - \langle \alpha j | \hat{H}\hat{G}(E) | \alpha' k \rangle = \langle \alpha | \alpha' \rangle \delta_{jk}. \quad \text{(I.B.19)}$$

Using the relation (Klauder and Sudarshan, 1968)

$$\langle \alpha | W(\hat{a}^\dagger, \hat{a}) | \psi \rangle = \hat{W}(\alpha^*, \alpha/2 + \partial/\partial\alpha^*)\langle \alpha | \psi \rangle, \quad \text{(I.B.20)}$$

we obtain

$$\langle \alpha j | \hat{H}\hat{G}(E) | \alpha' k \rangle = \langle \alpha j | \hat{H}(\hat{a}^+, \hat{a})\hat{G}(E) | \alpha' k \rangle$$

$$= \sum_l \left\langle j \left| \hat{H}\left(\alpha^*, \frac{\alpha}{2} + \frac{\partial}{\partial\alpha^*} \right) \right| l \right\rangle \langle \alpha l | \hat{G}(E) | \alpha' k \rangle \quad \text{(I.B.21)}$$

and

$$E\langle \alpha j | \hat{G}(E) | \alpha' k \rangle - \sum_l \left\langle j \left| \hat{H}\left(\alpha^*, \frac{\alpha}{2} + \frac{\partial}{\partial\alpha^*} \right) \right| l \right\rangle \langle \alpha l | \hat{G}(E) | \alpha' k \rangle$$

$$= \langle \alpha | \alpha' \rangle \delta_{jk}. \quad \text{(I.B.22)}$$

Thus for $j = k = g$, Eq. (I.B.22) becomes

$$E\langle \alpha g | \hat{G}(E) | \alpha' g \rangle - \langle g | \hat{H}(\alpha^*, \alpha/2 + \partial/\partial\alpha^*) | g \rangle \langle \alpha g | \hat{G}(E) | \alpha' g \rangle$$

$$- \langle g | \hat{H}(\alpha^*, \alpha/2 + \partial/\partial\alpha^*) | e \rangle \langle \alpha e | \hat{G}(E) | \alpha' g \rangle = \langle \alpha | \alpha' \rangle, \quad \text{(I.B.23)}$$

or, using the relations

$$\langle g | \hat{H}(\alpha^*, \alpha/2 + \partial/\partial\alpha^*) | g \rangle = \hbar\omega\alpha^*(\alpha/2 + \partial/\partial\alpha^*) \quad \text{(I.B.24)}$$

and

$$\langle g | \hat{H}(\alpha^*, \alpha/2 + \partial/\partial\alpha^*) | e \rangle = \mu_{ge}\alpha^*, \quad \text{(I.B.25)}$$

we obtain

$$EG(E)_{\alpha g, \alpha' g} - \hbar\omega\alpha^*(\alpha/2 + \partial/\partial\alpha^*)G(E)_{\alpha g, \alpha' g} - \alpha^*\mu_{ge}G(E)_{\alpha e, \alpha' g} = \langle\alpha|\alpha'\rangle.$$

$$(I.B.26)$$

Defining the quantities

$$\varepsilon = E/\hbar\omega, \qquad \bar{\mu}_{ge} = \mu_{ge}/\hbar\omega,$$

$$k_{jk} = \hbar\omega\exp[\tfrac{1}{2}(|\alpha|^2 + |\alpha'|^2)]G(E)_{\alpha j, \alpha' k},$$

$$(I.B.27)$$

Eq. (I.B.26) becomes

$$(\varepsilon - \alpha^* \partial/\partial\alpha^*)k_{gg} - \alpha^*\bar{\mu}_{ge}k_{eg} = \exp(\alpha^*\alpha'),$$

$$(I.B.28)$$

where the following relation has been used:

$$(\partial/\partial\alpha^*)(e^{1/2|\alpha|^2}f) = \exp(\tfrac{1}{2}|\alpha|^2)(\alpha/2 + \partial/\partial\alpha^*)f.$$

$$(I.B.29)$$

Similarly, we obtain

$$(\varepsilon - \varepsilon_l - \alpha^* \partial/\partial\alpha^*)k_{eg} - \bar{\mu}_{eg}\partial k_{gg}/\partial\alpha^* = 0,$$

$$(I.B.30)$$

where

$$\varepsilon_e = E_e/\hbar\omega, \qquad \bar{\mu}_{eg} = \mu_{eg}/\hbar\omega.$$

$$(I.B.31)$$

Next we have to solve Eqs. (I.B.28) and (I.B.30) for k_{gg} and k_{eg}. This can usually be accomplished by employing the power-series method,

$$k_{gg} = \sum_n G_n\alpha^{*n}, \qquad k_{eg} = \sum_n E_n\alpha^{*n}.$$

$$(I.B.32)$$

Substituting Eq. (I.B.32) into Eqs. (I.B.28) and (I.B.30) yields

$$(\Delta\varepsilon_e - n)E_n - (n + 1)\bar{\mu}_{eg}G_{n+1} = 0$$

$$(I.B.33)$$

and

$$(\varepsilon - n)G_n - \bar{\mu}_{ge}E_{n-1} = (\alpha')^n/n!,$$

$$(I.B.34)$$

where $\Delta\varepsilon_l = \varepsilon - \varepsilon_l$. From Eqs. (I.B.33) and (I.B.34) we obtain

$$G_n = \frac{(\alpha')^n/n!}{\varepsilon - n - n|\bar{\mu}_{eg}|^2/(\Delta\varepsilon_l - n + 1)},$$

$$(I.B.35)$$

$$k_{gg} = \sum_{n=0}^{\infty} \frac{(\alpha'\alpha^*)^n/n!}{\varepsilon - n - n|\bar{\mu}_{eg}|^2/(\Delta\varepsilon_l - n + 1)},$$

$$(I.B.36)$$

$$E_n = \frac{\bar{\mu}_{eg}}{\Delta\varepsilon_e - n} \cdot \frac{(\alpha')^{n+1}/n!}{\varepsilon - n - 1 - (n + 1)|\bar{\mu}_{eg}|^2/(\Delta\varepsilon_l - n)},$$

$$(I.B.37)$$

$$k_{eg} = \alpha'\bar{\mu}_{eg} \sum_{n=0}^{\infty} \frac{(\alpha'\alpha^*)^n/n!}{(\varepsilon - n - 1)(\Delta\varepsilon_l - n) - (n + 1)|\bar{\mu}_{eg}|^2}.$$

$$(I.B.38)$$

In order to obtain the matrix elements of the time evolution operator, we
use Eqs. (I.B.16), (I.B.36), and (I.B.38); for example,

$$\left\langle \alpha j \left| \exp\left(-\frac{it\hat{H}}{\hbar}\right) \right| \alpha' k \right\rangle = \frac{i}{2\pi} \int_c d\varepsilon \exp(-it\varepsilon\omega)$$

$$\times k_{jk}(\varepsilon, \alpha^*, \alpha') \exp\left[-\frac{1}{2}(|\alpha|^2 + |\alpha'|^2)\right]. \quad \text{(I.B.39)}$$

For the case $j = k = g$, Eq. (I.B.39) becomes

$$\left\langle \alpha g \left| \exp\left(-\frac{it\hat{H}}{\hbar}\right) \right| \alpha' g \right\rangle = \frac{i}{2\pi} \exp\left[-\frac{1}{2}(|\alpha|^2 + |\alpha'|^2)\right]$$

$$\times \int_c d\varepsilon \, e^{-it\varepsilon\omega} \sum_{n=0}^{\infty} \frac{(\Delta\varepsilon_l - n + 1)(\alpha'\alpha^*)^n/n!}{(\varepsilon - \varepsilon_{n_1})(\varepsilon - \varepsilon_{n_2})}$$

$$= \exp\left[-\frac{1}{2}(|\alpha|^2 + |\alpha'|^2)\right] \sum_{n=0}^{\infty} \frac{(\alpha'\alpha^*)^n}{n!(\varepsilon_{n_2} - \varepsilon_{n_1})}$$

$$\times \left[(\varepsilon_{n_1} - \varepsilon_l - n + 1)\exp(-it\omega\varepsilon_{n_1})\right.$$

$$\left. - (\varepsilon_{n_2} - \varepsilon_l - n + 1)\exp(-it\omega\varepsilon_{n_2})\right], \quad \text{(I.B.40)}$$

where ε_{n_1} and ε_{n_2} are given by

$$\varepsilon_{n_1} = \tfrac{1}{2}(\varepsilon_l + 2n - 1 + S_n), \qquad \varepsilon_{n_2} = \tfrac{1}{2}(\varepsilon_e + 2n - 1 - S_n), \quad \text{(I.B.41)}$$

with

$$S_n = [(\varepsilon_l - 1)^2 + 4n|\bar{\mu}_{eg}|^2]^{1/2} \quad \text{(I.B.42)}$$

The probability of finding the molecule in state g at time t, given that it
is initially in state g with the field in $|\alpha'\rangle$, is obtained from the absolute mag-
nitude of Eq. (I.B.40). Summed over all possible final states of the field,

$$P_{gg}(t|\alpha') = \int \frac{d^2\alpha}{\pi} \left| \left\langle \alpha g \left| \exp\left(-\frac{it\hat{H}}{\hbar}\right) \right| \alpha' g \right\rangle \right|^2, \quad \text{(I.B.43)}$$

where the coherent state algebra (Louisell, 1973; Glauber, 1963; Klauder and
Sudarshan, 1968) requires that the integration in Eq. (I.B.43) be carried out
over the real and imaginary parts of α. Substituting Eq. (I.B.40) into Eq.
(I.B.43) and using the basic integral

$$\int \frac{d^2\alpha}{\pi} (\alpha^*)^n \alpha^m \exp(-|\alpha|^2) = n!\delta_{nm} \quad \text{(I.B.44)}$$

leads to the general result (Freed and Villaeys, 1979)

$$P_{gg}(t|\alpha') = \sum_{n=0}^{\infty} \exp(-|\alpha'|^2) \frac{|\alpha'|^{2n}}{n!} \left(1 - \frac{\Delta n|\bar{\mu}_{eg}|^2}{(\varepsilon_e - 1)^2 + 4n|\mu_{eg}|^2}\right.$$

$$\left. \times \sin^2 \frac{\omega t}{2} [(\varepsilon_l - 1)^2 + 4n|\bar{\mu}_{eg}|^2]^{1/2}\right). \tag{I.B.45}$$

Next we shall compare the result, given by Eq. (I.B.45), obtained from the coherent state method with that obtained from the photon number (or Fock state) method. For this purpose, we define

$$|1\rangle \equiv |ng\rangle, \qquad |2\rangle \equiv |(n-1)e\rangle, \tag{I.B.46}$$

where n represents the number of photons. Using the relation

$$E\langle j|\hat{G}(E)|k\rangle - \langle j|\hat{H}\hat{G}(E)|K\rangle = \delta_{jk} \tag{I.B.47}$$

for the Green's function, we obtain

$$(E - H_{11})G_{11} - H_{12}G_{21} = 1,$$
$$-H_{21}G_{11} + (E - H_{22})G_{21} = 0. \tag{I.B.48}$$

It follows that

$$G_{11}(E) = \frac{E - H_{22}}{(E - H_{11})(E - H_{22}) - |H_{12}|^2}$$

$$= \frac{1}{E - H_{11} - |H_{12}|^2/(E - H_{22})} \tag{I.B.49}$$

and

$$G_{21}(E) = \frac{H_{21}}{(E - H_{11})(E - H_{22}) - |H_{12}|^2}$$

$$= \frac{H_{21}}{(E - H_{11})[(E - H_{22}) - |H_{12}|^2/(E - H_{11})]}, \tag{I.B.50}$$

where

$$H_{11} = n\hbar\omega, \qquad H_{22} = E_e + (n-1)\hbar\omega, \qquad H_{12} = \mu_{ge}\sqrt{n}. \tag{I.B.51}$$

Thus the probability of finding the system at $|g\rangle$ with n photons is given by

$$P_{11}(t) = \left|\left\langle 1 \left| \exp\left[-\frac{it}{\hbar}\hat{H}\right]\right|1\right\rangle\right|^2$$

$$= \left|\frac{i}{2\pi}\int_z \exp\left(-\frac{itE}{\hbar}\right)G_{11}(E)\,dE\right|^2. \tag{I.B.52}$$

Substituting Eq. (I.B.49) into Eq. (I.B.57) yields

$$P_{11}(t) = 1 - \frac{4n|\bar{\mu}_{eg}|^2}{(\varepsilon_l - 1)^2 + 4n|\bar{\mu}_{eg}|^2} \sin^2 \frac{\omega t}{2}$$
$$\times \left[(\varepsilon_l - 1)^2 + 4n|\bar{\mu}_{eg}|^2\right]^{1/2}. \tag{I.B.53}$$

Comparing Eqs. (I.B.45) and (I.B.53), we can see that the expression given by Eq. (I.B.45) is just a Poisson average of the semiclassical expression given by Eq. (I.B.53) with an average photon number of $n_{av} \equiv |\alpha'|^2$. Thus when

$$n_{av} \gg \sqrt{n_{av}}, \qquad \sqrt{n_{av}} \gg 1, \tag{I.B.54}$$

the terms in Eq. (I.B.45) vary negligibly over the dominant range of the Poisson distribution, so Eq. (I.B.45) reduces to Eq. (I.B.53), showing that the semiclassical approximation becomes exact when the condition given by Eq. (I.B.54) is satisfied.

Next we shall discuss how to use the coherent state method to treat multi-photon processes. For simplicity of demonstration, we choose the following simple three-level model:

$$\hat{H} = \hat{H}_0 + \hat{H}', \tag{I.B.55}$$

$$\hat{H}_0 = |e\rangle E_l \langle e| + |m\rangle E_m \langle m| + \hbar \omega \hat{a}^\dagger \hat{a}, \tag{I.B.56}$$

$$\hat{H}' = (\mu_{mg} + \mu_{em})\hat{a} + (\mu_{mg}^* + \mu_{em}^*)\hat{a}^\dagger. \tag{I.B.57}$$

Repeating the same type of calculation that was made for the two-level case, we obtain

$$\left[(\varepsilon - \varepsilon_l) - \alpha^* \frac{\partial}{\partial \alpha^*}\right] k_{eg} - \bar{\mu}_{em} \frac{\partial k_{mg}}{\partial \alpha^*} = 0, \tag{I.B.58}$$

$$\left[(\varepsilon - \varepsilon_m) - \alpha^* \frac{\partial}{\partial \alpha^*}\right] k_{mg} - \bar{\mu}_{mg} \frac{\partial k_{gg}}{\partial \alpha^*} = \bar{\mu}_{em}^* \alpha^* k_{eg}, \tag{I.B.59}$$

$$\left(\varepsilon - \alpha^* \frac{\partial}{\partial \alpha^*}\right) k_{gg} - \bar{\mu}_{mg}^* \alpha^* k_{mg} = e^{\alpha'\alpha^*}. \tag{I.B.60}$$

Equations (I.B.58)–(I.B.60) can again be solved by using the power-series method:

$$k_{gg} = \sum_n G_n(\alpha^*)^n, \tag{I.B.61}$$

$$k_{mg} = \sum_m M_n(\alpha^*)^n, \tag{I.B.62}$$

$$k_{eg} = \sum_n E_n(\alpha^*)^n. \tag{I.B.63}$$

Substituting Eqs. (I.B.61)–(I.B.63) into Eqs. (I.B.58)–(I.B.60) yields

$$(\Delta\varepsilon_l - n)E_n - (n + 1)\bar{\mu}_{em}M_{n+1} = 0, \qquad (\text{I.B.64})$$

$$(\Delta\varepsilon_m - n)M_n - (n + 1)\bar{\mu}_{mg}G_{n+1} = \bar{\mu}_{em}^* E_{n-1}, \qquad (\text{I.B.65})$$

$$(\varepsilon - n)G_n - \bar{\mu}_{mg}^* M_{n-1} = (\alpha')^n/n!, \qquad (\text{I.B.66})$$

where $\Delta\varepsilon_l = \varepsilon - \varepsilon_l$ and $\Delta\varepsilon_m = \varepsilon - \varepsilon_m$. We can solve Eqs. (I.B.64)–(I.B.66) for G_n, M_n, and E_n as

$$G_n = \cfrac{(\alpha')^n/n!}{\varepsilon - n - \cfrac{n|\bar{\mu}_{mg}|^2}{[\Delta\varepsilon_m - n + 1 - (n - 1)|\bar{\mu}_{em}|^2/(\Delta\varepsilon_l - n + 2)]}}, \qquad (\text{I.B.67})$$

$$M_n = \frac{(n + 1)\bar{\mu}_{mg}}{\Delta\varepsilon_m - n - n|\bar{\mu}_{em}|^2/(\Delta\varepsilon_l - n + 1)} G_{n+1}, \qquad (\text{I.B.68})$$

$$E_n = \frac{(n + 1)\bar{\mu}_{em}}{\Delta\varepsilon_l - n} M_{n+1}. \qquad (\text{I.B.69})$$

It follows that

$$k_{gg} = \sum_{n=0}^{\infty} \frac{(\alpha'\alpha^*)^n/n!}{\varepsilon - n - n|\bar{\mu}_{mg}|^2/[\Delta\varepsilon_m - n + 1 - (n - 1)|\bar{\mu}_{em}|^2/(\Delta\varepsilon_e - n + 2)]}, \qquad (\text{I.B.70})$$

$$k_{mg} = \alpha'\bar{\mu}_{mg} \sum_{n=0}^{\infty} \left\{ \frac{(\alpha'\alpha^*)^n/n!}{\Delta\varepsilon_m - n - n|\bar{\mu}_{em}|^2/(\Delta\varepsilon_e - n + 1)} \right.$$
$$\left. \times \frac{1}{\varepsilon - n + 1 - (n + 1)|\bar{\mu}_{mg}|^2/[\Delta\varepsilon_m - n - n|\bar{\mu}_{em}|^2/(\Delta\varepsilon_e - n + 1)]} \right\}, \qquad (\text{I.B.71})$$

and

$$k_{eg} = \bar{\mu}_{em}\bar{\mu}_{mg}\alpha' \sum_{n=0}^{\infty} \left\{ \frac{(\alpha'\alpha^*)^n/n!\,(\Delta\varepsilon_e - n)}{\Delta\varepsilon_m - n + 1 - (n + 1)|\bar{\mu}_{em}|^2/(\Delta\varepsilon_1 - n)} \right.$$
$$\left. \times \frac{1}{\varepsilon - n - 2 - (n + 2)|\bar{\mu}_{mg}|^2/[\Delta\varepsilon_m - n - 1 - (n + 1)|\bar{\mu}_{em}|^2/(\Delta\varepsilon_n - n)]} \right\}. \qquad (\text{I.B.72})$$

Next, for comparison we solve the three-level model given by Eqs. (I.B.55)–(I.B.57) by the photon number (or Fock state) method. For convenience, we define

$$|1\rangle \equiv |gn\rangle, \qquad |2\rangle \equiv |mn - 1\rangle, \qquad |3\rangle \equiv |en - 2\rangle. \qquad (\text{I.B.73})$$

To find the matrix elements of $\hat{G}(E)$ we have to solve the simultaneous equations [see Eq. (I.B.47)]

$$(E - H_{11})G_{11} - H_{12}G_{21} = 1,$$

$$-H_{21}G_{11} + (E - H_{22})G_{21} - H_{23}G_{31} = 0, \qquad \text{(I.B.74)}$$

$$-H_{32}G_{21} + (E - H_{33})G_{31} = 0.$$

We obtain

$$G_{11}(E) = \frac{1}{E - H_{11} - |H_{12}|^2/[E - H_{22} - |H_{23}|^2/(E - H_{33})]}, \qquad \text{(I.B.75)}$$

$$G_{21}(E) = \frac{H_{21}}{E - H_{22} - |H_{23}|^2/(E - H_{33})}$$

$$\times \frac{1}{E - H_{11} - |H_{12}|^2/[E - H_{22} - |H_{23}|^2/(E - H_{33})]}, \qquad \text{(I.B.76)}$$

$$G_{31}(E) = \frac{H_{32}H_{21}}{(E - H_{33})\,[E - H_{22} - |H_{23}|^2/(E - H_{33})]}$$

$$\times \frac{1}{E - H_{11} - |H_{12}|^2/[E - H_{22} - |H_{23}|^2/(E - H_{33})]} \qquad \text{(I.B.77)}$$

where

$$H_{11} = n\hbar\omega, \qquad H_{22} = E_m + (n - 1)\hbar\omega, \qquad H_{33} = E_e + (n - 2)\hbar\omega,$$

$$H_{21} = \sqrt{n}\,\mu_{mg}, \qquad H_{23} = \sqrt{n-1}\,\mu_{em}, \qquad H_{31} = 0. \qquad \text{(I.B.78)}$$

We can see that $G_{11}(E)$, $G_{21}(E)$, and $G_{31}(E)$ are related to k_{gg}, k_{mg}, and k_{eg} by the Poisson distribution.

Next we shall consider the case in which there are many intermediate states. Since the coherent state method and the photon number state method are closely related, we shall employ the latter for this purpose. Suppose we have the Hamiltonian given by

$$\hat{H} = \hat{H}_0 + \hat{H}', \qquad \text{(I.B.79)}$$

with

$$\hat{H}_0 = |e\rangle E_e \langle e| + \sum_i |m_i\rangle E_{m_i} \langle m_i| + \hbar\omega \hat{a}^\dagger \hat{a} \qquad \text{(I.B.80)}$$

and

$$\hat{H}' = \sum_i [(\mu_{m_ig} + \mu_{em_i})\hat{a} + (\mu_{m_ig}^* + \mu_{em_i}^*)\hat{a}^\dagger]. \qquad \text{(I.B.81)}$$

If we define the wave functions

$$|1\rangle \equiv |gn\rangle, \qquad |2_i\rangle \equiv |m_in - 1\rangle, \qquad |3\rangle \equiv |en - 2\rangle, \qquad \text{(I.B.82)}$$

the matrix elements of $\hat{G}(E)$ are given by

$$(E - H_{11})G_{11} - \sum_i H_{12_i}G_{2_i1} = 0,$$

$$-H_{2_i1}G_{11} + (E - H_{2_i2_i})G_{2_i1} - H_{2_i3}G_{31} = 0, \qquad \text{(I.B.83)}$$

$$-H_{32_i}G_{2_i1} + (E - H_{33})G_{31} = 0.$$

It follows that

$$G_{11}(E) = \frac{1}{E - H_{11} - \sum_i\{|H_{12_i}|^2/[E - H_{2_i2_i} - |H_{2_i3}|^2/(E - H_{33})]\}}, \qquad \text{(I.B.84)}$$

$$G_{2_i1}(E) = \frac{H_{2_i1}}{E - H_{2_i2_i} - |H_{2_i3}|^2/(E - H_{33})} G_{11}(E) \qquad \text{(I.B.85)}$$

$$G_{31}(E) = \frac{H_{32_i}}{E - H_{33}} G_{2_i1}(E), \qquad \text{(I.B.86)}$$

where

$$H_{11} = n\hbar\omega; \qquad H_{2_i2_i} = E_{m_i} + (n - 1)\hbar\omega; \qquad H_{33} = E_e + (n - 2)\hbar\omega,$$

$$H_{2_i1} = \sqrt{n}H_{m_ig}; \qquad H_{32_i} = \sqrt{n - 1}\mu_{em_i}; \qquad H_{2_i2_j} = 0 \quad (i \neq j),$$

$$H_{2_i3} = 0$$

$$\text{(I.B.87)}$$

We now discuss the application of the coherent state method to the case of two different photons:

$$\hat{H} = \hat{H}_0 + \hat{H}', \qquad \text{(I.B.88)}$$

$$\hat{H}_0 = |e\rangle E_e\langle e| + |m\rangle E_m\langle m| + \hbar\omega_1\hat{a}_1^\dagger\hat{a}_1 + \hbar\omega_2\hat{a}_2^\dagger\hat{a}_2, \qquad \text{(I.B.89)}$$

$$\hat{H}' = |e\rangle\mu_{em}\langle m|\hat{a}_2 + |m\rangle\mu_{mg}\langle g|\hat{a}_1 + \text{c.c.} \qquad \text{(I.B.90)}$$

In this case Eq. (I.B.18) becomes

$$E\langle\alpha_1\alpha_2 j|\hat{G}|\alpha_1'\alpha_2'k\rangle - \sum_l \left\langle j\left|\hat{H}\left(\alpha_1^*, \frac{\alpha_i}{2} + \frac{\partial}{\partial\alpha_i^*}\right)\right|l\right\rangle\langle\alpha_1\alpha_2|\hat{G}|\alpha_1'\alpha_2'k\rangle$$

$$= \langle\alpha_1\alpha_2|\alpha_1'\alpha_2'\rangle\delta_{jk}. \qquad \text{(I.B.91)}$$

We obtain

$$\left(\varepsilon - \omega_1\alpha_1^* \frac{\partial}{\partial\alpha_1^*} - \omega_2\alpha_2^* \frac{\partial}{\partial\alpha_2^*}\right)k_{gg} - \bar{\mu}_{mg}^*\alpha_1^* k_{mg} = \exp(\alpha_1^*\alpha_1' + \alpha_2^*\alpha_2'), \quad \text{(I.B.92)}$$

$$\left(\Delta\varepsilon_m - \omega_1\alpha_1^* \frac{\partial}{\partial\alpha_1^*} - \omega_2\alpha^* \frac{\partial}{\partial\alpha_2^*}\right)k_{mg} - \bar{\mu}_{mg}\frac{\delta k_{gg}}{\partial\alpha_1^*} - \alpha_2^*\bar{\mu}_{em}^* k_{eg} = 0, \quad \text{(I.B.93)}$$

$$\left(\Delta\varepsilon_e - \omega_1\alpha_1^* \frac{\partial}{\partial\alpha_1^*} - \omega_2\alpha_2^* \frac{\partial}{\partial\alpha_2^*}\right)k_{eg} - \bar{\mu}_{em}\frac{\partial k_{mg}}{\partial\alpha_2^*} = 0. \quad \text{(I.B.94)}$$

To solve this set of differential equations, we set

$$k_{gg} = \sum_{n_1}\sum_{n_2} G_{n_1 n_2}(\alpha_1^*)^{n_1}(\alpha_2^*)^{n_2}, \qquad \text{(I.B.95)}$$

$$k_{eg} = \sum_{n_1}\sum_{n_2} E_{n_1 n_2}(\alpha_1^*)^{n_1}(\alpha_2^*)^{n_2}, \qquad \text{(I.B.96)}$$

$$k_{mg} = \sum_{n_1}\sum_{n_2} M_{n_1 n_2}(\alpha_1^*)^{n_1}(\alpha_2^*)^{n_2}. \qquad \text{(I.B.97)}$$

Substituting Eqs. (I.B.95)–(I.B.97) into Eqs. (I.B.92)–(I.B.94) yields

$$G_{n_1 n_2} = \cfrac{(\alpha_1')^{n_1}(\alpha_2')^{n_2}/n_1! n_2!}{\varepsilon - n_1\omega_1 - n_2\omega_2}$$

$$-\cfrac{n_1|\bar{\mu}_{mg}|^2}{\Delta\varepsilon_m - (n_1 - 1)\omega_1 - n_2\omega_2 - \cfrac{n_2|\bar{\mu}_{em}|^2}{\Delta\varepsilon_e - (n_1 - 1)\omega_1 - (n_2 - 1)\omega_2}}$$

$$\text{(I.B.98)}$$

$$M_{n_1 n_2} = \cfrac{\bar{\mu}_{mg}(n_1 + 1)}{\Delta\varepsilon_m - n_1\omega_1 - n_2\omega_2 - \cfrac{n_2|\bar{\mu}_{em}|^2}{\Delta\varepsilon_e - n_1\omega_1 - (n_2 - 1)\omega_2}} G_{n_1 + n_2},$$

$$\text{(I.B.99)}$$

$$E_{n_1 n_2} = \frac{(n_2 + 1)\bar{\mu}_{em}}{\Delta\varepsilon_e - n_1\omega_1 - n_2\omega_2} M_{n_1 n_2 + 1}, \qquad \text{(I.B.100)}$$

where, for example,

$$\varepsilon = E/\hbar, \qquad \bar{\mu}_{mg} = (1/\hbar)\bar{\mu}_{mg},$$

$$k_{gg} = \hbar G_{gg}\exp[\tfrac{1}{2}(|\alpha_1|^2 + |\alpha_2|^2 + |\alpha_1|^2 + |\alpha_2'|^2)]. \qquad \text{(I.B.101)}$$

This model can also be treated by using the photon number state method.

I.C   THE SUSCEPTIBILITY METHOD

In this section, we shall derive an expression for the linear susceptibility taking into account the saturation effect (Bloembergen, 1965; Klauder and Sudarshan, 1968; Shimoda et al., 1972; Sargent et al., 1974; Yariv, 1975). For simplicity, we shall assume that the molecular system has only two levels with energies E_m and E_n with $E_m > E_n$. In this case, the master equations can be expressed as

$$d\rho_{nn}/dt = -(i/\hbar)(H'_{nm}\rho_{mn} - \rho_{nm}H'_{mn}) - \Gamma_{n:n}\rho_{nn} - \Gamma_{n:m}\rho_{mm}, \quad \text{(I.C.1)}$$

$$d\rho_{nm}/dt = -(i\omega_{nm} + \Gamma_{nm})\rho_{nm} - (i/\hbar)H'_{nm}(\rho_{mm} - \rho_{nn}), \quad \text{(I.C.2)}$$

and so on, where $\Gamma_{n:n} = \Gamma_{nn:nn}$, $\Gamma_{n:m} = \Gamma_{nn:mm}$, and $\Gamma_{mn} = \Gamma_{mn:mn}$, for example, and

$$H'_{nm} = -\boldsymbol{\mu}_{nm} \cdot \mathbf{E}(t), \quad \text{(I.C.3)}$$

with

$$\mathbf{E}(t) = \tfrac{1}{2}\mathbf{E}_0[\exp(it\omega) + \exp(-it\omega)]. \quad \text{(I.C.4)}$$

Letting

$$\rho_{nm}(t) = \sigma_{nm}(t)\exp(it\omega), \qquad \rho_{mn}(t) = \sigma_{mn}(t)\exp(-it\omega), \quad \text{(I.C.5)}$$

Eqs. (I.C.1) and (I.C.2) become

$$\frac{d\rho_{nn}}{dt} = \frac{i}{2\hbar}\{(\boldsymbol{\mu}_{nm} \cdot \mathbf{E}_0)[1 + \exp(-2it\omega)]\sigma_{mn}$$

$$- \sigma_{mn}[1 + \exp(2it\omega)](\boldsymbol{\mu}_{mn} \cdot \mathbf{E}_0)\} - \Gamma_{n:n}\rho_{nn} - \Gamma_{n:m}\rho_{mm} \quad \text{(I.C.6)}$$

and

$$\frac{d\sigma_{nm}}{dt} = -[i(\omega - \omega_{mn}) + \Gamma_{mn}]\sigma_{nm}$$

$$+ \frac{i}{2\hbar}(\boldsymbol{\mu}_{nm} \cdot \mathbf{E}_0)[1 + \exp(-2it\omega)](\rho_{mm} - \rho_{nn}). \quad \text{(I.C.7)}$$

Ignoring factors with time dependence $\exp(2it\omega)$ and $\exp(-2it\omega)$ (i.e, the rotating wave approximation), we obtain

$$\frac{d\rho_{nn}}{dt} = \frac{i}{2\hbar}[(\boldsymbol{\mu}_{nm} \cdot \mathbf{E}_0)\sigma_{mn} - \sigma_{nm}(\boldsymbol{\mu}_{mn} \cdot \mathbf{E}_0)] - \Gamma_{n:n}\rho_{nn} - \Gamma_{n:m}\rho_{mm}, \quad \text{(I.C.8)}$$

$$\frac{d\rho_{mm}}{dt} = \frac{i}{2\hbar}[(\boldsymbol{\mu}_{mn} \cdot \mathbf{E}_0)\sigma_{nm} - \sigma_{mn}(\boldsymbol{\mu}_{mn} \cdot \mathbf{E}_0)] - \Gamma_{m:m}\rho_{mm} - \Gamma_{m:n}\rho_{nn}, \quad \text{(I.C.9)}$$

$$\frac{d\sigma_{nm}}{dt} = -[i(\omega - \omega_{mn}) + \Gamma_{mn}]\sigma_{nm} + \frac{i}{2\hbar}(\mathbf{\mu}_{nm} \cdot \mathbf{E}_0)(\rho_{mm} - \rho_{nn}). \quad \text{(I.C.10)}$$

Notice that for a two-level system we have

$$\rho_{nn}(t) + \rho_{mm}(t) = \rho_{nn}^0 + \rho_{mm}^0, \quad \text{(I.C.11)}$$

$$\Gamma_{n:n} = -\Gamma_{m:n} \quad \text{and} \quad \Gamma_{m:m} = -\Gamma_{n:m}, \quad \text{(I.C.12)}$$

recalling that, for example, $-\Gamma_{m:n}$ denotes the rate constant for $n \to m$, and ρ_{nn}^0 and ρ_{mm}^0 denote the equilibrium distributions, that is,

$$\rho_{nn}^0\Gamma_{n:n} = \rho_{mm}^0\Gamma_{m:m}. \quad \text{(I.C.13)}$$

Applying Eqs. (I.C.11) and (I.C.13) to Eqs. (I.C.8) and (I.C.9) yields

$$\frac{d\,\Delta\rho_{mn}}{dt} = \frac{i}{2\hbar}(\mathbf{\mu}_{nm} \cdot \mathbf{E}_0\sigma_{mn} - \sigma_{nm}\mathbf{\mu}_{mn} \cdot \mathbf{E}_0) - \frac{1}{\tau_{mn}}(\Delta\rho_{mn} - \Delta\rho_{mn}^0), \quad \text{(I.C.14)}$$

where

$$\Delta\rho_{mn} = \rho_{nn}(t) - \rho_{mm}(t), \quad \text{(I.C.15)}$$

$$\Delta\rho_{mn}^0 = \rho_{nn}^0 - \rho_{mm}^0, \quad \text{(I.C.16)}$$

$$1/\tau_{mn} = \Gamma_{n:n} + \Gamma_{m:m}, \quad \text{(I.C.17)}$$

A solution of Eq. (I.C.10) is given by

$$\sigma_{nm}(t) = \frac{i\mathbf{\mu}_{nm} \cdot \mathbf{E}_0}{2\hbar} \int_0^t \exp\{[i(\omega - \omega_{mn}) + \Gamma_{mn}](t' - t)\}\,\Delta\rho_{nm}(t')\,dt'. \quad \text{(I.C.18)}$$

For $t > \Gamma_{mn}$, and if $\Delta\rho_{mn}(t)$ is constant during the time duration, Eq. (I.C.18) can be expressed as

$$\sigma_{nm}(t) = \frac{i\mathbf{\mu}_{nm} \cdot \mathbf{E}_0}{2\hbar}\,\frac{\Delta\rho_{nm}(t)}{i(\omega - \omega_{mn}) + \Gamma_{mn}}. \quad \text{(I.C.19)}$$

This is equivalent to the solution of Eq. (I.C.10) in the adiabatic (steady-state) approximation. We can thus see under what condition the adiabatic approximation is valid. Substituting Eq. (I.C.19) into Eq. (I.C.14), we obtain

$$\frac{d\,\Delta\rho_{mn}}{dt} = -\frac{\|\mathbf{\mu}_{nm} \cdot \mathbf{E}_0\|^2\Gamma_{nm}}{\hbar^2[(\omega - \omega_{mn})^2 + \Gamma_{nm}^2]}\Delta\rho_{mn} - \frac{1}{\tau_{mn}}(\Delta\rho_{mn} - \Delta\rho_{mn}^0). \quad \text{(I.C.20)}$$

Applying the steady-state approximation to Eq. (I.C.20) yields

$$\Delta\rho_{mn} = \frac{[(\omega - \omega_{mn})^2 + \Gamma_{mn}^2]\Delta\rho_{mn}^0}{(\omega - \omega_{mn})^2 + \Gamma_{mn}^2 + 4\Gamma_{mn}\tau_{mn}\Omega_{mn}^2}, \quad \text{(I.C.21)}$$

where Ω_{mn} is the precession frequency defined by

$$\Omega_{mn}^2 = \frac{|\boldsymbol{\mu}_{mn} \cdot \mathbf{E}_0|^2}{4\hbar^2} \tag{I.C.22}$$

We note from Eq. (I.C.21) that a saturation of the population takes place as the laser intensity becomes strong: $\rho_{mm} = \rho_{nn}$ for $\Omega_{mn} \to \infty$.

As shown in Section 2.4, we can derive susceptibility as

$$\bar{\bar{\chi}} = \frac{i\,\Delta N^0(\boldsymbol{\mu}_{mn}\boldsymbol{\mu}_{nm})[i(\omega - \omega_{mn}) - \Gamma_{mn}]}{\hbar[(\omega - \omega_{mn})^2 + \Gamma_{mn}^2 + 4\Gamma_{mn}\tau_{mn}\Omega_{mn}^2]}, \tag{I.C.23}$$

where $\Delta N^0 = N\Delta\rho^0$ represents the population difference in the absence of an electromagnetic field. Defining

$$\bar{\bar{\chi}} = \bar{\bar{\chi}}' - i\bar{\bar{\chi}}'', \tag{I.C.24}$$

we obtain

$$\bar{\bar{\chi}}' = \frac{\Delta N^0(\boldsymbol{\mu}_{mn}\boldsymbol{\mu}_{nm})(\omega_{mn} - \omega)}{\hbar[(\omega - \omega_{mn})^2 + \Gamma_{mn}^2 + 4\Gamma_{mn}\tau_{mn}\Omega_{mn}^2]}, \tag{I.C.25}$$

and

$$\bar{\bar{\chi}}'' = \frac{\Delta N^0(\boldsymbol{\mu}_{mn}\boldsymbol{\mu}_{nm})\Gamma_{mn}}{\hbar[(\omega - \omega_{mn})^2 + \Gamma_{mn}^2 + 4\Gamma_{mn}\tau_{mn}\Omega_{mn}^2]}. \tag{I.C.26}$$

If the molecules in the system are randomly oriented, then a spatial average over the molecular orientation has to be performed for the macroscopic polarization. In this case we have

$$\chi' = \frac{\Delta N^0|\boldsymbol{\mu}_{mn}|^2}{3\hbar} \frac{(\omega_{mn} - \omega)}{(\omega - \omega_{mn})^2 + \Gamma_{mn}^2 + 4\Gamma_{mn}\tau_{mn}\Omega_{mn}^2} \tag{I.C.27}$$

and

$$\chi'' = \frac{\Delta N^0|\boldsymbol{\mu}_{mn}|^2}{3\hbar} \frac{\Gamma_{mn}}{(\omega - \omega_{mn})^2 + \Gamma_{mn}^2 + 4\Gamma_{mn}\tau_{mn}\Omega_{mn}^2}. \tag{I.C.28}$$

It should be noted that $\chi'(\omega)$ is related to $\chi''(\omega)$ by the Kramers–Krönig relation

$$\chi'(\omega) = \frac{1}{\pi} P \int_{-\infty}^{\infty} \frac{\chi''(\omega')}{\omega' - \omega}\,d\omega' \tag{I.C.29}$$

and

$$\chi''(\omega) = -\frac{1}{\pi} P \int_{-\infty}^{\infty} \frac{\chi'(\omega')}{\omega' - \omega}\,d\omega'. \tag{I.C.30}$$

REFERENCES

Bloembergen, N. (1965). "Nonlinear Optics." Benjamin, New York.

Freed, K. F., and Villaeys, A. A. (1979). *J. Chem. Phys.* **70**, 3071.

Glauber, R. J. (1963), *Phys. Rev.* **131**, 2766.

Heitler, W. (1954). "The Quantum Theory of Radiation." Oxford Univ. Press, London and New York.

Klauder, J. R., and Sudarshan, E. C. G. (1968). "Fundamentals of Quantum Optics." Benjamin, New York.

Lin, S. H. (1967). *In* "Physical Chemistry" (H. Eyring, ed.), Vol. 2, pp. 109–168. Academic Press, New York.

Louisell, W. H. (1973). "Quantum Statistical Properties of Radiation." Wiley (Interscience), New York.

Sargent, M., III, Scully, M. O., and Lamb, W. E., Jr. (1974). "Laser Physics." Addison-Wesley, Reading, Massachusetts.

Shimoda, K., Yajima, T., Ueda, Y., Shimizu, T., and Kasuya, T. (1972). "Quantum Electronics." Syokabo, Tokyo.

Yariv, A. (1975). "Quantum Electronics." Wiley, New York.

APPENDIX

II

In this appendix to Chapter 4 we shall review the symmetric properties of $D_{MK}^{(J)}(\Omega)$ functions and the Wigner $3j$ symbols.

II.A SYMMETRIC PROPERTIES OF $D_{MK}^{(J)}(\Omega)$ FUNCTIONS

In this section some of the symmetric properties of D functions are presented. Detailed treatments of D functions can be found in many textbooks [see, for example, Edmonds (1960) and Davydov (1966)].

Let $D(\alpha, \beta, \gamma)$ be an operator of the successive rotations that are associated with the Euler angles and defined with respect to the original coordinates (X, Y, Z). This rotation operator is represented by the angular momentum operators $\hat{J}_x$, $\hat{J}_y$, and $\hat{J}_z$ as

$$D(\alpha, \beta, \gamma) = \exp\left(\frac{i\alpha}{\hbar}\,\hat{J}_X\right)\exp\left(\frac{i\beta}{\hbar}\,\hat{J}_Y\right)\exp\left(\frac{i\gamma}{\hbar}\,\hat{J}_Z\right) \qquad \text{(II.A.1)}$$

$$\equiv D(\Omega).$$

The D function is defined by

$$D_{MK}^{(J)}(\Omega) = \langle JM|D^{(J)}(\Omega)|JK\rangle, \qquad \text{(II.A.2)}$$

which is an eigenfunction of the square of the total angular momentum $\hat{J}^2$ with eigenvalue $\hbar^2 J(J + 1)$ and that of $\hat{J}_Z$ with eigenvalue $\hbar M$. It is also an eigenfunction of the angular momentum operator of rotation, $-i\hbar\,\partial/\partial\gamma$, with eigenvalue $\hbar K$. The complex conjugate of the D functions is given by

$$D_{MK}^{(J)*}(\Omega) = (-1)^{M-K} D_{-M-K}^{(J)}(\Omega). \tag{II.A.3}$$

The D functions satisfy the unitary relation

$$\sum_M D_{MK}^{(J)*} D_{MK'}^{(J)} = \delta_{KK'}. \tag{II.A.4}$$

The products of the D functions of different orders are expressed as

$$D_{M_1K_1}^{(J_1)}(\Omega) D_{M_2K_2}^{(J_2)}(\Omega) = \sum_J \sum_M \sum_K (2J+1) \begin{pmatrix} J_1 & J_2 & J \\ M_1 & M_2 & M \end{pmatrix}$$

$$\times \begin{pmatrix} J_1 & J_2 & J \\ K_1 & K_2 & K \end{pmatrix} D_{MK}^{(J)*}(\Omega), \tag{II.A.5}$$

with $|J_1 - J_2| \leqslant J \leqslant |J_1 + J_2|$ and $M_1 + M_2 + M = 0$, where

$$\begin{pmatrix} J_1 & J_2 & J \\ M_1 & M_2 & M \end{pmatrix}$$

are the Wigner $3j$ symbols and their definition and properties are presented in Appendix II.B. Another product formula is expressed as

$$\sum_{M_1} \sum_{M_2} \sum_M D_{M_1K_1}^{(J_1)}(\Omega) D_{M_2K_2}^{(J_2)}(\Omega) D_{MK}^{(J)}(\Omega) \begin{pmatrix} J_1 & J_2 & J \\ M_1 & M_2 & M \end{pmatrix}$$

$$= \begin{pmatrix} J_1 & J_2 & J \\ K_1 & K_2 & K \end{pmatrix}. \tag{II.A.6}$$

The orthonormalization of D functions with respect to integration over the Euler angles is given by

$$\frac{1}{8\pi^2} \int d\Omega \, D_{M_1K_1}^{(J_1)}(\Omega) D_{M_2K_2}^{(J_2)}(\Omega)$$

$$= \frac{1}{8\pi^2} \int_0^{2\pi} d\alpha \int_0^\pi d\beta \int_0^{2\pi} d\gamma \sin\beta \, D_{M_1K_1}^{(J_1)}(\Omega) D_{M_2K_2}^{(J_2)}(\Omega)$$

$$= (2J_1 + 1)^{-1} \delta_{M_1M_2} \delta_{K_1K_2} \delta_{J_1J_2}. \tag{II.A.7}$$

Applying Eq. (II.A.7) to Eq. (II.A.5) yields

$$\frac{1}{8\pi^2} \int_0^{2\pi} d\alpha \int_0^\pi d\beta \int_0^{2\pi} d\gamma \sin\beta \, D_{M_1K_1}^{(J_1)}(\Omega) D_{M_2K_2}^{(J_2)}(\Omega) D_{MK}^{(J)}(\Omega)$$

$$= \begin{pmatrix} J_1 & J_2 & J \\ M_1 & M_2 & M \end{pmatrix} \begin{pmatrix} J_1 & J_2 & J \\ K_1 & K_2 & K \end{pmatrix}. \tag{II.A.8}$$

II.B   SYMMETRIC PROPERTIES OF
THE WIGNER 3j SYMBOLS

In this section, the symmetric properties and the orthogonality of the Wigner $3j$ symbols are summarized. These are frequently utilized in the derivation of the two-photon transition probability of rotating molecules in Section 4.2.

The Wigner $3j$ symbol

$$\begin{pmatrix} J_1 & J_2 & J \\ M_1 & M_2 & M \end{pmatrix}$$

is originally defined in terms of the vector-coupling coefficients (Edmonds, 1960; Landau and Lifshitz, 1964). It can be expressed as

$$\begin{pmatrix} J_1 & J_2 & J \\ M_1 & M_2 & M \end{pmatrix} = \Delta[J_1 \quad J_2 \quad J]\sum_Z (-1)^{J_1 - J_2 - M + Z}$$

$$\times \frac{[(J_1 + M_1)!(J_1 - M_1)!(J_2 + M_2)!(J_2 - M_2)!(J + M)!(J - M)!]^{1/2}}{Z!(J_1 + J_2 - J - Z)!(J_1 - M_1 - Z)!(J_2 + M_2 - Z)!}$$
$$\times (J - J_2 + M_1 + Z)!(J - J_1 - M_2 + Z)!,$$

$$\text{(II.B.1)}$$

where

$$\Delta[J_1 \quad J_2 \quad J]$$
$$= [(J_1 + J_2 - J)!(J_1 + J - J_2)!(J_2 + J - J_1)!/(J_1 + J_2 + J + 1)!]^{1/2}$$

and $M_1 + M_2 + M = 0$ and where $|J_1 - J_2| \leqslant J \leqslant |J_1 + J_2|$ are satisfied. The permutation rules of the Wigner $3j$ symbols are written as

$$\begin{pmatrix} J_1 & J_2 & J \\ M_1 & M_2 & M \end{pmatrix} = \begin{pmatrix} J_2 & J & J_1 \\ M_2 & M & M_1 \end{pmatrix} = \begin{pmatrix} J & J_1 & J_2 \\ M & M_1 & M_2 \end{pmatrix}, \quad \text{(II.B.2)}$$

$$\begin{pmatrix} J_2 & J_1 & J \\ M_2 & M_1 & M \end{pmatrix} = (-1)^{J_1 + J_2 + J}\begin{pmatrix} J_1 & J_2 & J \\ M_1 & M_2 & M \end{pmatrix}, \quad \text{(II.B.3)}$$

$$\begin{pmatrix} J_1 & J_2 & J \\ M_1 & M_2 & M \end{pmatrix} = (-1)^{J_1 + J_2 + J}\begin{pmatrix} J_1 & J_2 & J \\ -M_1 & -M_2 & -M \end{pmatrix}. \quad \text{(II.B.4)}$$

The orthogonality property is expressed as

$$\sum_J \sum_M (2J + 1)\begin{pmatrix} J_1 & J_2 & J \\ M_1 & M_2 & M \end{pmatrix}\begin{pmatrix} J_1 & J_2 & J \\ M_1' & M_2' & M \end{pmatrix} = \delta_{M_1 M_1'}\delta_{M_2 M_2'} \quad \text{(II.B.5)}$$

and

$$\sum_{M_1} \sum_{M_2} \begin{pmatrix} J_1 & J_2 & J \\ M_1 & M_2 & M \end{pmatrix} \begin{pmatrix} J_1 & J_2 & J' \\ M_1 & M_2 & M' \end{pmatrix} = (2J+1)^{-1} \delta_{JJ'} \delta_{MM'}. \quad \text{(II.B.6)}$$

Some of the simple formulas of the Wigner $3j$ symbols are:

$$\begin{pmatrix} J & J & 0 \\ M & -M & 0 \end{pmatrix} = (-1)^{J-M} (2J+1)^{-1/2}, \quad \text{(II.B.7)}$$

$$\begin{pmatrix} J & L & J \\ -J & 0 & J \end{pmatrix} = (2J+L+1)^{-1/2} \frac{(2J)!}{[(2J+L)!(2J-L)!]^{1/2}}. \quad \text{(II.B.8)}$$

APPENDIX

III

In this appendix to chapter 5 we shall derive Eq. (5.2.20), outline the theory of intramolecular vibrational relaxation, and present the evaluation of a generating function.

III.A DERIVATION OF EQUATION (5.2.20)

In this section we shall present a brief derivation of the expression for the simultaneous TPA probability given in Eq. (5.2.20) using the Feynman disentangle and boson operator techniques (Louisell, 1973).

Let us first summarize these techniques.

The Feynman Disentangle Technique

$$\hat{T} \exp\left\{ \int_0^t d\tau \left[f(\tau) + g(\tau) \right] \right\} = \exp\left[\int_0^t d\tau f(\tau) \right] \hat{T} \exp\left[\int_0^t d\tau\, G(\tau) \right], \quad \text{(III.A.1)}$$

where

$$G(\tau) = \exp\left[-\int_0^\tau d\tau' f(\tau') \right] g(\tau) \exp\left[\int_0^\tau d\tau' f(\tau') \right]. \quad \text{(III.A.2)}$$

Here $f(\tau)$ and $g(\tau)$ are functions of τ involving noncommuting operators, and $\hat{T}$ denotes the time-ordering operator.

The Boson Operator Technique

$$\exp(\alpha\hat{b}^\dagger\hat{b})f(\hat{b}, \hat{b}^\dagger) \exp(-\alpha\hat{b}^\dagger\hat{b}) = f[\hat{b} \exp(-\alpha), \hat{b}^\dagger \exp(\alpha)], \qquad \text{(III.A.3)}$$

$$\exp(\alpha\hat{b})f(\hat{b},\hat{b}^\dagger) \exp(-\alpha\hat{b}) = f(\hat{b},\hat{b}^\dagger + \alpha), \qquad \text{(III.A.4)}$$

$$\exp(-\alpha\hat{b}^\dagger)f(\hat{b}, \hat{b}^\dagger) \exp(\alpha\hat{b}^\dagger) = f(\hat{b} + \alpha, \hat{b}^\dagger), \qquad \text{(III.A.5)}$$

$$\exp(\alpha\hat{b}^\dagger + \beta\hat{b}^\dagger) = \exp(\alpha\hat{b}^\dagger) \exp(\beta\hat{b}) \exp(\alpha\beta/2)$$

$$= \exp(\beta\hat{b}) \exp(\alpha\hat{b}^\dagger) \exp(-\alpha\beta/2), \qquad \text{(III.A.6)}$$

where $\hat{b}$ and $\hat{b}^\dagger$, respectively, denote the boson annihilation and creation operators, which satisfy $[\hat{b}, \hat{b}^\dagger] = \hat{b}\hat{b}^\dagger - \hat{b}^\dagger\hat{b} = 1$, α and β are C numbers, and $f(\hat{b},\hat{b}^\dagger)$ is a function which may be expanded in a power series in $\hat{b}$ and $\hat{b}^\dagger$. The averaging $A_v \cdots = \Sigma_a$, $\langle X_a | \cdots \rho_a | X_a \rangle$, where X_a is the harmonic oscillator wave function $\chi_a = (\hat{b}^\dagger)^n / \sqrt{n!} |0\rangle$ and the density matrix $\rho_a = [1 - \exp(-\beta\hbar\omega)] \exp(-\beta\hbar\omega\hat{b}^\dagger\hat{b})$, with $\beta = 1/kT$, is evaluated as

$$A_v f(\hat{b}, \hat{b}^\dagger) = \langle 0, 0 | f(\sqrt{1 + \bar{n}}\,\hat{b} + \sqrt{\bar{n}}\,\hat{a}^\dagger, \sqrt{1 + \bar{n}}\,\hat{b}^\dagger + \sqrt{\bar{n}}\,\hat{a}) | 0, 0 \rangle. \qquad \text{(III.A.7)}$$

Here, $\bar{n}$, the occupation number, is given by $\bar{n} = [\exp(\beta\hbar\omega) - 1]^{-1}$. These techniques have been applied to the evaluation of nonradiative transition rate constants of molecules (Fong, 1975) and to the derivation of the cross section of the reasonance Raman scattering from molecules (Fujimura and Lin, 1979a, b).

Let us now derive Eq. (5.2.20). In the displaced harmonic oscillator model, the vibronic Hamiltonians of the initial, resonant, and final electronic states take the form

$$\hat{H}_a = \sum_{i=1}^N \hat{H}_{ai}, \qquad \hat{H}_{ai} = \hbar\omega_i \left(\hat{b}_i^\dagger \hat{b}_i + \frac{1}{2} \right), \qquad \text{(III.A.8)}$$

$$\hat{H}_m = \sum_{i=1}^N \hat{H}_{mi} + \varepsilon_m^0, \qquad \hat{H}_{mi} = \hat{H}_{ai} - \frac{\hbar\omega_i\Delta_{mai}}{\sqrt{2}} (\hat{b} + \hat{b}_i^\dagger) + \frac{\hbar\omega_i\Delta_{mai}^2}{2}, \qquad \text{(III.A.9)}$$

$$\hat{H}_n = \sum_{i=1}^N \hat{H}_{ni} + \varepsilon_n^0, \qquad \hat{H}_{ni} = \hat{H}_{ai} - \frac{\hbar\omega_i\Delta_{nai}}{\sqrt{2}} (\hat{b}_i + \hat{b}_i^\dagger) + \frac{\hbar\omega_i\Delta_{mai}^2}{2}. \qquad \text{(III.A.10)}$$

Noting that the vibrational wave function $|\theta_a\rangle$ is expressed in terms of the product of the harmonic wave functions, $|\theta_a\rangle = \Pi_i|\theta_{ai}\rangle$, and substituting Eqs. (III.A.8)–(III.A.10) into Eq. (5.2.15), we find that

$$G(\tau, \tau', t) = \exp\left[\frac{i\varepsilon_m^0(\tau - \tau')}{\hbar} + \frac{i\varepsilon_n^0 t}{\hbar} \right] \prod_i G_i(\tau, \tau', t), \qquad \text{(III.A.11)}$$

where G_i, the generating function for the ith mode, is given by

$$G_i(\tau, \tau', t) = A_{vi}\{\exp[-i\hat{H}_{ai}(t + \tau' - \tau)/\hbar]\exp(i\hat{H}_{mi}\tau'/\hbar)$$
$$\times \exp(i\hat{H}_{ni}t/\hbar)\exp(-i\hat{H}_{mi}\tau/\hbar)\}. \qquad \text{(III.A.12)}$$

By utilizing the Feynman disentangle technique, exponential terms involving the intermediate and final state Hamiltonians can be expressed in terms of the ground state Hamiltonian as, for example,

$$\exp\left(\frac{i\hat{H}_{mi}\tau'}{\hbar}\right) = \exp\left[\frac{\Delta^2_{mai}}{2}(e^{i\omega_i\tau'} - 1)\right]\exp\left(\frac{i\hat{H}_{ai}\tau'}{\hbar}\right)$$
$$\times \exp[-\lambda_{mai}(\tau')\hat{b}_i^\dagger]\exp[\lambda^*_{ma}(\tau')\hat{b}_i], \qquad \text{(III.A.13)}$$

where

$$\lambda_{mai}(\tau') = (\Delta_{mai}/\sqrt{2})(1 - e^{-i\omega_i\tau'}).$$

Substituting Eq. (III.A.13) and the similar reduced form for $\exp[i\hat{H}_{ni}t/\hbar]$ into Eq. (III.A.12) yields

$$G_i(\tau, \tau', t) = \exp[(\Delta^2_{mi}/2)(e^{i\omega_i t} - 1) + (\Delta^2_{mai}/2)(e^{i\omega_i\tau'} - 1)$$
$$+ (\Delta^2_{mai}/2)(e^{-i\omega_i\tau} - 1)]L_i(\tau, \tau', t), \qquad \text{(III.A.14)}$$

where

$$L_i(\tau, \tau', t) = A_{vi}\{\exp[-i\hat{H}_{ai}(t + \tau' - \tau)/\hbar]\exp(i\hat{H}_{ai}\tau'/\hbar)$$
$$\times \exp[-\lambda_{mai}(\tau')\hat{b}_i^\dagger]\exp[\lambda^*_{mai}(\tau')\hat{b}_i]\exp(i\hat{H}_{ai}t/\hbar)$$
$$\times \exp[-\lambda_{mai}(t)\hat{b}_i^\dagger]\exp[\lambda^*_{mai}(t)\hat{b}_i]\exp(-i\hat{H}_{ai}\tau/\hbar)$$
$$\times \exp[-\lambda^*_{mai}(\tau)\hat{b}_i^\dagger]\exp[\lambda_{mai}(\tau)\hat{b}_i]\}. \qquad \text{(III.A.15)}$$

The averaging procedure in Eq. (III.A.15) can be accomplished by using Eq. (III.A.7):

$$L_i(\tau, \tau', t) = \langle 0_i 0_i|\exp[-\lambda_{mai}(\tau')e^{-i\omega_i t}(\sqrt{1 + \bar{n}_i}\,\hat{b}_i^\dagger + \sqrt{\bar{n}_i}\,\hat{a}_i)]$$
$$\times \exp[\lambda^*_{mai}(\tau')e^{i\omega_i t}(\sqrt{1 + \bar{n}_i}\,\hat{b}_i + \sqrt{\bar{n}}\,\hat{a}_i^\dagger)]$$
$$\times \exp[-\lambda_{nai}(t)(\sqrt{1 + \bar{n}_i}\,\hat{b}_i^\dagger + \sqrt{\bar{n}_i}\,\hat{a}_i)]$$
$$\times \exp[\lambda^*_{nai}(t)(\sqrt{1 + \bar{n}_i}\,\hat{b}_i + \sqrt{\bar{n}}\,\hat{a}_i^\dagger)]$$
$$\times \exp[\lambda_{mai}(\tau)(\sqrt{1 + \bar{n}_i}\,\hat{b}_i^\dagger + \sqrt{\bar{n}_i}\,\hat{a}_i)]$$
$$\times \exp[-\lambda^*_{mai}(\tau)(\sqrt{1 + \bar{n}_i}\,\hat{b}_i + \sqrt{\bar{n}_i}\,\hat{a}_i^\dagger)]|0_i 0_i\rangle. \qquad \text{(III.A.16)}$$

By applying the boson operator technique, substituting Eqs. (III.A.3)–(III.A.6) into Eq. (III.A.16) yields

$L_i(\tau, \tau', t)$

$$
= \exp\left[-(2\bar{n}_i + 1)\left(\frac{\Delta_{mai}^2}{2} + \frac{\Delta_{nmi}^2}{2}\right) - \frac{\Delta_{nmi}\Delta_{mai}}{2}(\bar{n}_i + 1)(e^{i\omega_i\tau'} + e^{-i\omega_i\tau}) \right.
$$

$$
- \frac{\Delta_{nmi}\Delta_{mai}}{2}\bar{n}_i(e^{-i\omega_i\tau'} + e^{i\omega_i\tau}) + (\bar{n}_i + 1)\left(\frac{\Delta_{nmi}}{\sqrt{2}} + \frac{\Delta_{mai}}{\sqrt{2}}e^{i\omega_i\tau'}\right)
$$

$$
\times \left(\frac{\Delta_{nmi}}{\sqrt{2}} + \frac{\Delta_{mai}}{\sqrt{2}}e^{i\omega_i\tau}\right)e^{i\omega_i t} + \bar{n}_i\left(\frac{\Delta_{nmi}}{\sqrt{2}} + \frac{\Delta_{mai}}{\sqrt{2}}e^{-i\omega_i\tau'}\right)
$$

$$
\left. \times \left(\frac{\Delta_{nmi}}{\sqrt{2}} + \frac{\Delta_{mai}}{\sqrt{2}}e^{i\omega_i\tau}\right)e^{-i\omega_i t} \right], \tag{III.A.17}
$$

where $\Delta_{nmi} = \Delta_{nai} - \Delta_{mai}$. Substituting Eqs. (III.A.11), (III.A.14), and (III.A.17) into Eq. (5.2.14), we find

$W_{\text{sim}}(\omega_1, \omega_2)$

$$
= \frac{|M_{nm}M_{ma}|^2}{\hbar^4} \int_{-\infty}^{\infty} d\tau \int_0^{\infty} d\tau \int_0^{\infty} d\tau'
$$

$$
\times \exp\left[-\Gamma_{na}|t| - \Gamma_{ma}(\tau + \tau') - it(\omega_1 + \omega_2) - i\omega_1(\tau' - \tau) \right.
$$

$$
\left. + \frac{i\varepsilon_m^0\tau'}{\hbar} + \frac{i\varepsilon_n^0\tau}{\hbar} - \frac{i\varepsilon_m^0\tau}{\hbar} \right]
$$

$$
\times \exp\left[-\sum_i (2\bar{n}_i + 1)\left(\frac{\Delta_{mai}^2}{2} + \frac{\Delta_{nmi}^2}{2}\right) \right.
$$

$$
- \sum_i \frac{\Delta_{nmi}\Delta_{mai}}{2}(\bar{n}_i + 1)(e^{i\omega_i\tau'} + e^{-i\omega_i\tau})
$$

$$
- \sum_i \frac{\Delta_{nmi}\Delta_{mai}\bar{n}_i}{2}(e^{-i\omega_i\tau'} + e^{i\omega_i\tau})
$$

$$
+ \sum_i (\bar{n}_i + 1)\left(\frac{\Delta_{nmi}}{\sqrt{2}} + \frac{\Delta_{mai}}{\sqrt{2}}e^{i\omega_i\tau'}\right)\left(\frac{\Delta_{nmi}}{\sqrt{2}} + \frac{\Delta_{mai}}{\sqrt{2}}e^{-i\omega_i\tau}\right)e^{i\omega_i t}
$$

$$
\left. + \sum_i \bar{n}_i\left(\frac{\Delta_{nmi}}{\sqrt{2}} + \frac{\Delta_{mai}}{\sqrt{2}}e^{-i\omega_i\tau'}\right)\left(\frac{\Delta_{nmi}}{\sqrt{2}} + \frac{\Delta_{mai}}{\sqrt{2}}e^{i\omega_i\tau}\right)e^{-i\omega_i t} \right]. \tag{III.A.18}
$$

By expanding the exponents involving $\exp(i\omega_i t)$ and integrating over t, Eq. (III.A.18) can be expressed as

$$W_{\text{sim}}(\omega_1, \omega_2)$$

$$= \frac{2|M_{nm}M_{ma}|^2}{\hbar^4} \exp\left[-\sum_i (2\bar{n}_i + 1)\left(\frac{\Delta_{mai}^2}{2} + \frac{\Delta_{nmi}^2}{2} \right) \right]$$

$$\times \sum_{k_1=0}^{\infty} \sum_{k_2=0}^{\infty} \cdots \sum_{k_N=0}^{\infty} \sum_{l_1=0}^{\infty} \sum_{l_2=0}^{\infty}$$

$$\cdots \sum_{l_N=0}^{\infty} \frac{(\bar{n}_1 + 1)^{k_1}(\bar{n}_2 + 1)^{k_2} \cdots (\bar{n}_N + 1)^{k_N} \bar{n}_1^{l_1} \bar{n}_2^{l_2} \cdots \bar{n}_N^{l_N}}{k_1! k_2! \cdots k_N! l_1! l_2! \cdots l_N!}$$

$$\times \left| \int_0^{\infty} d\tau \exp\left(i\omega_1 \tau - \frac{i\varepsilon_m^0 \tau}{\hbar} - \Gamma_{ma}\tau \right) \left(\frac{\Delta_{nm1}}{\sqrt{2}} + \frac{\Delta_{ma1}}{\sqrt{2}} e^{-i\omega_1 \tau} \right)^{k_1} \right.$$

$$\times \left(\frac{\Delta_{nm2}}{\sqrt{2}} + \frac{\Delta_{ma2}}{\sqrt{2}} e^{-i\omega_2 \tau} \right)^{k_2} \cdots \left(\frac{\Delta_{nmN}}{\sqrt{2}} + \frac{\Delta_{maN}}{\sqrt{2}} e^{-i\omega_N \tau} \right)^{k_N}$$

$$\times \left(\frac{\Delta_{nmi}}{\sqrt{2}} + \frac{\Delta_{mai}}{\sqrt{2}} e^{i\omega_1 \tau} \right)^{l_1} \left(\frac{\Delta_{nm2}}{\sqrt{2}} + \frac{\Delta_{ma2}}{\sqrt{2}} e^{i\omega_2 \tau} \right)^{l_2}$$

$$\cdots \left(\frac{\Delta_{nmN}}{\sqrt{2}} + \frac{\Delta_{maN}}{\sqrt{2}} e^{i\omega_N \tau} \right)^{l_N}$$

$$\times \exp\left[-\sum_i \frac{\Delta_{nmi}\Delta_{mai}}{2} (\bar{n}_i + 1)e^{-i\omega_i \tau} - \sum_i \frac{\Delta_{nmi}\Delta_{mai}}{2} \bar{n}_i e^{i\omega_i \tau} \right] \Bigg|^2. \quad \text{(III.A.19)}$$

If one uses binomial and Taylor expansions in Eq. (III.A.19) and integrates over τ, the analytical expression for the transition probability of simultaneous TPA in the multimode case for nonzero temperature is finally given by

$$W_{\text{sim}}(\omega_1, \omega_2)$$

$$= \frac{2|M_{nm}M_{ma}|^2}{\hbar^4} \exp\left[-\sum_{i=1}^{N} (2\bar{n}_i + 1)\left(\frac{\Delta_{mai}^2}{2} + \frac{\Delta_{nmi}^2}{2} \right) \right] \sum_{k_1=0}^{\infty} \sum_{k_2=0}^{\infty}$$

$$\cdots \sum_{k_N=0}^{\infty} \sum_{l_1=0}^{\infty} \sum_{l_2=0}^{\infty} \cdots \sum_{l_N=0}^{\infty} \left[\prod_{i=1}^{N} \frac{(\bar{n}_i + 1)^{k_i} \bar{n}_i^{l_i}}{k_i! l_i!} \right]$$

$$\times \frac{\Gamma_{na}}{[\omega_1 + \omega_2 - (\varepsilon_n^0/\hbar) - \sum_{i=1}^{N} (k_i - l_i)\omega_i]^2 + \Gamma_{na}^2}$$

$$\times \left| \sum_{p_1=0}^{k_1} \sum_{p_2=0}^{k_2} \cdots \sum_{p_N=0}^{k_N} \sum_{q_1=0}^{l_1} \sum_{q_2=0}^{l_2} \right.$$

(*Equation continues*)

$$
\cdots \sum_{q_N=0}^{l_N} \sum_{r_1=0}^{\infty} \sum_{r_2=0}^{\infty} \cdots \sum_{r_N=0}^{\infty} \sum_{s_1=0}^{\infty} \sum_{s_2=0}^{\infty}
$$

$$
\cdots \sum_{s_N=0}^{\infty} \left\{ \prod_{i=1}^{N} \binom{k_i}{p_i}\binom{l_i}{q_i} [-(\Delta_{nmi}\Delta_{mai}/2)(\bar{n}_i + 1)]^{r_i} \right.
$$

$$
\times \frac{[-(\Delta_{nmi}\Delta_{mai}/2)\bar{n}_i]^{s_i}(\Delta_{nmi}/\sqrt{2})^{(k_i - p_i + l_i - q_i)}(\Delta_{mai}/\sqrt{2})^{(p_i + q_i)}}{r_i! \, s_i!}
$$

$$
\times \left. \frac{1}{i[\varepsilon_m^0/\hbar - \omega_1 + \Sigma_{i=1}^{N}(p_i - q_i + r_i - s_i)\omega_i] + \Gamma_{ma}} \right\}\Bigg|^2 . \tag{III.A.20}
$$

III.B INTRAMOLECULAR VIBRATIONAL RELAXATION

For convenience, we briefly outline here the theory of intramolecular vibrational relaxation (IVR) based on the use of the adiabatic approximation. The Hamiltonian of a polyatomic molecule can in general be expressed as

$$
\hat{H} = \hat{T}_Q + \hat{T}_{rv} + \hat{h}_q, \tag{III.B.1}
$$

where

$$
\hat{h}_q = \hat{T}_q + V(q, Q) \tag{III.B.2}
$$

$$
\hat{T}_{rv} = \frac{(\hat{M}_x - \hat{m}_x)^2}{2I_x} + \frac{(\hat{M}_y - \hat{m}_y)^2}{2I_y} + \frac{(\hat{M}_z - \hat{m}_z)^2}{2I_z} = \sum_u \frac{(\hat{M}_u - \hat{m}_u)^2}{2I_u}. \tag{III.B.3}
$$

Here q and Q represent the vibrational coordinates of high- and low-frequency modes, respectively, and $(\hat{M}_x, \hat{M}_y, \hat{M}_z)$ and $(\hat{m}_x, \hat{m}_y, \hat{m}_z)$ denote the rotational and vibrational angular momenta, respectively. In Eq. (III.B.2), $\hat{h}_q$ represents the Hamiltonian operator of q.

According to the adiabatic approximation, the solution of

$$
\hat{H}\Psi_{vnl} = E_{vnl}\Psi_{vnl} \tag{III.B.4}
$$

can be expressed as

$$
\Psi_{vnl} = \Psi_v(q, Q)\theta_{vnl}(Q, x), \tag{III.B.5}
$$

$$
\hat{h}_q\Psi_v(q, Q) = U_v(Q)\Psi_v(q, Q), \tag{III.B.6}
$$

and

$$
(\hat{T}_Q + \hat{T}_{rv} + U_v)\theta_{vnl} = E_{vnl}\theta_{vnl}, \tag{III.B.7}
$$

where n and l denote the sets of the quantum numbers of low-frequency modes and those of molecular rotation, respectively. Using the adiabatic approximation as a basis set, the IVR rate constant of a single rovibrational level (SRVL) can be written as

$$W_{vnl} = \frac{2\pi}{\hbar} \sum_{v'n'l'} |\langle \Psi_{v'n'l'} | \hat{H}' | \Psi_{vnl} \rangle|^2 D(E_{v'n'l'} - E_{vnl}), \qquad \text{(III.B.8)}$$

where $D(E_{v'n'l'} - E_{nvl})$ represents the line shape function:

$$D(E_{v'n'l'} - E_{vnl}) = \frac{1}{\pi} \frac{\Gamma_{v'n'l',vnl}}{(E_{v'n'l'} - E_{vnl})^2 + \Gamma^2_{v'n'l',vnl}}. \qquad \text{(III.B.9)}$$

Here $\Gamma_{v'n'l',vnl}$ denotes the dephasing (or damping) constant. In Eq. (III.B.8), $\hat{H}'$, which represents the perturbations inducing IVR, can be attributed to the Born–Oppenheimer (BO) coupling

$$\hat{H}'_{BO} = \frac{1}{2} \sum_j \hat{P}_j^2 = -\frac{\hbar^2}{2} \sum_j \frac{\partial^2}{\partial Q_j^2}, \qquad \text{(III.B.10)}$$

that is, the kinetic energy operator of low-frequency modes of vibration, and to the Coriolis coupling

$$\hat{H}'_{Cor} = -\sum_u \frac{\hat{M}_u \hat{m}_u}{I_u}. \qquad \text{(III.B.11)}$$

For example, if $V(q, Q)$ takes the form

$$V(q, Q) = \frac{1}{2} \sum_i \omega_i^2 q_i^2 + \frac{1}{2} \sum_j \omega_j^2 Q_j^2 + \frac{1}{3!} \sum_i \sum_j \sum_\alpha V_{ij\alpha} q_i q_j Q_\alpha$$

$$+ \frac{1}{3!} \sum_i \sum_\alpha \sum_\beta V_{i\alpha\beta} q_i Q_\alpha Q_\beta + \frac{1}{4!} \sum_i \sum_j \sum_\alpha \sum_\beta V_{ij\alpha\beta} q_i q_j Q_\alpha Q_\beta + \cdots,$$

$$\text{(III.B.12)}$$

that is, including cubic and quartic anharmonic potentials, then we obtain

$$U_v(Q) = \sum_i \left[\left(v_i + \frac{1}{2} \right) - \chi_i \left(v_i + \frac{1}{2} \right)^2 \right] \hbar\omega_i + \frac{1}{2} \sum_\alpha \omega_\alpha(\{v_i\})^2 Q_\alpha(\{v_i\})^2 + \cdots,$$

$$\text{(III.B.13)}$$

where

$$Q_\alpha(\{v_i\}) = Q_\alpha + \sum_i \frac{\hbar V_{ii\alpha}}{6\omega_i \omega_\alpha^2} \left(v_i + \frac{1}{2} \right) + \cdots \qquad \text{(III.B.14)}$$

and

$$
\omega_\alpha(\{v_i\})^2 = \omega_\alpha^2 + \sum_i \frac{\hbar V_{ii\alpha\alpha}}{12\omega_i}\left(v_i + \frac{1}{2}\right)
$$

$$
+ \sum_i \sum_j \frac{\hbar V_{ij\alpha}^2}{36\omega_i\omega_j(\omega_i^2 - \omega_j^2)}\left[\omega_j\left(v_i + \frac{1}{2}\right) - \omega_i\left(v_j + \frac{1}{2}\right)\right] + \cdots .
$$

$$(\text{III.B.15})$$

It has been shown that

$$
\langle \Psi_{v'n'l'}|\hat{H}'_{\text{BO}}|\Psi_{vnl}\rangle = \sum_i \sum_\alpha \sum_\beta \frac{1}{3!}\, V_{i\alpha\beta}\, \langle \Psi_{v'n'l'}|q_i Q_\alpha Q_\beta|\Psi_{vnl}\rangle + \cdots . \quad (\text{III.B.16})
$$

From Eqs. (III.B.14)–(III.B.16), we can see that different types of anharmonic potentials play different roles in IVR. Whereas the type $(1/3!)V_{i\alpha\beta}q_i Q_\alpha Q_\beta$ plays the promoting role of the IVR of q_i, the type $(1/3!)V_{ii\alpha}q_i^2 Q_\alpha$, which introduces a vibrational coordinate shift in a totally symmetric mode so that the Q_α mode can accept more than one vibrational quantum in IVR, determines the effectiveness of Q_α in accepting the vibrational excitation energy during IVR. Notice that the types $(1/4!)V_{ii\alpha\alpha}q_i^2 Q_\alpha^2$ and $(1/3!)V_{ij\alpha}q_i q_j Q_\alpha$ can introduce the frequency shift in the Q_α mode so that it can accept more than one vibrational quantum in IVR.

In this section, we are concerned only with the calculation of the SRVL IVR rate constant; this is usually sufficient for the purpose of interpreting the experimental IVR data for molecules in the low energy range. For other purposes, a distribution function for the low-frequency modes and molecular rotation may be required; in these cases, the desired IVR rate constant can be obtained by multiplying the above-mentioned distribution function with the corresponding SRVL IVR rate constants.

III.C EVALUATION OF A GENERATING FUNCTION

In this section we are concerned with the calculation of $G_{nj}(t)$ for the case of a vibrational frequency shift. Notice that by applying the Mehler formula

$$
\sum_{n=0}^{\infty} \frac{\exp[-(n + \frac{1}{2})t]}{\sqrt{\pi}\, 2^n n!}\, H_n(x)H_n(x')\exp\left[-\frac{1}{2}(x^2 + x'^2)\right]
$$

$$
= (2\pi \sinh t)^{-1/2}\exp\left[-\frac{1}{4}(x + x')^2 \tanh\frac{t}{2}\right.
$$

$$
\left. -\frac{1}{4}(x - x')^2 \coth\frac{t}{2}\right]
$$

$$(\text{III.C.1})$$

to Eq. (5.3.6), we obtain

$$G_{n_j}(t) = E_{n_j}(t) \frac{\beta'_j}{(2\pi \sinh \lambda'_j)^{1/2}} \int_{-\infty}^{\infty} \int_{-\infty}^{\infty} dQ_j \, d\bar{Q}_j \, \chi_{vn_j}(Q_j)\chi_{v'n'_j}(\bar{Q}_j)$$

$$\times \exp\left\{-\frac{\beta'^2_j}{4}\left[(Q_j + \bar{Q}_j)^2 \tanh \frac{\lambda'_j}{2} + (Q_j - \bar{Q}_j)^2 \coth \frac{\lambda'_j}{2}\right]\right\},$$

$$\text{(III.C.2)}$$

where

$$\beta'^2_j = \omega'_j/\hbar, \tag{III.C.3}$$

$$\lambda'_j = t(\Gamma_j - i\omega'_j), \tag{III.C.4}$$

$$E_{n_j}(t) = \exp[-it\omega_j(n_j + \tfrac{1}{2}) - (n_j - \tfrac{1}{2})\Gamma_j t]. \tag{III.C.5}$$

By using the contour integral representation for the Hermite polynomial

$$H_n(x) = (-1)^n \frac{n!}{2\pi i} \int_c \frac{dz}{z^{n+1}} \exp(-z^2 - 2xz), \tag{III.C.6}$$

we can carry out the integrations over Q_j and $\bar{Q}_j$ in Eq. (III.C.2) as

$$G_{n_j}(t) = G_{n_j=0}(t)\frac{E_{n_j}(t)}{E_{n_j=0}(t)}\frac{n_j!}{2^{n_j}}\left(\frac{1}{2\pi i}\right)^2 \int_c \int_c \frac{dz_1 dz_2}{z_1^{n_j+1}z_2^{n_j+1}}$$

$$\times \exp(-a_j z_1^2 - a_j z_2^2 + b_j z_1 z_2). \tag{III.C.7}$$

Here a_j and b_j are given in Eqs. (5.3.17) and (5.3.18), respectively. Performing the contour integrations in Eq. (III.C.7), we obtain $G_{n_j}(t)$ as given by Eq. (5.3.12).

REFERENCES

Fong, F. K. (1975). " Theory of Molecular Relaxation." Wiley, New York.
Fujimura, Y., and Lin, S. H. (1979a). *J. Chem. Phys.* **70**, 247. Fujimura, Y., and Lin, S. H. (1979b). *J. Chem. Phys.* **71**, 3733.
Louisell, W. H. (1973). "Quantum Statistical Properties of Radiation." Wiley (Interscience), New York.

APPENDIX
IV

In this appendix to Chapter 7, a brief derivation of the unimolecular rate constant $k(E)$ is outlined by using quasi-equilibrium theory (Eyring et $al.$, 1980).

IV.A QUASI-EQUILIBRIUM THEORY

From the absolute reaction rate theory, the behavior of an isolated system can be described in terms of a microcanonical ensemble with systems uniformly distributed over all states having energies in the region E to $E + dE$ (Eyring et $al.$, 1964). The unimolecular rate of an isolated molecule having a fixed energy E can be obtained by evaluating the rate at which systems of the ensemble pass through saddle points of the potential energy curve in configuration space.

Let $dW(E)$ represent the number of states of the system having energy between E and $E + dE$. If the number of states is large, it can be expressed in terms of a density of state function $\rho(E)$ as

$$dW(E) = \rho(E)\,dE. \qquad (\text{IV.A.1})$$

The activated state is defined by the total energy between E and $E + dE$ with activation energy $E_0^{\ddagger}$ and kinetic energy between ε_t and $\varepsilon_t + d\varepsilon_t$ in the reaction coordinate. The number of activated states is given by $\rho^{\ddagger}(E - E_0^{\ddagger} - \varepsilon_t)\,dE\rho_t(\varepsilon_t)\,d\varepsilon_t$. The unimolecular rate constant $k(E)$ is then given by half the ratio of activated complexes to initial normal molecules, multiplied by the frequency of crossing the barrier, and by integrating over all possible values of the translational energy in the reaction coordinates as

$$k(E) = \int_0^{E-E_0^\ddagger} \frac{\gamma^\ddagger}{2} \frac{\rho^\ddagger(E - E_0^\ddagger - \varepsilon_t)\, dE\, \rho_t(\varepsilon_t)\, d\varepsilon_t}{\rho(E)\, dE}, \qquad \text{(IV.A.2)}$$

where $\gamma^\ddagger$, the frequency of crossing the barrier, with v_t the velocity along the reaction coordinate, is given by

$$\gamma^\ddagger = v_t/l = (1/l)(2\varepsilon_t/\mu^\ddagger)^{1/2}. \qquad \text{(IV.A.3)}$$

Here the equation $\varepsilon_t = \frac{1}{2}\mu^\ddagger v_t^2$ has been used, $\mu^\ddagger$ and l are the effective reduced mass for the translation and the length in the reaction coordinate, respectively. Noting that the translational energy is defined as

$$\varepsilon_t = n^2 h^2 / 8\mu^\ddagger l^2 \qquad \text{(IV.A.4)}$$

and that the state density $\rho_t(\varepsilon_t)$ of translational states in the reaction coordinate is given by

$$\rho_t(\varepsilon_t) = dn/d\varepsilon_t = (l/h)(2\mu^\ddagger/\varepsilon_t)^{1/2}, \qquad \text{(IV.A.5)}$$

Eq. (IV.A.2) can be rewritten as

$$k(E) = \frac{1}{h\rho(E)} \int_0^{E-E_0^\ddagger} d\varepsilon_t\, \rho^\ddagger(E - E_0^\ddagger - \varepsilon_t). \qquad \text{(IV.A.6)}$$

From Eq. (IV.A.1), $\rho^\ddagger(E - E_0^\ddagger - \varepsilon_t) = dW^\ddagger(E - E_0^\ddagger - \varepsilon_t)\,d(E - E_0^\ddagger - \varepsilon_t)$, and Eq. (IV.A.5) can be expressed as

$$k(E) = (1/h)[W^\ddagger(E - E_0^\ddagger)/\rho(E)], \qquad \text{(IV.A.7)}$$

where $W^\ddagger(E - E_0^\ddagger)$ is the total number of states of the activated complex for the system for all energies up to $E - E_0^\ddagger$.

References

Eyring, H., Henderson, D., Stover, B. J., and Eyring, E. M. (1964). "Statistical Mechanics and Dynamics." Wiley (Interscience), New York.
Eyring, H., Lin, S. H., and Lin, S. M. (1980). "Basic Chemical Kinetics." Wiley, New York.

Index